Topics in Theoretical and Experimental Gravitation Physics

NATO ADVANCED STUDY INSTITUTES SERIES

A series of edited volumes comprising multifaceted studies of contemporary scientific issues by some of the best scientific minds in the world, assembled in cooperation with NATO Scientific Affairs Division.

Series B: Physics

RECENT VOLUMES IN THIS SERIES

Volume 19 – Defects and Their Structure in Nonmetallic Solids
edited by B. Henderson and A. E. Hughes

Volume 20 – Physics of Structurally Disordered Solids
edited by Shashanka S. Mitra

Volume 21 – Superconductor Applications: SQUIDs and Machines
edited by Brian B. Schwartz and Simon Foner

Volume 22 – Nuclear Magnetic Resonance in Solids
edited by Lieven Van Gerven

Volume 23 – Photon Correlation Spectroscopy and Velocimetry
edited by E. R. Pike and H. Z. Cummins

Volume 24 – Electrons in Finite and Infinite Structures
edited by P. Phariseau and L. Scheire

Volume 25 – Chemistry and Physics of One-Dimensional Metals
edited by Heimo J. Keller

Volume 26 – New Developments in Quantum Field Theory and Statistical Mechanics–Cargèse 1976
edited by Maurice Lévy and Pronob Mitter

Volume 27 – Topics in Theoretical and Experimental Gravitation Physics
Edited by V. De Sabbata and J. Weber

Volume 28 – Material Characterization Using Ion Beams
Edited by J. P. Thomas and A. Cachard

Volume 29 – Electron–Phonon Interactions and Phase Transitions
Edited by Tormod Riste

The series is published by an international board of publishers in conjunction with NATO Scientific Affairs Division

A	Life Sciences	Plenum Publishing Corporation
B	Physics	New York and London
C	Mathematical and Physical Sciences	D. Reidel Publishing Company Dordrecht and Boston
D	Behavioral and Social Sciences	Sijthoff International Publishing Company Leiden
E	Applied Sciences	Noordhoff International Publishing Leiden

Topics in Theoretical and Experimental Gravitation Physics

Edited by

V. De Sabbata

Institute of Physics
University of Bologna
Bologna, Italy
and
Institute of Physics
University of Ferrara
Ferrara, Italy

and

J. Weber

University of Maryland
College Park, Maryland
and
University of California at Irvine
Irvine, California

PLENUM PRESS • LONDON AND NEW YORK
Published in cooperation with NATO Scientific Affairs Division

Library of Congress Cataloging in Publication Data

International School of Cosmology and Gravitation, Erice, Italy, 1975.
Topics in theoretical and experimental gravitation physics.

(NATO advanced study institutes series: Series B, Physics; 27)
"Proceedings of the International School of Cosmology and Gravitation held in Erice, Trapani, Sicily, March 13-25, 1975."
Includes indexes.
1. Gravitation–Congresses. I. De Sabbata, V. II. Weber, Joseph, 1919- III. Title. IV. Series.
QC178.I57 1975 531'.14 77-14029
ISBN 0-306-35727-5

Proceedings of the International School of Cosmology and Gravitation held in Erice, Trapani, Sicily, March 13–25, 1975

"Ettore Majorana" International Centre for Scientific Culture
Antonino Zichichi – *Director*

International School of Cosmology and Gravitation
Venzo De Sabbata – *Director*

Fourth Course: "Gravitational Waves"
Joe Weber – *Director*

Printed in the United States of America

Preface

During the period March 13-25, 1975, the IV course of the International School of Cosmology and Gravitation was held at Erice, Sicily and devoted primarily to gravitational waves. A number of participants have prepared their lectures for publication here. These include topics in Gravitational Radiation , Gravitational Radiation Antennas and Data Analyses, Singularities in General Relativity Theory, the Bimetric Theory of Gravitation, Tachyons, and General Relativity and Quantum Theory.

V. De Sabbata and J. Weber

Contents

OPENING REMARKS

J. Weber

University of California at Irvine and

University of Maryland, College Park

This appears to be the first school course devoted primarily to gravitational radiation. It is very appropriate to have a school of relativity and cosmology here in Erice. Probably the first work on the stress tensor of a fluid was done here by Archimedes. The great Italian physicists of modern times, Galileo, Fermi and Majorana, continue to be an inspiration to all of us.

We are very fortunate to be working in the field of General Relativity. Landau remarked that Einstein's achievement was the most perfect and most beautiful of all theories. The difficult mathematics has been a challenge to the greatest theoretical physicists in the world. The great weakness of the gravitational interaction -- 25 orders weaker than the weak interactions, makes possible experiments difficult, interesting and challenging.

The other lecturers are truly great physicists who are skilled in communicating their ideas to students. Each lecturer has a unique philosophy of doing physics and this school provides a splendid opportunity for students to become familiar with the different philosophies of doing physics.

The early work of Einstein, and Eddington predicted gravitational radiation, using the linearized field equations. For many years, there appeared to be no hope of doing meaningful experiments. We can admire the work of research workers during the period 1920-1960. This was a field pursued purely out of intellectual curiosity.

To test a theory one first thinks in terms of laboratory experiments. For electrons the quantity

$$\frac{Gm^2}{e^2} \rightarrow 10^{-43}$$

and this implies that in lowest order only one graviton is emitted for 10^{43} photons. Experiments of the type done by Hertz are therefore very difficult.

Serious work began on antennas to receive astronomical sources of radiation, in 1959. It still appeared that no sources existed with sufficient intensity to make direct observations possible. The displacement sensitivity x of an antenna with mass m, angular normal mode frequency ω is given by

$$x_{Rms} \sim \sqrt{\frac{kT}{m\omega^2 [\beta Q_1 Q_2]^{1/2}}}$$

and for $h\omega > kT$

$$x_{Rms} \sim \sqrt{\frac{\hbar\omega}{m\omega^2 [\beta Q_1 Q_2]^{1/2}}}$$

Here β is a coupling factor, Q_1 is the quality factor of the normal mode of an antenna, and Q_2 is the quality factor of some device which couples the energy out. By going to very low temperatures great increases in sensitivity are possible, and this makes more sources of gravitational radiation accessible.

The Einstein Eddington formula for the radiation from a spinning rod is

$$P = \frac{32GI^2\omega^6}{5c^5} \tag{1}$$

where P is the radiated power
G is the constant of gravitation
I is the amplitude of the quadrupole moment
ω is the angular frequency
c is the speed of light.

This formula is very discouraging because the very small quantity G -- a coupling constant -- appears in the numerator, while the very large quantity c^5 appears in the denominator.

For a binary star (1) may be written

$$P = \frac{G m^2 r^4 \omega^6}{c^5} \qquad (2)$$

where m is the mass. F.J. Dyson* was the first to point out that (2) transforms in a very interesting way for an extreme relativistic system. In this case

$$\frac{Gm^2}{4r^2} - \frac{mc^2}{r} \quad , \quad \frac{m}{r} = \frac{4c^2}{G} , \qquad c = \omega r$$

Substituting these gives us

$$P \Rightarrow G \frac{m^2 c}{r^2} \simeq \frac{c^5}{G}$$

In fact Dyson's more precise result is **

$$P = 128 \frac{c^5}{5G}$$

This stupendous amount of radiation might be expected in say a gravitational collapse, and was part of the motivation for the coincidence experiments search for radiation starting in 1967.

Finally, the very important discoveries of astronomy--the quasars, pulsars, and x ray sources refocussed attention on very early work of Chandresekhar, Oppenheimer and Volkcoff and Landau on neutron stars, black holes, and related very important gravitational radiation.

We thank Professors De Sabbata, Zichichi, NATO, NSF, Gravity Research Foundation, the Italian Ministry of Public Education, the Italian Ministry of Scientific and Technological Research, the National Research Council of Italy, the Regional Sicilian Government and many countries for making this school possible. Thank you.

*F.J. Dyson in Interstellar Communication p. 118, edited by AGW Cameron, W.A. Benjamin Inc. 1963

**L. Smarr's Computer Analyses suggest that the gravitational radiation from relativistic sources may be much smaller

THE GENERATION OF GRAVITATIONAL WAVES:

A REVIEW OF COMPUTATIONAL TECNIQUES *†

Kip S. Thorne

California Institute of Technology

Pasadena, California

TABLE OF CONTENTS

*Sections 1-6 of these lectures were presented at the International School of Cosmology and Gravitation, Erice, Sicily, 13-25 March 1975. Section 7, which is included here for completeness, was presented at the Second Latin-American Symposium on Relativity and Gravitation, Caracas, Venezuela, December 8-13, 1975. These are the lecture notes passed out at those conferences, except for a modest (and inadequate!) amount of updating as of January 1977.

†Supported in part by the National Aeronautics and Space Administration (NGR 05-002-256) and the National Science Foundation (AST 75-01398 A01). The updating and finalizing of the manuscript was carried out at Cornell University with partial support from the National Science Foundation (AST 75-21153).

1. INTRODUCTION

1.1 Nature of these lectures

In these lectures I shall review techniques for calculating gravitational-wave generation. My emphasis will be on the techniques themselves, on their realms of validity, and on typical applications of them. Most derivations will be omitted or sketched only briefly, but I shall give references to places where the derivations can be found.

I shall presume that the reader is familiar with general relativity at the level of "track one" of Misner, Thorne, and Wheeler (1973) -- cited henceforth as "MTW." My sign conventions and notation will be the same as in MTW.

1.2 Regions of Spacetime Around a Source

One can characterize a source of gravitational waves, semi-quantitatively, by the following length scales:

$$L \equiv \text{"Size of source"} \equiv \begin{Bmatrix}\text{radius of region inside which} \\ \text{the stress-energy } T^{\alpha\beta} \text{ is contained}\end{Bmatrix},$$

$$2M \equiv \text{"Gravitational radius of source"} \equiv \begin{Bmatrix}2 \times \text{mass of source in} \\ \text{units where } G = c = 1\end{Bmatrix},$$

$$\lambdabar \equiv \begin{matrix}\text{"reduced wavelength} \\ \text{of waves"}\end{matrix} \equiv \begin{Bmatrix}1/2\pi \times \text{characteristic wave-} \\ \text{length of gravitational} \\ \text{waves emitted by source}\end{Bmatrix}, \qquad (1.2.1)$$

$$\left.\begin{matrix} r_I \equiv \text{"inner radius of local wave zone"} \\ r_0 \equiv \text{"outer radius of local wave zone"}\end{matrix}\right\} \text{(see below).}$$

Corresponding to these length scales, one can divide space around a source into the following regions (See Fig. 1):

<u>Source</u>: $r \equiv \text{radius} \lesssim L$

<u>Strong-field region</u>: $r \lesssim 10\ M$ if $10\ M \gtrsim L$
typically does not exist if $L >> 10\ M$

<u>Weak-field near zone</u>: $L<r$, $10\ M<<r$, $r<<\lambdabar$ (1.2.2)

<u>Wave generation region</u>: $r \lesssim r_I$ (includes source, strong-field region, and weak-field near zone)

<u>Local wave zone</u>: $r_I \lesssim r \lesssim r_0$

<u>Distant wave zone</u>: $r_0 \lesssim r$.

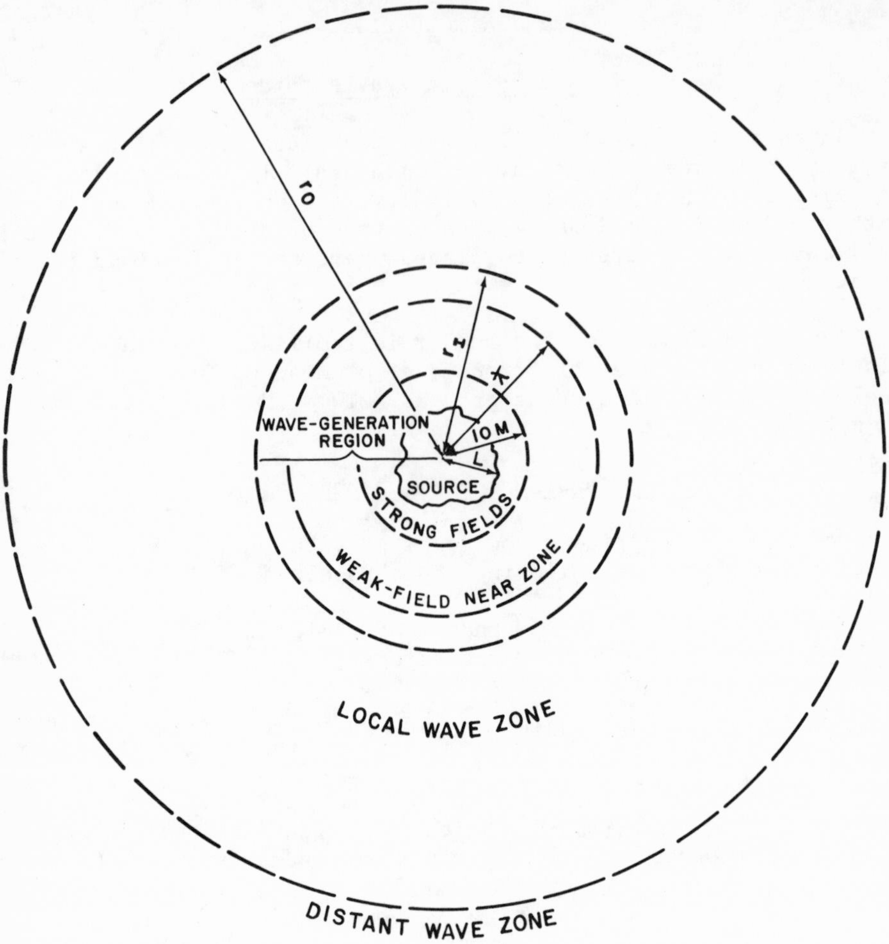

Figure 1. Regions of spacetime surrounding a source.

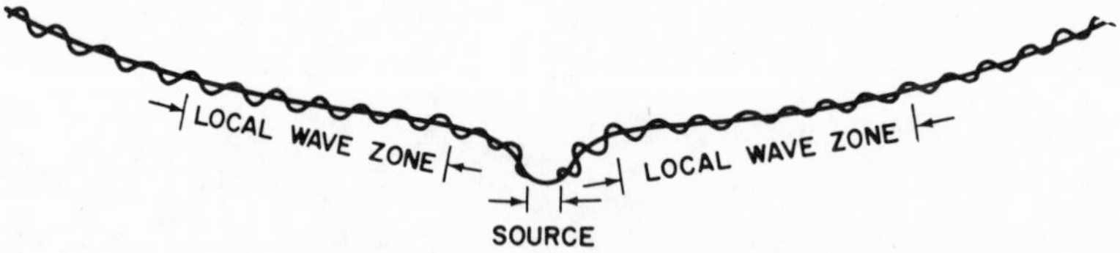

Figure 2. The local wave zone (schematic diagram). The smooth curve depicts the curvature of the background spacetime on which the waves (rippled curve) propagate. Near the source one cannot necessarily split the spacetime geometry into background plus waves; but in the local wave zone one can.

The "local wave zone" is the region in which (i) the source's waves are weak, outgoing ripples on a background spacetime; and (ii) the effects of the background curvature on the wave propagation are totally negligible (see Fig. 2).

The inner edge of the local wave zone r_I is the location at which one or more of the following effects becomes important: (i) the waves cease to be waves and become near-zone field, i.e., r becomes $\lesssim ƛ$; (ii) the gravitational pull of the source produces a significant redshift, i.e., r becomes $\backsim$ 2M = (Schwarzschild radius of source); (iii) the background curvature produced by the source distorts the wave fronts and backscatters the waves, i.e., $(r^3/M)^{1/2}$ becomes $\lesssim ƛ$; (iv) the outer limits of the source itself are encountered, i.e., r becomes $\lesssim$ L = (size of source). Thus, the inner edge of the local wave zone is given by

$$r_I = \alpha \times \text{Maximum } \{ƛ, 2M, (Mƛ^2)^{1/3}, L\}, \tag{1.2.3}$$

α $\equiv$ (some suitable number large compared to unity).

The outer edge of the local wave zone r_0 is the location at which one or more of the following effects becomes important: (i) a significant phase shift has been produced by the "M/r" gravitational field of the source, i.e., $(M/ƛ)\cdot \ln(r/r_I)$ is no longer $<< \pi$; (ii) the background curvature due to nearby masses or due to the external universe perturbs the propagation of the waves, i.e., r is no longer $<< R_B$ = (background radius of curvature). Thus, the outer edge of the local wave zone is given by

$$r_0 = \text{Minimum } \{r_I \cdot \exp(ƛ/\beta M), R_B/\gamma\}, \tag{1.2.4}$$

β, γ $\equiv$ (some suitable numbers large compared to unity).

Of course, we require that our large numbers $\alpha, \beta, \gamma,$ be adjusted so that the thickness of the local wave zone is very large compared to the reduced wavelength,

$$r_0 - r_I >> ƛ. \tag{1.2.5}$$

Throughout these lectures I shall confine attention to sources which possess a local wave zone -- and I shall call such sources "isolated." It seems likely that every source of gravitational waves in the Universe today is "isolated." However, in the very early Universe the background curvature, $1/R_B{}^2$, was so large that sources might not have been isolated.

In complex situations the location of the local wave zone might not be obvious. Consider, for example, a neutron star passing very near a super-massive black hole. The tidal pull of the hole sets the neutron star into oscillation, and the star's oscillations produce gravitational waves (Mashoon 1973; Turner 1977). If the hole is large enough, or if the star is far enough from it, there may exist a local wave zone around the star which does not also enclose the entire hole. Of greater interest - because more radiation will be produced - is the case where the star is very near the hole and the hole is small enough ($M_h \lesssim 100 M_\odot$) to produce large-amplitude oscillations, and perhaps even disrupt the star. In this case, before the waves can escape the influence of the star, they get perturbed by the background curvature of the hole. One must then consider the entire star-hole system as the source, and construct a local wave zone that surrounds them both.

The local wave zone acts as a buffer between the wave-generation region $r < r_I$ and the distant wave zone $r > r_0$. The existence of this buffer enables one to separate cleanly the theory of wave generation (applicable for $r < r_0$; treated in §§ 2-6 of these lectures) from the theory of wave propagation (applicable for $r > r_I$; treated in § 7 of these lectures).

1.3 The Gravitational-Wave Field

In the local wave zone where gravity is weak we shall use coordinate systems $(t,x,y,z) \equiv (x^0,x^1,x^2,x^3)$ which are centered on the source and are very nearly Minkowskiian; and we shall sometimes introduce the corresponding spherical coordinates (t,r,θ,ϕ) with

$$x = r \sin\theta \cos\phi \ , \ y = r \sin\theta \sin\phi \ , \quad z = r \cos\theta. \qquad (1.3.1)$$

The components of the metric then differ only slightly from Minkowskii form

$$g_{\alpha\beta} \equiv \eta_{\alpha\beta} + h_{\alpha\beta}, \ \eta_{\alpha\beta} = \text{diag}(-1,1,1,1), \ |h_{\alpha\beta}| \ll 1. \qquad (1.3.2)$$

Throughout these lectures it will be adequate, in the local wave zone, to treat $h_{\alpha\beta}$ as a linearized field residing in flat space-time. If one knows $h_{\alpha\beta}$ in any such coordinate system (i.e. in any "gauge"), one can compute from it the "gravitational wave field" (§35.4 of MTW)

$$h^{TT}_{jk} \equiv (P_{ja}h_{ab}P_{bk} - \ 1/2 \ P_{jk}P_{ab}h_{ab}). \qquad (1.3.3)$$

Here Latin indices run from 1 to 3; repeated Latin indices must be summed even if they are both "down"; and "TT" means "transverse, traceless part". The projection tensor used in this computation is

$$P_{jk} \equiv \delta_{jk} - n_j n_k \ ; \ n_j \equiv x^j/r. \qquad (1.3.4)$$

The task of wave-generation theory (§§2-6) is to determine h_{jk}^{TT} in the local wave zone. Once that has been done, wave-propagation theory (§7) can be used to calculate h_{jk}^{TT} in the distant wave zone.

1.4. Formalisms for Calculating Wave Generation

To calculate the generation of gravitational waves one must solve simultaneously the Einstein field equations and the equations of motion of the source. These equations are so complicated that one cannot hope to solve them exactly in any realistic situation. Therefore, one must resort to approximation schemes.

Several different approximation schemes have been devised for handling different types of sources. Most of these approximation schemes have been written in the form of "plug-in-and-grind" formalisms (i.e., computational algorithms). In these lectures I shall describe the following formalisms:

Weak-field formalisms (§2). These formalisms are valid for sources with weak internal gravitational fields ($L \gg 2M$; sources without any strong-field region; cf. Fig. 1). Section 2.1 will present a classification of weak-field formalisms and their realms of validity; and §§2.2-2.7 will present specific weak-field formalisms, each with its own realm of validity. A catalogue of the weak-field formalisms is given in Table 1.

Multipole analysis of the radiation field (§3). The gravitational waves from any source can be resolved into multipole fields, and that multipole resolution yields simple formulas for the energy, linear momentum, and angular momentum carried off by the waves. Section 3 presents such a multipole analysis, restricted to the local wave zone. That analysis can be used fruitfully in conjunction with the wave-generation formulas of §§2,4, and 5.

Slow-motion formalisms (§4). These formalisms are valid for sources with slow internal motions ($L/\lambda\!\!\!^{-} \ll 1$). They make use of near-zone multipole-moment expansions, and consequently they tie in tightly with the multipole analysis of the radiation field (§3). The slow-motion formalisms are catalogued in Table 1 along with the weak-field formalisms.

Small-perturbation formalisms (§5). These formalisms are applicable to sources consisting of small dynamical perturbations of a nonradiative "background" system (e.g., small particles falling into black holes, and small-amplitude pulsations of stars).

TABLE 1

A CATALOGUE OF WEAK-FIELD FORMALISMS AND SLOW-MOTION FORMALISMS

——— SLOW-MOTION EXPANSION ———→ $\left(\delta h^{TT}_{jk}\right)_{\ell\text{-pole}} \Big/ \left(h^{TT}_{jk}\right)_{\ell\text{-pole}}$

←WEAK-FIELD EXPANSION→ $\left(\frac{\delta T^{\alpha\beta}}{T^{00}}, \frac{\delta \bar{h}^{\alpha\beta}}{\bar{h}^{00}}\right)$

$\frac{\delta T^{\alpha\beta}}{T^{00}}$	$\frac{\delta \bar{h}^{\alpha\beta}}{\bar{h}^{00}}$	$\left(\frac{L}{ƛ}\right)^2$	$\left(\frac{L}{ƛ}\right)^4$	Complete Accuracy
ε	1			Special Relativity (§2.2)
ε	ε			Linearized Theory (§2.3)
ε^2	ε	Newtonian theory (§2.6)		Post-Linear Theory (§2.4)
ε^2	ε^2	Quadrupole-Moment Formalism (§2.6); also, weak-field limit of Slow-Motion Formalism (§4.4)		Post-Linear Wave-Generation Formalism (§2.5)
ε^3	ε^2			
ε^3	ε^3		Post-Newtonian Wave-Generation Formalism (§§2.7 and 4.5)	
Complete Accuracy		Slow Motion, Multipole-Moment Formalism (§4.3); errors $M/ƛ$		

Notation:

δ means "errors in."

$\bar{h}^{\alpha\beta}$ is the gravitational field (including both nonradiative and radiative parts)

h^{TT}_{jk} is the radiation field.

ε is the magnitude of the source's internal gravitational fields.

L is the source's linear size.

M is the source's mass.

$ƛ$ is the reduced wavelength of the gravitational waves.

Formalisms for studying systems with strong internal fields and fast, large-amplitude internal motions (§6). Unfortunately, no analytic formalisms now exist for studying such systems. However one can study their evolution and the waves they emit by numerical solution of the Einstein field equations on a large computer. The necessary numerical techniques are now under development.

2. WEAK-FIELD FORMALISMS

2.1 Classification of Weak-Field Formalisms

Weak-field formalisms are applicable to systems with weak internal gravitational fields.

To compute a characteristic dimensionless strength ε of the internal field of a source, one can analyze the source as though spacetime were flat, using globally Minkowskii coordinates in which the center-of-mass is at rest. In these coordinates one can compute ε from the retarded integral

$$\varepsilon = \begin{array}{l}\text{Maximum over all}\\ \text{"relevant" values}\\ \text{of the field point}\\ (t, x^j)\end{array} \left\{ \int \frac{T^{00}(t - |\underset{\sim}{x} - \underset{\sim}{x}'|, \underset{\sim}{x}')}{|\underset{\sim}{x} - \underset{\sim}{x}'|} d^3x' \right\} . \qquad (2.1.1)$$

Here T^{00} is the time-time component of the stress-energy tensor, and the "relevant" field points are those points at which one portion of the source interacts with fields produced by other portions of the source. One need not know ε with high precision; errors as large as a factor 3, say, are perfectly allowable.

For a source consisting of a single coherent body (e.g., a pulsating star) with mass M and linear size L,

$$\varepsilon \sim M/L. \qquad (2.1.2a)$$

For a source consisting of several lumps, each with mass m and size ℓ, separated by distance $b >> \ell$,

$\varepsilon \sim m/b$ if one is interested only in waves generated by relative motions of the lumps; (2.1.2b)

$\varepsilon \sim m/\ell$ if one is interested in waves generated by internal motions of the lumps. (2.1.2c)

Weak-field formalisms are valid only for sources that have $\varepsilon << 1$.

All weak-field formalisms have roughly the same structure: They utilize a single coordinate system $\{x^\alpha\}$ that covers the entire wave-generation region and local wave zone ($r \lesssim r_0$) and that is as nearly globally Minkowskiian as possible. In this coordinate system they define a "gravitational field" $\bar{h}^{\alpha\beta}$ (second-rank symmetric tensor) by

$$\bar{h}^{\alpha\beta} \equiv - [(-g)^{1/2} g^{\alpha\beta} - \eta^{\alpha\beta}], \tag{2.1.3}$$

where $g^{\alpha\beta}$ are the contravariant components of the metric tensor, where $g \equiv \det||g_{\alpha\beta}||$, and where $\eta^{\alpha\beta}$ are the components of the Minkowskii metric tensor [diag (-1,1,1,1)]. The formalisms then consist of field equations by which the stress-energy tensor $T^{\alpha\beta}$ (associated with matter and non-gravitational fields) generates the gravitational field $\bar{h}^{\alpha\beta}$, and equations of motion by which the gravitational field and internal stresses regulate the time evolution of the stress-energy tensor.

Thorne and Kovacs (1975; cited henceforth as TK) have devised a classification scheme for weak-field wave-generation formalisms--a scheme resembling that by which Havas and Goldberg (1962) classify "point-mass equation-of-motion formalisms." This scheme characterizes wave-generation formalisms by two integers n_T and n_h. These "order indices" tell one the magnitude of the errors made by the formalism--i.e., the amount by which the formalism's predictions should differ from those of exact general relativity theory:[1]

$$|(\text{errors in } T^{\mu\nu})/T^{00}| \sim \varepsilon^{n_T}, \tag{2.1.4a}$$

$$|(\text{errors in } \bar{h}^{\mu\nu})/\bar{h}^{00}| \sim \varepsilon^{n_h}. \tag{2.1.4b}$$

For example, a formalism of order $(n_T, n_h) = (1,1)$ makes fractional errors of order ε in both the stress-energy tensor and the gravitational field, while a formalism of order (2,1) makes fractional errors ε^2 in $T^{\mu\nu}$ and ε in $\bar{h}^{\mu\nu}$.

Errors in $\bar{h}^{\mu\nu}$, when fed into the equation of motion, produce errors in $T^{\mu\nu}$; and similarly, errors in $T^{\mu\nu}$, when fed into the field equations, produce errors in $\bar{h}^{\mu\nu}$. This feeding process places constraints on the order indices (n_T, n_h) of any self-consistent, weak-field wave-generation formalism:

$$n_h = n_T \quad \text{or} \quad n_h = n_T - 1. \tag{2.1.5}$$

---

[1] Note that all of the $|T^{\mu\nu}|$ are $\lesssim T^{00}$, and consequently all of the $|\bar{h}^{\mu\nu}|$ are $\lesssim \bar{h}^{00}$. This fact dictates the form of equations (2.1.4).

Thus, every weak-field formalism has order indices of the form (n,n-1) or (n,n) for some integer n.

TK devise a scheme for constructing formalisms of any desired order. They also show that any formalism of order (n,n-1) can readily be "strengthened" by augmenting onto it a higher-accuracy field-generation equation. The resulting, augmented formalism will have order (n,n).

In these lectures I shall not describe the general analysis of TK; I shall merely present examples of weak-field formalisms of various orders, and describe the augmentation process which improves their accuracy.

2.2 Special Relativity: A Formalism of Order (1,0)

Special Relativity is characterized by the equations of motion

$$T^{\mu\nu}{}_{,\nu} = 0, \tag{2.2.1}$$

which are oblivious of all gravitational effects. The largest of the individual terms that occur in these equations are of order

$$T^{00}/\ell, \tag{2.2.2}$$

where ℓ is a characteristic length scale inside the source. By contrast, the gravitational forces ignored by these equations of motion are

$$\Gamma^{\mu}{}_{\alpha\nu}T^{\alpha\nu}+\Gamma^{\nu}{}_{\alpha\nu}T^{\mu\alpha}\sim(\bar{h}^{00}/\ell)T^{00}\sim\varepsilon(T^{00}/\ell). \tag{2.2.3}$$

Comparison of the forces ignored (eq. 2.2.3) with the terms included (eq. 2.2.2) shows that special relativity makes fractional errors of order ε in the evolution of the stress-energy--and hence fractional errors of order ε in the stress-energy tensor itself. Evidently, the stress-energy tensor has order index $n_T = 1$.

The gravitational field, by contrast, has order index $n_h=0$, since special relativity is totally oblivious of gravity and therefore makes fractional errors $\delta\bar{h}^{\mu\nu}/\bar{h}^{00}\sim 1 = \varepsilon^0$.

Conclusion: Special relativity (eq. 2.2.1) is a weak-field formalism of order (1,0).

2.3. Linearized Theory: A Formalism of Order (1,1)

One can obtain Linearized Theory from Special Relativity very easily: One leaves unchanged the stress-energy tensor and its equations of motion

$$T^{\mu\nu}{}_{,\nu} = 0, \qquad (2.3.1a)$$

but one postulates that this stress-energy generates a gravitational field $\bar{h}^{\mu\nu}$ by the relation

$$\bar{h}^{\mu\nu}(x) = 16\pi \int G(x,x')T^{\mu\nu}(x')d^4x'. \qquad (2.3.1b)$$

Here d^4x' is the special relativistic volume element

$$d^4x' \equiv dx^{0'}dx^{1'}dx^{2'}dx^{3'}, \qquad (2.3.2)$$

and $G(x,x')$ is the flat-space propagator (Green's function)

$$G(x,x') = \frac{1}{4\pi}\,\delta_{ret}[1/2(x^{\alpha}-x^{\alpha'})(x^{\beta}-x^{\beta'})\eta_{\alpha\beta}], \qquad (2.3.3a)$$

$$\delta_{ret} \equiv \begin{cases} \text{Dirac delta function for } x^{\alpha} \text{ in future of } x^{\alpha'} \\ 0 \qquad\qquad\qquad\qquad \text{for } x^{\alpha} \text{ in past of } x^{\alpha'}. \end{cases} \qquad (2.3.3b)$$

[Notice: (1) in the arguments of $\bar{h}^{\mu\nu}$, G, and $T^{\mu\nu}$ we omit the indices of the coordinates x^{α} and x^{α} ; also (2) by integrating over time, $x^{0'}$, in eq. (2.3.1b) one obtains the expression

$$\bar{h}^{\mu\nu}(x^0,x^j) = 4\int \frac{T^{\mu\nu}(x^{0'}=x^0-|\underset{\sim}{x}-\underset{\sim}{x}'|,\ x^{j'})}{|\underset{\sim}{x}-\underset{\sim}{x}'|}\,d^3x', \qquad (2.3.4)$$

which is familiar from elementary treatises on Linearized Theory; e.g., chapter 18 of MTW.]

The linearized gravitational field (2.3.1b) is a fairly good approximation to the exact general relativistic gravitational field, $\bar{h}^{\alpha\beta} \equiv -[(-g)^{1/2}g^{\alpha\beta} - \eta^{\alpha\beta}]$. It makes fractional errors of order ε. Hence, Linearized Theory has a gravitational order index $n_h = 1$, which is one order "better" than Special Relativity; but its stress-energy order index is the same as that of Special Relativity, $n_T = 1$. The total order of Linearized Theory is $(n_T, n_h) = (1,1)$.

To within the accuracy of Linearized Theory the metric perturbation $h_{\mu\nu}$ is the trace-reversal of $\bar{h}_{\mu\nu}$:

$$h_{\mu\nu} = \bar{h}_{\mu\nu} - 1/2\eta_{\mu\nu}\bar{h}; \quad \bar{h} \equiv \bar{h}_{\alpha}{}^{\alpha}. \qquad (2.3.5a)$$

(Indices in Linearized Theory are raised and lowered with $\eta_{\mu\nu}$.) Since $h_{\mu\nu}$ and $\bar{h}_{\mu\nu}$ differ only by a trace, the gravitational-wave field in the local wave zone is (cf. eq. 1.3.3)

$$h^{TT}_{jk} = \bar{h}^{TT}_{jk} = P_{ja}P_{kb}\bar{h}_{ab} - 1/2P_{jk}(P_{ab}\bar{h}_{ab}). \qquad (2.3.5b)$$

The fractional errors in this gravitational-wave field are

$$(\text{fractional errors in } \bar{h}^{TT}_{jk}) \equiv \left| \frac{\delta \bar{h}^{TT}_{jk}}{\bar{h}^{TT}_{jk}} \right| = \left| \frac{\delta \bar{h}^{TT}_{jk}}{\bar{h}^{00}} \frac{\bar{h}^{00}}{\bar{h}^{TT}_{jk}} \right|$$

$$\sim \frac{\varepsilon}{|\bar{h}^{TT}_{jk}/\bar{h}^{00}|} \sim \frac{\varepsilon}{|\bar{h}^{TT}_{jk}/Mr^{-1}|} . \qquad (2.3.6)$$

Here $\bar{h}^{00} \backsim$ (Newtonian potential) $\backsim M/r$ is the largest of the components of the gravitational field; cf. equation (2.1.4b) and footnote 1.

The "rules" for using Linearized Theory to calculate gravitational-wave generation are as follows: (1) Express the special relativistic stress-energy tensor $T^{\mu\nu}$ in terms of the non-gravitational variables of the specific system being analyzed. (2) Solve the special-relativistic equations of motion (2.3.1a) to determine the evolution of the stress-energy tensor. (3) Evaluate the integral (2.3.1b) to determine the first-order gravitational field $\bar{h}^{\mu\nu}$ in the local wave zone. (4) Project out the gravitational-wave field $\bar{h}^{TT}_{jk}$ using equation (2.3.5b). (5) Check, using equation (2.3.6), that the errors in the wave field are acceptably small.

This set of rules shows very clearly the sense in which "Linearized Theory is the theory of order (1,1) obtained by augmenting onto Special Relativity [theory of order (1,0)] field generation equations"; cf. next to last paragraph of § 2.1. In particular, when working in Linearized Theory one at first (Rules 1 and 2) pretends that one is in Special Relativity. Only when one starts evaluating the radiation field (Rules 3 and 4) and its errors (Rule 5) does one depart from Special Relativity.

Recently Press (1977) has transformed the gravitational-wave field (2.3.4) and (2.3.5) of Linearized Theory into the form

$$\bar{h}^{TT}_{jk} = \frac{2}{r} \left\{ \frac{d^2}{dt^2} \int [T_{00} + 2T_{0p}n_p + T_{pq}n_p n_q]_{ret} x_{j'} x_{k'} d^3x' \right\}^{TT} . \qquad (2.3.7a)$$

Here "ret" means "evaluated at the retarded time"

$$[T_{00}]_{ret} \equiv T_{00}(t - |\underset{\sim}{x} - \underset{\sim}{x}'|, \underset{\sim}{x}') \qquad (2.3.7b)$$

and n_p is the unit radial vector pointing from the source toward the distant observer,

$$n_p \equiv x^p/r \ . \tag{2.3.7c}$$

Press's expression (2.3.7a) is particularly useful for systems in which $|T_{0j}| \ll T_{00}$ and $|T_{jk}| \ll T_{00}$, since then it involves only the second moment of the retarded energy distribution, T_{00ret}. Similar expressions are encountered in the "quadrupole-moment formalism" (eqs. 2.6.4 and 2.6.5 below). However there one requires that the source be confined deep within its own near zone ("slow-motion assumption"), whereas Press's expression requires no such constraint.

Press's expression, appropriately modified, has wider validity than just Linearized Theory: Whenever one has a formalism in which the local-wave-zone field $\bar{h}^{\mu\nu}$ can be written as:

$$\bar{h}^{\mu\nu}(x^0,\underset{\sim}{x})=\frac{4}{r}\int \tau^{\mu\nu}{}_{ret}\, d^3x', \text{ with } \tau^{\mu\nu}{}_{,\nu}=0. \tag{2.3.8}$$

then (as Press emphasizes) expression (2.3.7a) is valid with $T^{\mu\nu}$ replaced by $\tau^{\mu\nu}$ --whatever that animal may be. The weak-field formalisms classified by TK (§2.1 above) all have this property; cf.§II of TK.

2.3.1. A Valid Application of Linearized Theory

Linearized Theory is applicable whenever one can ignore self-gravitational forces inside the source -- i.e., whenever fractional errors

$$\begin{aligned} &|\delta T^{\mu\nu}/T^{00}| \sim \varepsilon \sim \text{(internal gravitational potential)} \\ &|\delta \bar{h}^{TT}_{ij}/\bar{h}^{TT}_{ij}| \sim \frac{\varepsilon}{|\bar{h}^{TT}_{ij}/\bar{h}^{00}|} \end{aligned} \tag{2.3.9}$$

are acceptable. The following is an example of a valid application of Linearized Theory:

A steel bar of mass $M \sim 500$ tons, length $L \sim 20$ meters, and diameter $D \sim 2$ meters generates gravitational waves by rotating end-over-end with angular velocity $\omega \sim 28$ radians per second [Einstein (1918); Eddington (1922); §36.3 of MTW]. (Faster rotation would tear the bar apart.) For such a bar the internal gravitational field is

$$\varepsilon \sim \frac{M}{D} \sim \left(\frac{5\times10^{8}\,\mathrm{g}}{2\times10^{2}\,\mathrm{cm}}\right)\times\left(\frac{0.7\times10^{-28}\,\mathrm{cm}}{\mathrm{g}}\right) \sim 2\times10^{-22}. \tag{2.3.10}$$

The radiation field and the gravitational potential $\bar{h}^{00}$, as calculated from Linearized Theory, turn out to be

$$|\bar{h}^{TT}_{ij}| \sim (ML^2\omega^2/r)\cos[2\omega(t-r) + \text{phase}] \tag{2.3.11}$$

$$\bar{h}^{00} = 4M/r; \tag{2.3.12}$$

and the fractional errors in the radiation field (eq. 2.3.6) are thus

$$\left|\frac{\delta\bar{h}^{TT}_{jk}}{\bar{h}^{TT}_{jk}}\right| \sim \frac{\varepsilon}{|\bar{h}^{TT}_{jk}/\bar{h}^{00}|} \sim \frac{\varepsilon}{(\omega L)^2}$$

$$\sim \frac{10^{-22}}{[(2\times10^{3}\,\mathrm{cm})(28\,\mathrm{sec}^{-1})(3\times10^{10}\,\mathrm{cm/sec})^{-1}]^2} \sim 10^{-11} \tag{2.3.13}$$

2.3.2. Invalid Applications of Linearized Theory

For most astrophysical systems internal gravity is important, and Linearized Theory is thus invalid. Examples are as follows:

(1) Gravitational waves from nonradial pulsations of a star. Let the star have mass M and radius R. Then its mean density ρ and its internal gravitational field strength ε are

$$\rho \sim M/R^3, \qquad \varepsilon \sim M/R; \tag{2.3.14}$$

and its mean pressure [calculated from hydrostatic equilibrium, $dp/dr \sim \rho M/R^2$] is

$$p \sim \rho(M/R) \sim \varepsilon\rho\ . \tag{2.3.15}$$

Since Linearized Theory makes fractional errors $|\delta T^{\mu\nu}/T^{00}| \sim \varepsilon$, its fractional errors in the pressure are

$$\frac{\delta p}{p} \sim \left|\frac{\delta T^{ij}}{T^{00}}\ \frac{T^{00}}{T^{ij}}\right| \sim \left|\varepsilon\,\frac{\rho}{p}\right| \sim 1. \tag{2.3.16}$$

Thus, Linearized Theory makes unacceptable errors in the star's internal pressure forces, as well as in its internal gravity.

One should not be surprised to learn that its errors in the gravitational waves due to stellar pulsation are also unacceptably large

$$\left| \frac{\delta \bar{h}^{TT}_{jk}}{\bar{h}^{TT}_{jk}} \right| \gtrsim 1. \tag{2.3.17}$$

(2) Gravitational waves from a near-encounter between two fast-moving stars. Linearized theory predicts no gravitational interaction between the stars. Therefore, according to Linearized Theory, each star proceeds undisturbed along its straight path through flat space, oblivious of the other star. As a result, no gravitational waves are produced -- an obviously incorrect prediction.

2.4 Post-Linear Theory: A Formalism of Order (2,1)

When analyzing the internal structure and motion of systems with significant self gravity, one needs a formalism which makes fractional errors $|\delta T^{\mu\nu}/T^{00}| \lesssim \varepsilon^2$ ---i.e., which has a stress-energy order index $n_T \geq 2$. The least accurate and least complex such formalism is Post-Linear Theory. It has order $(n_T, n_h) = (2,1)$.

For a derivation of Post-Linear Theory from general relativity theory, see Thorne and Kovács (1975; "TK").

In Post-Linear Theory one describes gravity by a gravitational field ${}_1\bar{h}^{\mu\nu}$, which is a good approximation to the general relativistic field (eq. 2.1.3):

$${}_1\bar{h}^{\mu\nu} = -[\sqrt{-g}\, g^{\mu\nu} - \eta^{\mu\nu}][1 + O(\varepsilon)]. \tag{2.4.1}$$

Post-Linear Theory, like Special Relativity and Linearized Theory, utilizes the language of flat spacetime. For example, one raises and lowers indices on the gravitational field by means of the flat Lorentz metric $\eta_{\mu\nu}$; and one performs "trace reversals" in the familiar manner of Linearized Theory:

$$\begin{gathered} {}_1h^{\mu\nu} = {}_1\bar{h}^{\mu\nu} - 1/2\eta^{\mu\nu}\, {}_1\bar{h} \; ; \quad {}_1\bar{h}^{\mu\nu} = {}_1h^{\mu\nu} - 1/2\eta^{\mu\nu}\, {}_1h; \\ {}_1\bar{h} = -{}_1h = {}_1\bar{h}^{\alpha}_{\alpha} = -\, {}_1h^{\alpha}_{\alpha} \end{gathered} \tag{2.4.2}$$

(cf. chapter 18 of MTW). The metric of general relativity is approximated by

$$g_{\mu\nu} = \eta_{\mu\nu} + {}_1h_{\mu\nu}\, [1 + O(\varepsilon)]. \tag{2.4.3}$$

The stress-energy tensor in Post-Linear Theory will typically be expressed in terms of the electromagnetic field $F_{\mu\nu}$, the matter variables (density, pressure, velocity, viscosity,...), and the gravitational field ${}_1\bar{h}^{\mu\nu}$. It must be constructed from these variables in accordance with all the rules of general relativity, to within fractional errors of $O(\varepsilon^2)$.

The equations of motion of Post-Linear Theory are formally the same as those of general relativity

$$T^{\mu\nu}{}_{,\nu} + {}_1\Gamma^{\mu}{}_{\alpha\nu}\, T^{\alpha\nu} + {}_1\Gamma^{\nu}{}_{\alpha\nu}\, T^{\mu\alpha} = 0; \qquad (2.4.4a)$$

however, the Christoffel symbols ${}_1\Gamma^{\mu}{}_{\alpha\beta}$ are given by the "first-order approximation"

$${}_1\Gamma^{\mu}{}_{\alpha\beta} = \eta^{\mu\nu}\, {}_1\Gamma_{\nu\alpha\beta}\ ;\ {}_1\Gamma_{\nu\alpha\beta} = ({}_1h_{\nu\alpha,\beta} + {}_1h_{\nu\beta,\alpha} - {}_1h_{\alpha\beta,\nu}). \qquad (2.4.4b)$$

The field equations of Post-Linear Theory are the same as those of Linearized Theory

$${}_1\bar{h}^{\mu\nu,\alpha}{}_{\alpha} = -\,16\pi\, T^{\mu\nu}; \qquad (2.4.4c)$$

and their solution can be written, using the flat-space propagator (eq. 2.3.3), as

$${}_1\bar{h}^{\mu\nu}(x) = 16\pi \int G(x,x')T^{\mu\nu}(x')d^4x'. \qquad (2.4.5)$$

Notice that Post-Linear Theory differs from Linearized Theory in one crucial, but simple way: It allows the gravitational field ${}_1\bar{h}^{\mu\nu}$ to "push the matter around". [Christoffel symbols have been inserted into the equation of motion; compare eqs. (2.3.1a) and (2.4.4a).] This difference is crucial for astrophysics. It allows Post-Linear Theory to treat accurately the structure and evolution of stars, planets, planetary systems, star clusters, and near stellar encounters -- unless the stars are highly compact (i.e., unless they are neutron stars or black holes). Linearized Theory makes enormous errors on all such systems.

2.5. Post-Linear Wave Generation: A Formalism of Order (2,2)

Despite its fine ability to analyze the internal dynamics of astrophysical systems, Post-Linear Theory does a bad job of predicting their gravitational-wave generation. For wave generation it has no better accuracy than Linearized Theory.

Fortunately, there is a simple way to improve its accuracy. One need only append to it a gravitational field ${}_2\bar{h}^{\mu\nu}$, which is more accurate than ${}_1\bar{h}^{\mu\nu}$:

$$ {}_2\bar{h}^{\mu\nu} = -[\sqrt{-g}\, g^{\mu\nu} - \eta^{\mu\nu}][1 + 0(\varepsilon^2)]. \tag{2.5.1} $$

By appending ${}_2\bar{h}^{\mu\nu}$ onto Post-Linear Theory, one boosts its accuracy from order (2,1) to order (2,2). The resulting formalism, called the "Post-Linear Wave-Generation Formalism," bears the same relationship to Post-Linear Theory as Linearized Theory does to Special Relativity.

TK have derived a formula for ${}_2\bar{h}^{\mu\nu}$ in terms of ${}_1\bar{h}^{\mu\nu}$ and $T^{\mu\nu}$. Their formula involves the flat-space propagator

$$ G(x,x') = \frac{1}{4\pi}\, \delta_{ret}\, [1/2(x^{\alpha}-x^{\alpha'})(x^{\beta}-x^{\beta'})\eta_{\alpha\beta}] \tag{2.5.2a} $$

[eq. (2.3.3)], and also its derivative with respect to the argument of the delta function

$$ G'(x,x') = \frac{1}{4\pi}\, \delta'_{ret}[1/2\ (x^{\alpha}-x^{\alpha'})(x^{\beta}-x^{\beta'})\eta_{\alpha\beta}]; \tag{2.5.2b} $$

$$ \delta'_{ret}(z) \equiv (d/dz)\, \delta_{ret}(z). $$

The TK formula splits ${}_2\bar{h}^{\mu\nu}$ into five pieces

$$ {}_2\bar{h}^{\mu\nu} = {}_2\bar{h}^{\mu\nu}_{D} + {}_2\bar{h}^{\mu\nu}_{F} + {}_2\bar{h}^{\mu\nu}_{TL} + {}_2\bar{h}^{\mu\nu}_{TR} + {}_2\bar{h}^{\mu\nu}_{W}. \tag{2.5.3} $$

Each piece has a particular physical interpretation. However, one must be careful not to take that interpretation too seriously -- for reasons discussed below. The five pieces are as follows:

Direct field, ${}_2\bar{h}^{\mu\nu}_{D}$. This field is produced directly by the Post-Linear stress-energy tensor $T^{\mu\nu}$; and it propagates from the source point $x^{\alpha'}$ to the field point x^{α} by means of the flat-space propagator. In other words, it propagates along the flat-space light cone with parallel-propagation of components and with a "1/r" fall-off of amplitude. The formula for this direct field is

$$ {}_2\bar{h}^{\mu\nu}_{D}(x) = 16\pi \int G(x,x')T^{\mu\nu}(x')[1-{}_1\bar{h}(x')]d^4x'. \tag{2.5.4} $$

This part of ${}_2\bar{h}^{\mu\nu}$ is $\sim M/r$, whereas the other four parts are typically $\lesssim M/r$.

Focussing field, ${}_2\bar{h}^{\mu\nu}_{F}$. When the gravitational field generated at $x^{\alpha'}$ propagates through regions of nonzero Ricci curvature -- i.e., through matter --, it gets focussed. This focussing increases the amplitude of ${}_2\bar{h}^{\mu\nu}$, without changing its directionality (i.e., without changing the relative magnitude of its components). The amount of focussing between a source point x^{α} and a field point x^{α} is described by the "focussing function"

$$\alpha(x,x') \equiv 1/2\; X^{\alpha}X^{\beta}\int_0^1 {}_1R_{\alpha\beta}(x^{\mu'} + \lambda X^{\mu})\lambda(1-\lambda)d\lambda, \tag{2.5.5a}$$

$$X^{\alpha} \equiv x^{\alpha} - x^{\alpha'}.$$

Here ${}_1R_{\alpha\beta}(x^{\mu'} + \lambda X^{\mu})$ is the first-order Ricci tensor, (calculated from $g_{\alpha\beta} = \eta_{\alpha\beta} + {}_1h_{\alpha\beta}$), evaluated at the event $x^{\mu'} + \lambda X^{\mu}$, which lies a fraction λ of the way along the straight line between source point and field point. A formula for ${}_1R_{\alpha\beta}$ is

$${}_1R_{\alpha\beta} = -1/2\; {}_1h_{\alpha\beta,\rho}{}^{\rho} = 8\pi(T_{\alpha\beta} - 1/2\eta_{\alpha\beta}T). \tag{2.5.5b}$$

For intuition into the focussing function, see Figure 3. In terms of the focussing function α, the flat-space propagator G, and the stress-energy tensor $T^{\mu\nu}$, the focussing field is

$${}_2\bar{h}_F^{\mu\nu}(x) = 16\pi\int \alpha(x,x')G(x,x')T^{\mu\nu}(x')d^4x'. \tag{2.5.5c}$$

Tail Field, ${}_2\bar{h}_{TL}^{\mu\nu}$. Consider the gravitational field ${}_2\bar{h}^{\mu\nu}$ generated at an event $x^{\alpha'}$. It has a "wave front" that propagates outward, initially spherically and initially along the future light cone of $x^{\alpha'}$. However, focussing produces dimples in the wave front; and dimpling, when analyzed from the viewpoint of Huygens' principle, produces waves that radiate outward from the dimpled region in all directions. (See Figure 4.) The result is a "tail" of the wave field. Let a wave originating at $x^{\alpha'}$ arrive at an event $x^{\alpha''}$, with a dimple in its wave front due to focussing. The amplitude $\beta(x'', x')$ for this dimple to produce a tail is given by

$$\beta(x'',x') = X^{\alpha''}X^{\beta''}\int_0^1 {}_1R_{\alpha\beta}(x^{\mu'} + \lambda X^{\mu''})\lambda^2 d\lambda, \tag{2.5.6a}$$

$$X^{\alpha''} \equiv x^{\alpha''} - x^{\alpha'}. \tag{2.5.6b}$$

This amplitude is called the "tail-generating function." The tail which it generates is given by

$$\bar{h}_{TL}^{\mu\nu}(x) = -16\pi \iint_{x'\in I^-(x)} G(x,x'')\beta(x'',x')G'(x'',x')T^{\mu\nu}(x')d^4x''d^4x'. \tag{2.5.6c}$$

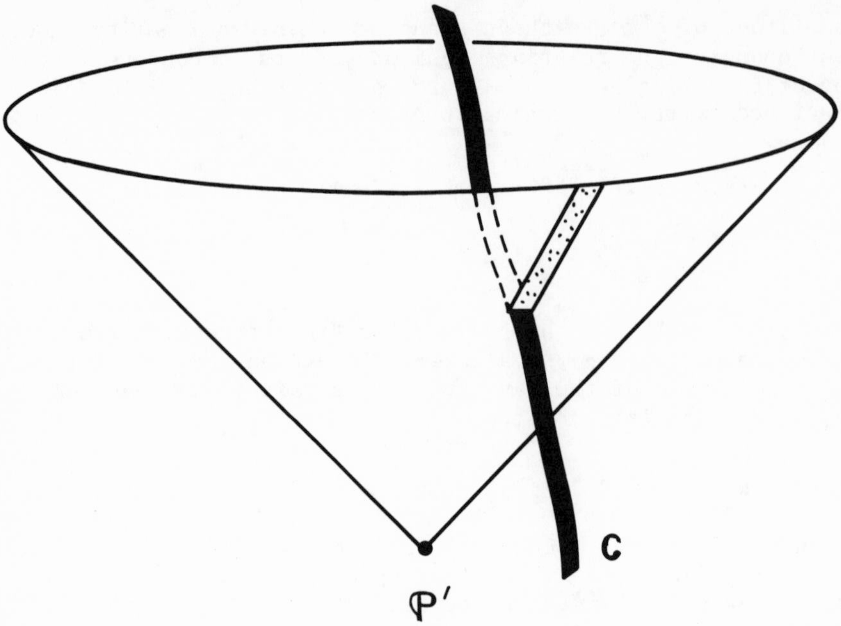

Figure 3. The region of focussing for a gravitational field generated at $\mathcal{P}'$, which propagates through spacetime containing a lump of matter (world tube $\mathbf{C}$). Focussing occurs, and α and $\bar{h}_F^{\mu\nu}$ are nonzero, in the stippled region--i.e., in the region containing rays that have passed through the lump.

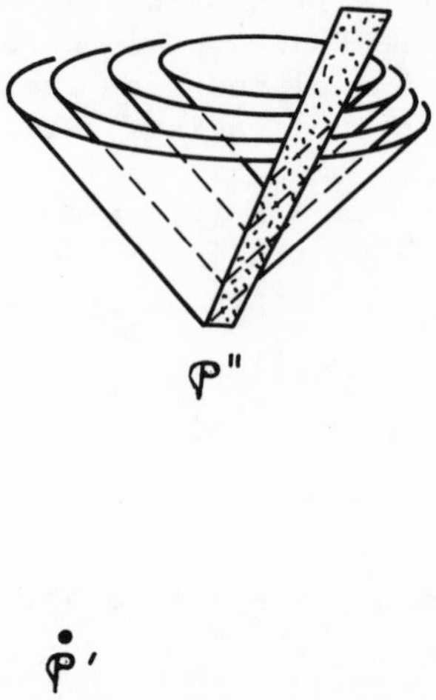

Figure 4. The tail of the gravitational field described in Figure 3. In the region of focussing (stippled region) the tail-generating function β is non-zero. Thus, each point $\mathcal{P}''$ in the stippled region is a source of tail -- and the tail propagates outward from each such $\mathcal{P}''$ along its future light cone.

Here the expression $x' \epsilon I^-(x)$ means that the integration must extend over source points $x^{\alpha'}$ that lie inside but not on the past (flat-space) light cone of x^α.

Transition Field, ${}_2\bar{h}^{\mu\nu}_{TR}$. The gravitational field ${}_2\bar{h}^{\mu\nu}$ generated at an event $x^{\alpha'}$ does not really propagate along light cones of the (fictitious) flat space of Post-Linear Theory. Rather, it propagates along light cones of the slightly curved metric $g_{\mu\nu} = \eta_{\mu\nu} + {}_1h_{\mu\nu}$. The result is a time delay relative to flat-space propagation -- a delay measured, for example, in the "Shapiro radar time-delay experiment" (Shapiro 1964;§40.4 of MTW). The direct field (eq. 2.5.4) fails to take this time delay into account. Therefore, one must correct it for the effects of the time delay. The "transition field" ${}_2\bar{h}^{\mu\nu}_{TR}$ does that correcting. The correction is embodied in a "time-delay function"

$$\gamma(x,x') = 1/2X^\alpha X^\beta \int_0^1 {}_1h_{\alpha\beta}(x^{\mu'} + \lambda X^\mu)d\lambda, \qquad (2.5.7a)$$

$$X^\mu \equiv x^\mu - x^{\mu'}. \qquad (2.5.7b)$$

For two events x and x' separated by a distance ℓ in the laboratory frame, this time-delay function is

$$\gamma(x,x') = \ell\Delta t_s, \qquad (2.5.7c)$$

where Δt_s is the "Shapiro time delay" (difference between curved-space and flat-space propagation times). In terms of this time-delay function, the transition field is given by[2]

$$\bar{h}^{\mu\nu}_{TR}(x) = 16\pi \int \gamma(x,x')G'(x,x')T^{\mu\nu}(x')d^4x'. \qquad (2.5.7d)$$

The name"transition field" is taken from electromagnetic theory: When a charged particle moves, with uniform velocity, through a medium of variable index of refraction, it radiates. The radiation (called "electromagnetic transition radiation") is caused by variations in the speed of propagation of the particle's Coulomb field.

[2]The expression for $\bar{h}^{\mu\nu}_{TR}$ given by TK differs slightly from eq.(2.5.7d): TK have to "truncate the external time delay" because they do not restrict attention to the local wave zone, wher the external delay is negligible.

The "gravitational transition field" (2.5.7d) has a similar origin: It is caused by variations in the "flat-space" speed of propagation of the particle's direct ("Coulomb") gravitational field.

Whump field, ${}_2\bar{h}^{\mu\nu}_W$. The gravitational field ${}_2\bar{h}^{\mu\nu}$ is generated not only by the material stress-energy tensor $T^{\mu\nu}$, but also by "gravitational stresses." For example, when two stars go flying past each other at high speed, their relative gravitational potential energy builds up quickly, then dies out quickly (i.e., goes "whump"), and in doing so it produces a burst of second-order gravitational field. TK call this field the "whump field", not only in the case of stellar fly-by, but also in general. The "gravitational stresses" $W^{\mu\nu}$ which generate it are a certain sum of products of first derivatives of ${}_1\bar{h}^{\alpha\beta}$; in particular,

$$W^{\mu\nu} \equiv {}_1t^{\mu\nu}_{LL} + (16\pi)^{-1}\, {}_1\bar{h}^{\mu\rho}{}_{,\sigma}\, {}_1\bar{h}^{\nu\sigma}{}_{,\rho} \tag{2.5.8a}$$

where ${}_1t^{\mu\nu}$ is the "Landau-Lifshitz pseudotensor," accurate up to fractional errors $\sim \varepsilon$:

$$\begin{aligned} {}_1t^{\alpha\beta}_{LL} = (16\pi)^{-1} \Big\{ & \tfrac{1}{2}\eta^{\alpha\beta}\, {}_1\bar{h}_{\lambda\nu,\rho}\, {}_1\bar{h}^{\lambda\rho,\nu} + {}_1\bar{h}^{\alpha}{}_{\lambda,\nu}\, {}_1\bar{h}^{\beta\lambda,\nu} \\ & - ({}_1\bar{h}_{\nu\rho}{}^{,\alpha}\, {}_1\bar{h}^{\beta\nu,\rho} + {}_1\bar{h}_{\nu\rho}{}^{,\beta}\, {}_1\bar{h}^{\alpha\nu,\rho}) + \tfrac{1}{2}\, {}_1\bar{h}_{\lambda\nu}{}^{,\alpha}\, {}_1\bar{h}^{\lambda\nu,\beta} \\ & - \tfrac{1}{4}\eta^{\alpha\beta}\, {}_1\bar{h}_{\lambda\nu,\rho}\, {}_1\bar{h}^{\lambda\nu,\rho} - \tfrac{1}{4}\, {}_1\bar{h}^{,\alpha}\, {}_1\bar{h}^{,\beta} + \tfrac{1}{8}\eta^{\alpha\beta}\, {}_1\bar{h}_{,\lambda}\, {}_1\bar{h}^{,\lambda} \Big\}. \end{aligned} \tag{2.5.8b}$$

In terms of these stresses, the whump field is

$${}_2\bar{h}^{\mu\nu}_W(x) = 16\pi \int G(x,x')\, W^{\mu\nu}(x')\, d^4x'. \tag{2.5.8c}$$

The gravitational waves emitted by a Post-Linear source are described by the "tranverse, traceless part" of ${}_2\bar{h}^{\mu\nu}$, evaluated in the local wave zone

$$h^{TT}_{jk} = \bar{h}^{TT}_{jk} = P_{ja}\, {}_2\bar{h}_{ab}\, P_{bk} - \tfrac{1}{2} P_{jk}(P_{ab}\, {}_2\bar{h}_{ab}). \tag{2.5.9}$$

Here P_{jk} is the usual transverse projection tensor(eq. 1.3.4).

In the local wave zone ${}_2\bar{h}^{\mu\nu}$ satisfies the flat-space Einstein field equations ${}_2\bar{h}^{\mu\nu}{}_{,\alpha\beta}\eta^{\alpha\beta} = 0$ and the flat-space "Lorentz gauge condition" ${}_2\bar{h}^{\mu\nu}{}_{,\nu} = 0$, except for fractional errors of $O(\varepsilon^2)$. As a consequence, the gravitational-wave field $\bar{h}^{TT}_{jk}$

transforms, under Lorentz transformations, in the usual manner for a field of spin two and zero rest mass.

Unfortunately, each of the five individual pieces of ${}_2\bar{h}^{\mu\nu}$, by itself, fails to satisfy the gauge condition. As a consequence, the transverse-traceless part of each piece (e.g., $[{}_2\bar{h}^{jk}_{TL}]^{TT}$) fails to transform as a spin-two, zero-rest-mass field. This fact prevents one from attributing a rigorous physical significance to each of the five pieces by itself. The physical interpretations given above are therefore somewhat heuristic. For further discussion, see TK, and also §VIc of Kovács and Thorne (1977a).

Crowley and Thorne (1977; their eq. 37) have derived several potentially useful formulas for the sum of the focussing, tail, and transition fields. An example is:

$$
\begin{aligned}
&{}_2\bar{h}^{\mu\nu}_{F}(x) + {}_2\bar{h}^{\mu\nu}_{TL}(x) + {}_2\bar{h}^{\mu\nu}_{TR}(x) = \\
&= -\frac{\partial}{\partial x^\alpha}\int G(x,x')\, {}_1\bar{h}^{\alpha\beta}(x')\frac{\partial}{\partial x^\beta}\, {}_1\bar{h}^{\mu\nu}(x')\,d^4x' . \qquad (2.5.10)
\end{aligned}
$$

Crowley and Thorne also examine the issue of when one can combine the Post-Linear Wave-Generation Formalism with a point-particle description of gravitating bodies. The answer is non-trivial: Some sets of post-linear formulas are compatible with point-particle descriptions of matter; others are not.

The accuracy of the Post-Linear Wave-Generation Formalism is a slightly delicate issue. Until one looks closely, one expects

$$\text{(fractional errors in } {}_2\bar{h}^{\mu\nu}) \sim \varepsilon^2 ; \qquad (2.5.11a)$$

$$\text{(fractional errors in } {}_2\bar{h}^{TT}_{jk}) \sim \frac{\varepsilon^2}{|\bar{h}^{TT}_{jk}/\bar{h}^{00}|} . \qquad (2.5.11b)$$

However, the formalism as expounded above relies on two approximations which can sometimes produce larger errors than (2.5.11): First, it assumes that focussing is small -- i.e., that the focussing function satisfies

[3] See, e.g., §III.B of Eardley, Lee, and Lightman (1974), with the change of notation $e_{\hat{z}} \to \underset{\sim}{n} = \underset{\sim}{e}_r$ = (unit radial vector), $\underset{\sim}{e}_{\hat{x}} \to \underset{\sim}{e}_{\hat{\theta}}$ = (unit vector in θ direction), $\underset{\sim}{e}_{\hat{y}} \to \underset{\sim}{e}_{\hat{\varphi}}$ = (unit vector in φ direction), and with specialization to the spin-2 case, $\Psi_2 = \Psi_3 = \Phi_{22} = 0$; and also Thorne (1977d).

$$|\alpha(x,x')| << 1 \text{ for all } x' \text{ inside source and } x \text{ in local wave zone.} \qquad (2.5.12)$$

Second, it assumes that the Shapiro time delays inside the source are short compared to the characteristic timescale for the source's motion -- i.e., that Δt_s (eq. 2.5.7c) and the reduced wavelength of the radiation, $\lambda\!\!\!^{-}$, satisfy

$$\Delta t_s \quad << \quad \lambda\!\!\!^{-} \ . \qquad (2.5.13)$$

These assumptions could be dropped, but only at the cost of making the formalism somewhat more complicated than it already is.

For realistic sources the "short-Shapiro-time-delay" assumption (2.5.13) is generally satisfied. However, one can readily imagine interesting weak-field sources ($\varepsilon << 1$) which violate the "small-focussing assumption." For example, for two stars of mass m and size ℓ separated by a distance $b >> \ell$, the focussing function for rays originating in one star and passing through the other is

$$\alpha \sim (b/\ell)(m/\ell) \sim 10^{-6} \ (b/\ell)(m/M_{\odot})(\ell/R_{\odot})^{-1} \ . \qquad (2.5.14)$$

This can be $\gtrsim 1$ for sufficiently large separations b.

The small-focussing and short-Shapiro-time-delay assumption force one to modify the error estimates (2.5.11). The correct error estimates are

$$(\text{fractional errors in } {}_2\bar{h}^{\mu\nu}) \sim \text{maximum } (\varepsilon^2, \varepsilon\alpha, \varepsilon\Delta t_s/\lambda\!\!\!^{-}), \qquad (2.5.15a)$$

$$(\text{fractional errors in } {}_2\bar{h}^{TT}_{jk}) \sim \frac{\text{maximum } (\varepsilon^2, \ \varepsilon\alpha, \ \varepsilon\Delta t_s/\lambda\!\!\!^{-})}{|\bar{h}^{TT}_{jk}/\bar{h}^{00}|} \ . \qquad (2.5.15b)$$

The "rules" for using the Post-Linear Wave-Generation Formalism are as follows: (1) Express the Post-Linear stress-energy tensor $T^{\mu\nu}$ in terms of the nongravitational variables of the specific system being analyzed, and in terms of the first-order gravitational field ${}_1\bar{h}^{\mu\nu}$. Do so in the manner of general relativity, up to fractional errors of $O(\varepsilon^2)$. (2) Solve the coupled Post-Linear equations of motion (2.4.4a) and Post-Linear gravitational field equations (2.4.4c) for the evolution of the source and its first-order gravitational field ${}_1\bar{h}^{\mu\nu}$. (3) Evaluate the integrals (2.5.3)-(2.5.8) for field points in the local wave zone to determine th five pieces of the second-order gravitational field ${}_2\bar{h}^{\mu\nu}$. Project out the gravitational-wave field $\bar{h}^{TT}_{jk}$ in the local wave zone using equation (1.3.3) (5) Check, using equation (2.5.15b), that the errors in the wave field are acceptably small.

2.5.1. Sample Applications

The Post-Linear Wave-Generation Formalism can be applied, with good accuracy, to most astrophysical systems. For example, it can be applied to the two systems of §2.3.2, in which Linearized Theory was a failure: pulsations of a noncompact star, and near-encounters between fast-moving stars.

Actually, for the stellar pulsation problem -- and for any other problem characterized by slow internal motions (speeds much less than that of light) -- one need not resort to the Post-Linear Formalism in its full complexity. Rather, one can apply a slow-motion variant of the formalism (§2.6 below).

However, for systems with fast motion and weak but significant internal gravity, the Post-Linear Formalism is the simplest formalism that will do the full job. An example is "gravitational Bremsstrahlung radiation" produced in a near-encounter between nearly Newtonian stars ($M/R \ll 1$), which fly past each other with a relative velocity $v \gtrsim 0.1$ x (speed of light). This example is treated in detail by Kovács and Thorne (1977a,b). [Limiting cases and special features of the Bremsstrahlung problem are actually amenable to other techniques, which I do not review in these lectures. These include the Feynman-diagram method (originally designed for quantum gravity problems, but also applicable to classical problems; see Feynman 1963); the method of virtual quanta (Matzner and Nutku 1974); the method of Green's functions for the linearized Schwarzschild metric (a special case of the perturbation methods of §5 below; Peters 1970); and a new colliding-plane-waves tecnique (D'Eath 1977)].

2.6. The Slow-Motion Limit of the Post-Linear Formalism: Newtonian Theory [Order (2,1)] and Quadrupole-Moment Formalism [Order (2,2)]

Most astrophysical systems have not only weak internal gravity, $\varepsilon \ll 1$, but also slow internal motions and weak internal stresses. For such systems the Post-Linear Formalism of §2.4, which has order (2,1), reduces to the Newtonian Theory of Gravity; and the Post-Linear Wave-Generation Formalism of §2.5, which has order (2,2), reduces to the Quadrupole-Moment Wave-Generation Formalism. I shall briefly summarize these formalisms and their realms of validity. For details of the slow-motion transition from the Post-Linear Formalism to these formalisms, see TK. For derivations of these formalisms assuming slow motion from the beginning see, e.g., § 104 of Landau and Lifshitz (1962) or chapter 36 of MTW.

Realm of validity: One can characterize a slow-motion, weak-field system by the following parameters:

$$M \equiv \text{(mass of system)}$$

$$L \equiv \text{(size of system)}$$

$$\bar{\lambda} \equiv \text{(characteristic timescale of system)} = \text{(reduced wavelength of radiation)}$$

$$v \equiv (|T^{0j}|/T^{00})_{max} = \text{(maximum internal velocity)} \qquad (2.6.1)$$

$$S \equiv (|T^{ij}|/T^{00})_{max} = \text{(maximum of stress/density)}$$

$$\varepsilon \equiv \text{(magnitude of internal gravity)} \sim M/L.$$

For a static system $\bar{\lambda}$ is not zero; rather, it is the characteristic timescale for perturbations against which one hopes the system is stable (e.g., for small-amplitude pulsations, if the system is a static star).

Newtonian Gravitation Theory and the Quadrupole-Moment Formalism require for their validity the following conditions:

Weak gravity: $\varepsilon \ll 1$ (2.6.2a)

Slow motion: $L/\bar{\lambda} \ll 1$, which implies $v \ll 1$ (2.6.2b)

Small Stresses: $S^2 \ll 1$, (2.6.2c)

System not near marginal stability: $\begin{cases} \omega^2 \equiv (1/\bar{\lambda})^2 \gg (M/L)(M/L^3) \\ \omega^2 \equiv (1/\bar{\lambda})^2 \gg S^2(S/L)^2 \end{cases}$ (2.6.2d)

[If the last condition is violated, then post-Newtonian gravitational forces may affect significantly the stability of the system and its motion; cf. Chandrasekhar (1964), or Boxes 24.2 and 26.2 of MTW.]

Newtonian Theory Summarized: Newtonian Theory describes gravity by a scalar gravitational potential U, and describes matter by its mass density ρ, its velocity v_j, and the stress t_{jk} measured in its rest frame. The equations of the theory are: (1) conservation of mass

$$\rho_{,t} + (\rho v_j)_{,j} = 0; \tag{2.6.3a}$$

(2) "Euler" equations of motion

$$\rho(v_{j,t}+v_{j,k}v_k) = \rho U_{,j} - t_{jk,k}; \tag{2.6.3b}$$

(3) gravitational field equation

$$U_{,jj} = -4\pi\rho, \tag{2.6.3c}$$

which has the solution

$$U(\mathbf{x},t) = \int \frac{\rho(\mathbf{x}',t)}{|\mathbf{x}-\mathbf{x}'|}\, d^3x'. \tag{2.6.3d}$$

For further details see, e.g., §39.7 of MTW.

Quadrupole-Moment Formalism Summarized: To calculate the gravitational waves from a nearly Newtonian system, one can proceed as follows: (1) Specify, by an equation of state, or viscosity, or some other method, the dependence of the stresses t_{jk} on ρ, v_j, and other variables of the system. (2) Solve the Newtonian equations (2.6.3) to determine the structure and evolution of the system. (3) Calculate the gravitational-wave amplitude $\bar{h}^{TT}_{jk}$ in the local wave zone using the following formula:

$$h^{TT}_{jk} = \frac{2}{r}\left[\ddot{\mathcal{I}}_{jk}(t-r)\right]^{TT}. \tag{2.6.4}$$

Here $\mathcal{I}_{jk}(t)$ is the "reduced quadrupole moment" of the source at time t

$$\mathcal{I}_{jk}(t) = \int \rho(t,\mathbf{x}')\left[x^{j'}x^{k'} - \frac{1}{3}\delta^{jk}(\mathbf{x}')^2\right] d^3x'; \tag{2.6.5}$$

$\ddot{\mathcal{I}}_{jk}$ is the second time derivative of $\mathcal{I}_{jk}$; r is distance from the source to the field point; and "TT" denotes the transverse-traceless projection process of equation (1.3.3). (4) Verify that the"conditions of validity" (2.6.2) are satisfied.

2.6.1. Sample Applications

The quadrupole-moment formalism has been used more widely in gravitational-wave calculations than any other wave-generation formalism. Examples of its application are: the waves emitted by quadrupole pulsations of a star [see, e.g., Wheeler (1966)]; the waves emitted by orbital motion of a binary star system [see, e.g., Peters and Mathews (1963)]; the gravitational bremsstrahlung radiation from low-velocity near-encounters of two stars [low velocity limit of the problem described in §2.5.1; see, e.g., Ruffini and Wheeler (1971)]; and the "tidal gravitational radiation" produced when a Newtonian star is tidally deformed by passage near a black hole or other star [Mashoon (1973), Turner (1977)].

2.7 The Post-Newtonian Wave-Generation Formalism [Order (3,3)]

For systems with slow, but not extremely slow motions (e.g., $v \sim 0.3$), and weak, but not extremely weak gravity (e.g., $\varepsilon \sim 0.1$), one needs a gravitational-wave formalism of higher accuracy than the Quadrupole-Moment Formalism. More specifically, one needs a formalism that takes account of Post-Newtonian effects in the structure and evolution of the system.

Epstein and Wagoner (1975) have devised such a formalism. Like all Post-Newtonian formalisms, it is valid only for gravitationally bound, or nearly gravitationally bound systems -- i.e., for systems with

$$v^2 \lesssim \varepsilon, \qquad S^2 \lesssim \varepsilon \tag{2.7.1a}$$

[cf. eq. (2.6.1)]; and it requires moderately weak internal gravity

$$\varepsilon \lesssim 0.1. \tag{2.7.1b}$$

The Epstein-Wagoner "Post-Newtonian Wave-Generation Formalism" has the following structure: (1) First one calculates the structure and evolution of the source using a Post-Newtonian Formalism of order $(n_T, n_h) = (3,2)$ -- e.g., the perfect-fluid Post-Newtonian formalism of Chandrasekhar (1965). (2) Then one evaluates the radiation field in the local wave zone using formulas to be given below (§4.5). [These formulas boost the order of the formalism up to $(n_T, n_h) = (3,3)$.] (3) Finally one verifies that the emitting system satisfies the Post-Newtonian validity conditions (2.7.1).

I delay presenting the Epstein-Wagoner equations until after I have treated multipole-moment formalisms, because Epstein and Wagoner utilize multipole moments in a nontrivial fashion.

2.7.1. Sample Applications

The Epstein-Wagoner formalism is being used in two different ways by the Stanford relativity group: (i) They use it to study deviations from the quadrupole-moment formalism for systems with moderately strong internal gravity or moderately high velocities or both [e.g. for moderately relativistic binary systems, and for stellar flybies at moderately high velocity (gravitational bremsstrahling); see Wagoner and Will (1976)]. (ii) They use it to seek insight into highly relativistic situations [e.g. waves from the collapse of a rotating star to form a black hole; Epstein (1976), Wagoner (1977)], where one can hope for, but is not guaranteed, more reliable results than from the quadrupole-moment formalism.

3. MULTIPOLE-MOMENT ANALYSIS OF THE RADIATION FIELD

3.1. Gravitational-Wave Field

I now wish to turn attention from weak-field formalisms to slow-motion formalisms. But before doing so, I must digress into a topic which is needed as an underpinning for the slow-motion discussion. This topic is the multipole structure of the gravitational-wave field for <u>any</u> isolated source. A number of researchers (not including me) have made major contributions to this subject; see end of §3.3 for references. Recently, I have taken the various contributions, have exhibited the relationships between some of their notations, and have tried to consolidate them into a single formalism in which (to me) the notation looks optimal. The following review of the subject is based on that work (Thorne 1977a--which I shall cite henceforth as TV, since it is Paper V in a series).

Consider the gravitational-wave field h_{jk}^{TT} in the local wave zone of an isolated source. The local wave zone is so constructed [eqs. (1.2.2)-(1.2.5)] that in it one can regard the background on which the waves propagate as completely flat. Consequently, the waves have the radially-propagating, flat-space form

$$h_{jk}^{TT} = r^{-1} A_{jk}(t - r,\theta,\phi) \qquad (3.1.1)$$

where A_{jk} is a symmetric, transverse, traceless amplitude that depends on retarded time $u = t - r$ and on angular location (θ,ϕ) around the source. The fact that A_{jk}^{TT} is symmetric, transverse, and traceless is embodied in the algebraic relations

Symmetry: $A_{jk} = A_{kj}$, (3.1.2a)

Transverse: $A_{jk}n_k = 0$, where $\underset{\sim}{n} \equiv \underset{\sim}{x}/r$
= unit radial vector, (3.1.2b)

Traceless: $A_{jj} = 0$. (3.1.2c)

It is often useful to expand the wave amplitude A_{jk} in spherical harmonics, and to identify the coefficients of the expansion as the multipole moments of the radiation field. In these lectures I shall follow the notation of Sachs (1961), Pirani (1965), MTW (Chap. 36), and TV (Thorne 1977a), which uses symmetric, trace-free tensors to represent spherical harmonics. The formalism of symmetric, trace-free tensors is described briefly in the next section.

3.2. Symmetric, Trace-Free Tensors and Spherical Harmonics

Let $\underline{A}$ be a symmetric, constant (position-independent) tensor of rank ℓ defined in 3-dimensional, flat space. We shall denote its components in a Cartesian coordinate system by $A_{k_1 \ldots k_\ell}$, and the components of its completely symmetric part by $A_{(k_1 \ldots k_\ell)}$:

$$A_{(k_1 \ldots k_\ell)} \equiv \frac{1}{\ell!} \sum_{\pi} A_{k_{\pi(1)} \ldots k_{\pi(\ell)}} . \qquad (3.2.1)$$

Here the summation goes over all $\ell!$ permutations, π, of $1 \ldots \ell$. We shall denote the trace-free, symmetric part of $\underline{A}$ by the corresponding capital script letter, $\mathcal{A}$:

$$\mathcal{A}_{k_1 \ldots k_\ell} = \sum_{n=0}^{[\ell/2]} (-1)^n \frac{\ell!(2\ell-2n-1)!!}{(\ell-2n)!(2\ell-n)!!(2n)!!} \times \qquad (3.2.2)$$

$$\times\ \delta_{(k_1 k_2} \ldots \delta_{k_{2n-1} k_{2n}} S_{k_{2n+1} \ldots k_\ell) j_1 j_1 \ldots j_n j_n} .$$

Here δ_{ab} = (Kronecker delta) are the components of the metric, $[\ell/2]$ means the largest integer less than or equal to $\ell/2$, $S_{k_1 \ldots k_\ell} \equiv A_{(k_1 \ldots k_\ell)}$, and the "double factorial" $(2n)!!$ is defined by (3.2.8) below.

Consider the set of all symmetric, trace-free tensors of rank ℓ ("STF-ℓ tensors"). The STF-ℓ tensors generate an irreducible representation of the rotation group, of weight ℓ. Hence, there exists a one-to-one mapping between them and the spherical harmonics of order ℓ. To exhibit that mapping we introduce the unit radial vector $\underset{\frown}{n}$ of our Euclidean space, and we express its Cartesian components in terms of spherical coordinates, θ and ϕ :

$$\underset{\sim}{n}=\underset{\sim}{x}/|\underset{\sim}{x}|, \quad n_1+in_2=\sin\theta\ e^{i\phi}, \quad n_3=\cos\theta. \tag{3.2.3}$$

From the usual expression for the spherical harmonic $Y^{\ell m}(\theta,\phi)$ in terms of cos θ and sin θ $e^{i\phi}$, it is easy to show [see pp. 289-290 of Pirani (1965) or see TV] that

$$Y^{\ell m}(\theta,\phi)=\mathcal{Y}^{\ell m}_{k_1\ldots k_\ell}\, n_{k_1}\ldots n_{k_\ell}. \tag{3.2.4}$$

Here $\mathcal{Y}^{\ell m}_{k_1\ldots k_\ell}$ is the following STF-ℓ tensor

$$\mathcal{Y}^{\ell m}_{k_1\ldots k_\ell}\equiv C^{\ell m}\sum_{j=0}^{[(\ell-m)/2]} a^{\ell mj}(\delta^1_{k_1}+i\delta^2_{k_1})\ldots(\delta^1_{k_m}+i\delta^2_{k_m})\,\cdot$$

$$\cdot\,\delta^3_{k_{m+1}}\ldots\delta^3_{k_{\ell-2j}}(\delta^{a_1}_{k_{\ell-2j+1}}\delta^{a_1}_{k_{\ell-2j+2}})\ldots(\delta^{a_j}_{k_{\ell-1}}\delta^{a_j}_{k_\ell}) \text{ for } m\geq 0, \tag{3.2.5}$$

$$\mathcal{Y}^{\ell m}_{k_1\ldots k_\ell}\equiv(-1)^m(\mathcal{Y}^{\ell\,-m}_{k_1\ldots k_\ell})^* \quad \text{for} \quad m<0.$$

$$C^{\ell m}\equiv(-1)^m\left[\frac{2\ell+1}{4\pi}\ \frac{(\ell-m)!}{(\ell+m)!}\right]^{1/2}, \quad a^{\ell mj}\equiv\frac{(-1)^j(2\ell-2j)!}{2^\ell j!(\ell-j)!(\ell-m-2j)!}.$$

Here $[(\ell-m)/2]$ means the largest integer less than or equal to $(\ell-m)/2$, and * means complex conjugate.

Henceforth, to simplify notation, when we encounter a sequence of many (say ℓ) indices on a tensor we shall denote it as follows

$$S_{A_\ell}\equiv S_{a_1\ldots a_\ell}. \tag{3.2.6a}$$

Similarly we shall abbreviate the tensor product of ℓ unit radial vectors by

$$N_{B_\ell}\equiv n_{b_1}\ldots n_{b_\ell}. \tag{3.2.6b}$$

In this abbreviated notation equation (3.2.4) says

$$Y^{\ell m}(\theta,\phi) = \mathcal{Y}^{\ell m}_{K_\ell} N_{K_\ell} . \tag{3.2.4'}$$

Thus, capital subscript letters denote sequences of lower-case subscript indices; and the number of indices in a sequence is denoted by a subscript to the capital subscript.

The tensors $\mathcal{Y}^{\ell m}_{K_\ell}$ with $-\ell \leq m \leq +\ell$ serve two roles: First, they generate the spherical harmonics of order ℓ (eq. 3.2.4'). Second, they form a basis for the $(2\ell+1)$-dimensional vector space of STF-ℓ tensors; i.e., any STF-ℓ tensor $\mathcal{F}$ can be expanded as

$$\mathcal{F}_{K_\ell} = \sum_{m=-\ell}^{\ell} F^{\ell m} \mathcal{Y}^{\ell m}_{K_\ell} . \tag{3.2.7a}$$

The tensor components $\mathcal{F}_{K_\ell}$ are real if and only if $F^{\ell -m} = (-1)^m F^{\ell m}$. The expansion (3.2.7a) can be inverted as follows (see, e.g., TV)

$$F^{\ell m} = 4\pi \frac{\ell!}{(2\ell+1)!!} \mathcal{F}_{K_\ell} \mathcal{Y}^{\ell m}_{K_\ell} \tag{3.2.7b}$$

where the "double factorial" is defined by

$$n!! \equiv n(n-2)(n-4)\ldots(2 \text{ or } 1). \tag{3.2.8}$$

In practical calculations one can use spherical harmonics and STF-ℓ tensors interchangeably: Consider any sphere centered on the coordinate origin, and on that sphere consider any scalar function $f(\theta,\phi)$. One can expand $f(\theta,\phi)$ in spherical harmonics with complex-number expansion coefficients

$$f(\theta,\phi) = \sum_{\ell=0}^{\infty} \sum_{m=-\ell}^{\ell} F^{\ell m} Y^{\ell m}(\theta,\phi); \tag{3.2.9a}$$

alternatively, one can expand it in powers of the unit radial vector $\underset{\sim}{n}$, with coefficients that are STF-ℓ tensors

$$f(\theta,\phi) = \sum_{\ell=0}^{\infty} \mathcal{F}_{K_\ell} N_{K_\ell} . \tag{3.2.9b}$$

The expansion coefficients of the two schemes are related by equations (3.2.7a,b).

Similar expansions can be made for vector fields and tensor fields. The resulting formalism, which relates vector and tensor spherical harmonics to STF-ℓ tensors, is presented in TV.

3.3. Multipole Expansion of Gravitational-Wave Field

The vacuum Einstein field equations permit the gravitational-wave amplitude A_{ij} (eq. 3.1.1) to be any symmetric, transverse, traceless function of retarded time $t-r$ and angle. One can show [see, e.g., TV] that this fact allows A_{ij} to be expanded in a multipole series of the form

$$h_{ij}^{TT} = r^{-1}A_{ij}$$

$$= \left[\sum_{\ell=2}^{\infty} (4/\ell!)r^{-1}\, {}^{(\ell)}\mathcal{I}_{ijA_{\ell-2}}(t-r)\, N_{A_{\ell-2}} \right. \tag{3.3.1}$$

$$\left. + \sum_{\ell=2}^{\infty} [8\ell/(\ell+1)!]r^{-1}\varepsilon_{pq(i}\, {}^{(\ell)}\mathcal{S}_{j)pA_{\ell-2}}(t-r)n_q N_{A_{\ell-2}}\right]^{TT} .$$

Here the normalizations (factors of 4, 8, ℓ, etc.) have been chosen to make equations (4.4.3) look simple; ε_{abc} is the Levi-Civita tensor; parentheses around tensor indices denote symmetrization; ${}^{(\ell)}\mathcal{I}_{A_\ell}(t-r)$ and ${}^{(\ell)}\mathcal{S}_{A_\ell}(t-r)$ are the trace-free, symmetric coefficients of the expansion evaluated at retarded time $t-r$; a prefix superscript in parentheses, e.g., ${}^{(k)}\mathcal{S}(u)$, means that the quality is to be differentiated k times with respect to its argument u

$${}^{(k)}\mathcal{S}(u) \equiv d^k\mathcal{S}/du^k; \tag{3.3.2}$$

and TT means that the transverse, traceless part is to be taken (eq. 1.3.3).

The coefficients ${}^{(\ell)}\mathcal{I}_{A_\ell}$ and ${}^{(\ell)}\mathcal{S}_{A_\ell}$ of the spherical harmonic expansions--integrated ℓ times (superscript ℓ removed)--are called the "mass" and "current" moments of the radiation field:

$\mathcal{I}_{ab} \equiv$ (mass quadrupole moment), $\mathcal{S}_{ab} \equiv$ (current quadrupole moment),

$\mathcal{I}_{abc} \equiv$ (mass octupole moment), $\mathcal{S}_{abc} \equiv$ (current octupole moment), (3.3.3)

$$\mathcal{I}_{A_\ell} \equiv \text{(mass } \ell\text{-pole moment)}, \quad \mathcal{S}_{A_\ell} \equiv \text{(current } \ell\text{-pole moment)}.$$

They are trace-free, symmetric tensors which depend on retarded time.

Notice that the mass quadrupole part of the radiation field (3.3.1) has a form familiar from the "Quadrupole Moment Formalism"

$$(h^{TT}_{jk})_{\text{mass quadrupole}} = 2r^{-1} \ddot{\mathcal{I}}^{TT}_{jk} \tag{3.3.4}$$

[cf. eq. (2.6.4)] . Also notice that the mass-multipole moments produce a radiation field with "electric-type parity" (also called "even-type parity"), $\pi = (-1)^\ell$; while the current-multipole moments produce a radiation field with "magnetic-type" ("odd-type") parity, $\pi = (-1)^{\ell+1}$

If the radiation field is known, one can project out its multipole moments using the following integrals over a sphere of constant (t-r):

$$^{(\ell)}\mathcal{I}_{B_\ell} = \begin{bmatrix}\text{Symmetric, trace-} \\ \text{free part of}\end{bmatrix} \left\{ \frac{\ell(\ell-1)(2\ell+1)!!}{2(\ell+1)(\ell+2)} \frac{r}{4\pi} \int h^{TT}_{b_1 b_2} n_{b_3} \cdots n_{b_\ell} d\Omega \right\}, \tag{3.3.5a}$$

$$^{(\ell)}\mathcal{S}_{B_\ell} = \begin{bmatrix}\text{Symmetric, trace-} \\ \text{free part of}\end{bmatrix} \left\{ \frac{(\ell-1)(2\ell+1)!!}{4(\ell+2)} \frac{r}{4\pi} \int \varepsilon_{b_1 jk} n_j h^{TT}_{kb_2} n_{b_3} \cdots n_{b_\ell} d\Omega \right\}. \tag{3.3.5b}$$

Multipole expansions of a generic radiation field have been given in various notations by Sachs (1961), Pirani (1965), Mathews (1962), Janis and Newman (1965), Bonnor and Rotenberg (1966), and Campbell and Morgan (1971). I like the above STF notation -- which is due to Sachs (1961), and Pirani (1965)-- because it ties in nicely with the theory of slow-motion sources (see below) and with the standard form of the quadrupole-moment formalism. Elsewhere (TV) I exhibit the relationship between the above STF expansions and expositions that use other conventions for spherical harmonics.

3.4. Multipole Expansion of Energy, Momentum, and Angular Momentum in Waves

The energy and linear momentum carried off by the radiation field (3.3.1) are most easily evaluated using the Isaacson (1968) stress-energy tensor for gravitational waves

$$T^{GW}_{\alpha\beta} = (1/32\pi) \langle h^{TT}_{jk,\alpha} h^{TT}_{jk,\beta} \rangle \tag{3.4.1}$$

(cf. MTW, §§35.7 and 35.15). Here the brackets, $\langle\ \rangle$, denote an average over several wavelengths. The power radiated into a unit solid angle about the radial, $\underset{\sim}{n}$, direction is

$$(dE/d\Omega dt) \equiv -r^2 n_j T^{GW}_{j0} = +r^2\ T^{GW}_{00}\ ; \tag{3.4.2}$$

and the total power radiated is the integral of this power over a sphere lying in the local wave zone:

$$dE/dt = \int (dE/d\Omega dt)d\Omega. \tag{3.4.3}$$

By inserting the gravitational-wave field (3.3.1) into equations (3.4.1)-(3.4.3) and integrating one obtains (see TV)

$$\frac{dE}{dt} = \sum_{\ell=2}^{\infty} \frac{(\ell+1)(\ell+2)}{(\ell-1)\ell}\ \frac{1}{\ell!(2\ell+1)!!} \left\langle {}^{(\ell+1)}\mathcal{I}_{A_\ell}\ {}^{(\ell+1)}\mathcal{I}_{A_\ell} \right\rangle$$
$$+ \sum_{\ell=2}^{\infty} \frac{4\ell(\ell+2)}{(\ell-1)}\ \frac{1}{(\ell+1)!(2\ell+1)!!} \left\langle {}^{(\ell+1)}\mathcal{S}_{A_\ell}\ {}^{(\ell+1)}\mathcal{S}_{A_\ell} \right\rangle . \tag{3.4.4}$$

The waves carry linear momentum out radially--and, as with any locally plane-fronted radiation field, the magnitude of their momentum flux is the same as that of their energy flux:

$$(dP_j/d\Omega dt) \equiv r^2\ n_k\ T^{GW}_{kj} = r^2\ n_j T^{GW}_{00} = n_j\ (dE/d\Omega dt) \tag{3.4.5}$$

(cf. eq. 35.77j of MTW). The total linear momentum carried off by the waves,

$$dP_j/dt = \int (dP_j/d\Omega dt)d\Omega, \tag{3.4.6}$$

can be evaluated by inserting the wave field (3.3.1) into the above equations and integrating:

$$\frac{dP_j}{dt} = \sum_{\ell=2}^{\infty} \left\{ \frac{2(\ell+2)(\ell+3)}{\ell(\ell+1)!(2\ell+3)!!} \left\langle {}^{(\ell+2)}\mathcal{I}_{jA_\ell}\ {}^{(\ell+1)}\mathcal{I}_{A_\ell} \right\rangle + \right.$$
$$+ \frac{8(\ell+3)}{(\ell+1)!(2\ell+3)!!} \left\langle {}^{(\ell+2)}\mathcal{S}_{jA_\ell}\ {}^{(\ell+1)}\mathcal{S}_{A_\ell} \right\rangle + \tag{3.4.7}$$
$$\left. + \frac{8(\ell+2)}{(\ell-1)(\ell+1)!(2\ell+1)!!} \left\langle \varepsilon_{jpq}\ {}^{(\ell+1)}\mathcal{I}_{pA_{\ell-1}}\ {}^{(\ell+1)}\mathcal{S}_{qA_{\ell-1}} \right\rangle \right\}.$$

Note, that, whereas the energy radiated (eq. 3.4.4) involves a beating of each multipole moment against itself, the linear momentum radiated (eq. 3.4.7) involves a beating of "adjacent" multipole moments.

The dependence of the wave amplitude, h^{TT}_{ij}, on angle causes its wave fronts to be not quite precisely spherical--and thereby enables the waves to carry off angular momentum. One might hope that the angular momentum loss could be calculated by integrating $\epsilon_{jab}\, x_a\, T^{GW}_{b0}$ over a sphere surrounding the source. Unfortunately, such a procedure fails--for this reason: the averaging process that underlies equation (3.4.1) for $T^{GW}_{\alpha\beta}$ treats as zero the "tiny corrections" which die out as $1/r^3$. However, it is precisely the $1/r^3$ part that carries off the angular momentum. I was vaguely aware of this fact when writing the relevant sections of MTW, but was not sufficiently certain to spell it out explicitly. Subsequently Bryce DeWitt (1971) derived a simple, correct expression for the flux of angular momentum:

$$\frac{dS_j}{dt} = \frac{1}{16\pi}\int \varepsilon_{jpq}\, x_p \left\langle (h^{TT}_{qa}\, {}^{(1)}h^{TT}_{ab})_{,b} - \frac{1}{2} h^{TT}_{ab,q}\, {}^{(1)}h^{TT}_{ab} \right\rangle r^2\, d\Omega. \tag{3.4.8}$$

A word of interpretation is needed. This equation is correct only if the integral is evaluated in the asymptotic rest frame of the source. (Similarly for all previous formulae in this section.) As the source's linear momentum changes (eq. 3.4.7), its asymptotic rest frame gradually changes; and one must gradually change the reference frame in which one evaluates (3.4.8) (and all previous integrals).

Return to equation (3.4.8). By inserting the wave field (3.3.1) and integrating, one obtains

$$\frac{dS_j}{dt} = \sum_{\ell=2}^{\infty} \frac{(\ell+1)(\ell+2)}{(\ell-1)\ell!\,(2\ell+1)!!} \left\langle \varepsilon_{jpq}\, {}^{(\ell)}\mathcal{I}_{pA_{\ell-1}}\, {}^{(\ell+1)}\mathcal{I}_{qA_{\ell-1}} \right\rangle + \sum_{\ell=2}^{\infty} \frac{4\ell^2(\ell+2)}{(\ell-1)(\ell+1)!\,(2\ell+1)!!} \left\langle \varepsilon_{jpq}\, {}^{(\ell)}\mathcal{S}_{pA_{\ell-1}}\, {}^{(\ell+1)}\mathcal{S}_{qA_{\ell-1}} \right\rangle. \tag{3.4.9}$$

Notice that this angular momentum radiated, like the energy radiated (eq. 3.4.4), involves a beating of each multipole with itself.

3.5. Discussion of the Multipole Formalism

Of what use are the above formulas? I view them as tools to to be used in studying the generation of gravitational waves from explicit sources. Given a source, one identifies the local wave zone. Using any technique one can dream up (and this is the tough part of the analysis!), one calculates the time-changing multipole moments of the source. One then plugs into the above formulas to get the radiation field and the rate it carries off energy, momentum, and intrinsic angular momentum.

As an alternative application, one can calculate the radiation field of a given source (the tough task; above formulas not necessarily useful); one can use equations (3.3.5) to resolve it into multipole pieces; and one can then use the other formulas above to read off the energy, momentum, and angular momentum radiated.

Once the radiation field h_{ij}^{TT} is known in the local wave zone, one can propagate it on outwards to Earth using the propagation equation of the geometric-optics formalism (§7 below). For typical situations the wave form (3.3.1) will remain highly accurate all the way to Earth, except for uninteresting phase shifts caused by the waves' self energy and by various masses present in the Universe, and except for changes in the weakest of the multipole fields caused by nonlinear beating together of stronger multipole fields.

4. SLOW-MOTION FORMALISMS

4.1. Overview

Turn attention now from weak-field systems (§2) and arbitrary systems (§3) to slow-motion systems--i.e., systems for which

$$\lambda\!\!\!^{-} \equiv (\text{reduced wavelength of radiation}) \gg L \equiv (\text{size of source}). \quad (4.1.1)$$

Thorne (1977b--cited henceforth as TVI) has derived a slow-motion multipole-moment formalism for calculating gravitational-wave generation by such systems. This slow-motion formalism generalizes the quadrupole-moment formalism of §2.6 to include all multipole moments, and to encompass sources with arbitrarily strong internal gravity; cf. Table 1.

The slow-motion formalism takes, as its starting point, a multipole analysis of the external gravitational field of a stationary source. That stationary multipole analysis is described in §4.2; and the slow-motion wave-generation formalism built on it is described in §4.3. Then, in §4.4 the weak-field limit of the slow-motion formalism is described. Throughout § 4 we shall use subscripted coordinate indices: $x_\alpha \equiv x^\alpha$.

4.2. Multipole Moments of a Stationary Source

Consider a stationary system residing in asymptotically flat spacetime and surrounded by vacuum. It is well known[4] that the vacuum geometry of spacetime surrounding such a stationary system is uniquely determined by two families of time-independent multipole moments: the mass moments, and the current moments. The object of this section is to give a definition of these moments (i.e., a set of conventions for them) that will tie in simply with radiation theory. More specifically, if the stationary source is set into very slow motion (§4.3 below), its moments as defined here (§4.2) will be identical to those which characterize the emitted radiation (§3 above).

4.2.1. ACMC-N Coordinate Systems

As a tool in defining the moments of a stationary source, we introduce a special class of coordinate systems: A coordinate system will be called "asymptotically Cartesian and mass centered to order N" ("ACMC-N") if and only if its metric coefficients have the following form. Expand them in inverse powers of "radius"

$$r \equiv [(x_1)^2 + (x_2)^2 + (x_3)^2]^{1/2} \quad . \tag{4.2.1}$$

The coefficients must be time-independent (so the Killing vector is $\partial/\partial x_0$); the leading term must be the Minkowskii metric; and the remaining terms must have the form

$$g_{\alpha\beta} = \eta_{\alpha\beta} + \frac{\overset{0}{A}_{\alpha\beta}(\underset{\sim}{n})}{r} + \frac{\overset{1}{A}_{\alpha\beta}(\underset{\sim}{n})}{r^2} + \ldots + \frac{\overset{N+1}{A}_{\alpha\beta}(\underset{\sim}{n})}{r^{N+2}}$$
$$+ \text{[terms that die out faster than } 1/r^{N+2}], \tag{4.2.2}$$

where

$$\underset{\sim}{n} = \underset{\sim}{x}/r = (x_j/r)\,(\partial/\partial x_j). \tag{4.2.3}$$

Expand each of the coefficients $\overset{P}{A}_{\alpha\beta}(\underset{\sim}{n})$ in spherical harmonics.

[4] See, e.g., van der Burg (1968), Geroch (1970), Clarke and Sciama (1971).

$A^0_{\alpha\beta}$ and A^1_{00} must involve only harmonics of order 0.

A^1_{0j} and A^1_{jk} must involve only harmonics of order 0 and 1. (4.2.4)

$A^2_{\alpha\beta}$ must involve only harmonics of order 0,1, and 2.

$A^3_{\alpha\beta}$ must involve only harmonics of order 0,1,2, and 3

.

. (4.2.4 con'd)

.

$A^N_{\alpha\beta}$ must involve only harmonics of order 0,1,...,N.

Note the absence of a dipole in A^1_{00}; that is what makes the coordinates "mass centered".

That such a coordinate system always exists is proved in TVI by showing that "deDonder coordinates," with appropriate specialization of the gauge, are ACMC-∞.

4.2.2. Multipole Moments

Given an ACMC-N coordinate system, we define the multipole moments of order $\ell \leq N+1$ in the following manner: The ℓ-order part of the $1/r^{\ell+1}$ piece of g_{00} is the mass ℓ-pole moment; and the ℓ-order, "odd-type parity" part of the $1/r^{\ell+1}$ piece of g_{0j} is the current ℓ-pole moment. More specifically,

$$A^\ell_{00} = [2(2\ell-1)!!/\ell!]\ \mathcal{J}_{A_\ell}\ N_{A_\ell} + (\text{parts of order} < \ell) \tag{4.2.5a}$$

$$A^\ell_{0j} = [-4\ell(2\ell-1)!!/(\ell+1)!]\ \varepsilon_{jpq}\ \mathcal{S}_{pA_{\ell-1}}\ n_q N_{A_{\ell-1}} \tag{4.2.5b}$$

$$+ (\text{"even-type" parts of order } \ell) + (\text{parts of order} < \ell)$$

(4.2.5b)

where $\mathcal{J}_{A_\ell}$ is the ℓ-pole mass moment, and $\mathcal{S}_{A_\ell}$ is the ℓ-pole current moment.

In TVI the self-consistency of this definition is proved; i.e., it is shown that every ACMC-N coordinate system gives the same result for the moments of order $\ell \leq N + 1$.

This definition of multipole moments is far less elegant than definitions given by other researchers.[4] However, the other definitions lead to moments which fail to tie in nicely with radiation theory. My moments must be nonlinear, algebraic functions of the moments defined elsewhere,[4] but I have not attempted to exhibit that relationship explicitly.

The above definition of multipole moments is embodied in the following formula for the metric coefficient of an ACMC-N coordinate system:

$$g_{00} = -1 + \frac{2\mathscr{I}}{r} + \frac{(0\text{-pole})}{r^2} + \frac{1}{r^3}[3\mathscr{I}_{ab}n_a n_b + (1\text{ pole}) + (0\text{ pole})$$

$$+\ldots+ \frac{1}{r^{\ell+1}}\left[\frac{2(2\ell-1)!!}{\ell!}\mathscr{I}_{A_\ell}N_{A_\ell} + (\ell\text{-1 pole})+\ldots+(0\text{ pole})\right]$$

$$+\ldots+ \frac{1}{r^{N+1}}\left[\frac{2(2N-1)!!}{N!}\mathscr{I}_{A_N}N_{A_N} + (N\text{-1 pole})+\ldots+(0\text{ pole})\right]$$

$$+\frac{1}{r^{N+2}}\left[\frac{2(2N+1)!!}{(N+1)!}\mathscr{I}_{A_{N+1}}N_{A_{N+1}} + (\text{poles with } \ell \neq N+1)\right]$$

$$+\left[\text{terms that die out faster than } 1/r^{N+2}\right]. \tag{4.2.6a}$$

$$g_{0j} = -\frac{1}{r^2}\left[2\varepsilon_{jpq}\mathscr{S}_p n_q + \begin{Bmatrix}\text{1 pole part with}\\ \text{parity } \pi = -\end{Bmatrix} + (0\text{ pole})\right]$$

$$-\ldots- \frac{1}{r^{\ell+1}}\left[\frac{4\ell(2\ell-1)!!}{(\ell+1)!}\varepsilon_{jpq}\mathscr{S}_{pA_{\ell-1}}n_q N_{A_{\ell-1}} + \right.$$

$$\left. + \begin{Bmatrix}\ell\text{ pole part with}\\ \text{parity } \pi = (-1)^\ell\end{Bmatrix} + (\ell\text{-1 pole}) + \ldots + (0\text{ pole})\right]$$

$$-\ldots- \frac{1}{r^{N+1}}\left[\frac{4N(2N-1)!!}{(N+1)!}\varepsilon_{jpq}\mathscr{S}_{pA_{N-1}}n_q N_{A_{N-1}} + \right.$$

$$\left. + \begin{Bmatrix}N\text{ pole part with}\\ \text{parity } \pi = (-1)^N\end{Bmatrix} + (N\text{-1 pole}) + \ldots + (0\text{ pole})\right] +$$

$$- \frac{1}{r^{N+2}} \left[\frac{4(N+1)(2N+1)!!}{(N+2)!} \varepsilon_{jpq} \mathcal{S}_{pA_N} n_q N_{A_N} + \begin{Bmatrix} \text{N+1 pole part with} \\ \text{parity } \pi = (-1)^{N+1} \end{Bmatrix} \right.$$

$$\left. + (\text{parts with } \ell \neq N+1) \right]$$

$$+ [\text{terms that die out faster than } 1/r^{N+2}] \quad . \qquad (4.2.6b)$$

$$g_{jk} = \delta_{jk} + \frac{(0 \text{ pole})}{r} + \frac{1}{r^2} [(1 \text{ pole}) + (0 \text{ pole})] + \ldots$$

$$+ \frac{1}{r^{\ell+1}} [(\ell \text{ pole}) + \ldots + (0 \text{ pole})] + \frac{1}{r^{N+1}} [(N \text{ pole})$$

$$+ \ldots + (0 \text{ pole})] + \frac{1}{r^{N+2}} \left[\begin{matrix} \text{any angular} \\ \text{dependence} \end{matrix} \right]$$

$$+ [\text{terms that die out faster than } 1/r^{N+2}]. \qquad (4.2.6c)$$

4.2.3. Example: The Kerr Metric

As an example, consider the Kerr metric (external field of a stationary black hole). In Boyer-Lindquist coordinates, made quasi-Cartesian by defining

$$x_1 = r \sin\theta \cos\phi, \; x_2 = r \sin\theta \sin\phi \, , \; x_3 = r \cos\theta,$$

the nonzero metric coefficients are (cf. page 877 of MTW)

$$g_{00} = - \frac{r^2 + a^2 \cos^2\theta - 2Mr}{r^2 + a^2 \cos^2\theta} = -1 + \frac{2M}{r} - \frac{2Ma^2 \cos^2\theta}{r^3} + 0\left(\frac{1}{r^5}\right) ,$$

$$g_{0j} = - \frac{2\varepsilon_{jpq} J_p n_q}{r^2 + a^2 \cos^2\theta} = - \frac{2\varepsilon_{jpq} J_p n_q}{r^2} + 0\left(\frac{1}{r^4}\right) , \qquad (4.2.7)$$

$$g_{jk} = \delta_{jk} + \frac{2M}{r} n_j n_k + \frac{1}{r^2} [(0 \text{ pole}) + (2 \text{ pole})] + 0\left(\frac{1}{r^3}\right),$$

where

$$\underset{\sim}{J} \equiv Ma(\partial/\partial x_3). \qquad (4.2.8)$$

The presence of a quadrupole $1/r^2$ term in g_{jk} prevents this coordinate system from being ACMC-1; it is only ACMC-0. Hence, from the metric coefficients we can read off only the moments of order 0 and 1:

$\mathcal{I} \equiv$ (mass monopole moment; i.e., total mass-energy) = M;

$\mathcal{I}_a \equiv$ (mass dipole moment) = 0 [guaranteed to vanish because coordinates are mass centered];

$\underset{\sim}{\mathcal{S}} \equiv$ (current dipole moment; i.e., total angular momentum)

$$= Ma(\partial/\partial x_3) \quad . \qquad (4.2.9)$$

In order to deduce the higher-order moments one must perform a coordinate transformation that gets rid of the offending quadrupole term in g_{jk}. Such a transformation is exhibited in TVI.

4.3. The Slow-Motion, Multipole-Moment Formalism For Strong-Field Sources

Turn attention now from stationary sources to the slow-motion, multipole-moment formalism.

Consider a specific (but arbitrary) slow-motion, isolated system. If the system were not changing at all ($\lambda\!\!\!^- = \infty$), the metric everywhere outside the source would be stationary -- i.e., it would be describable by the stationary multipole-moment formalism of § 4.2. We can pass from that stationary metric to the true metric by gradually turning on the time dependence of the system's multipole moments -- i.e., by gradually letting $\lambda\!\!\!^-$ decrease from infinity. Mathematically this is achieved by expanding the metric simultaneously in the two small dimensionless parameters R/r and $R/\lambda\!\!\!^-$, where R is the length scale that characterizes the weak-field, near-zone metric:

$$R \equiv \underset{\ell=1,2,\ldots}{\text{Maximum}} \left\{ \left|\frac{\mathcal{I}_{A_\ell}}{M}\right|^{1/\ell}, \left|\frac{\mathcal{S}_{A_\ell}}{M}\right|^{1/\ell} \right\} \qquad (4.3.1)$$

See TVI for details. For typical (not highly spherical) sources, R will be approximately equal to the size of the source L.

At zero order in $R/\lambda\!\!\!^-$, the R/r expansion of the metric must be formally identical to the general stationary expansion of § 4.2. But the time dependence of that "zero-order solution" will generate corrections or order $(R/\lambdabar)$, $(R/\lambdabar)^2$, ... In TVI it is shown that those corrections are determined uniquely by the (time dependent) multipole moments, plus the demand that the near-zone metric match onto outoing waves. TVI also performs the match onto outgoing waves in the local wave zone, with the following result: The matching requires that the moments $\mathcal{I}_{A_\ell}$, $\mathcal{S}_{A_\ell}$ that appear in the zero-order [i.e., $(R/\lambdabar)^0$] near-zone solutions [eqs. (4.2.6)] be identical to the moments that appear in the radiation field [eq. (3.3.1)] -- except for differences that are typically negligible:

$$|\delta\mathcal{I}_{A_\ell}| \lesssim MR^\ell(M/\lambdabar), \quad |\delta\mathcal{S}_{A_\ell}| \lesssim MR^\ell\,(M/\lambdabar). \qquad (4.3.2a)$$

For the most strongly radiating moment $\mathcal{M}_{A_\ell}$ (usually the mass quadrupole moment $\mathcal{I}_{A_2}$) the disagreement is

$$|\delta\mathcal{M}_{A_\ell}| \lesssim |\mathcal{M}_{A_\ell}|(M/\lambdabar). \qquad (4.3.2b)$$

This result leads to the following slow-motion formalism for calculating the generation of gravitational waves:
(1) Analyze the structure and evolution of the system in any convenient coordinates and by any fairly accurate approximation scheme. (2) From that analysis obtain an approximation to the external gravitational field which, at any instant, satisfies (to some degree of accuracy) the time-independent, vacuum Einstein field equations. (3) By transforming that external field to an ACMC coordinate system, read off its dominant multipole moments (the moments with the largest values of ${}^{(\ell)}\mathcal{I}_{A_\ell}$ or ${}^{(\ell)}\mathcal{S}_{A_\ell}$). (4) Plug those dominant moments into the gravitational-wave formulae of §3.

One attractive approximation scheme for use in steps (1) and (2) is the "instantaneous-gravity" approximation. In this scheme one sets to zero all time derivatives of the metric (but) not of the matter variables) when solving the Einstein field equations. This has the effect of removing all dynamical freedom from the gravitational field and making gravitational interactions within the source instantaneous rather than retarded. One automatically obtains an external gravitational field which satisfies the time-independent vacuum field equations; and, unless one has made a foolish choice of coordinates, the moments which one computes from that external field should contain errors no larger than $L/\lambdabar$. The radiation field computed from those moments will then have fractional errors $L/\lambdabar$ and $M/\lambdabar$. (Here L is the size of the source).

Typically the ℓ-pole mass and current moments will have magnitudes

$$\mathcal{I}_{A_\ell} \sim ML^\ell \ , \ \mathcal{S}_{A_\ell} \sim M(L/\lambda\!\!\!^{-})L^\ell \ . \tag{4.3.3}$$

In this case their contributions to the radiation field [eq.(3.3.1)] will be

$$(h^{TT}_{jk})_{\text{mass } \ell\text{-pole}} \sim (M/r)\ (L/\lambda\!\!\!^{-})^\ell , \tag{4.3.4a}$$

$$(h^{TT}_{jk})_{\text{current } \ell\text{-pole}} \sim (M/r)\ (L/\lambda\!\!\!^{-})^{\ell+1} \quad ; \tag{4.3.4b}$$

Since $L/\lambda\!\!\!^{-} \ll 1$ (slow-motion assumption), mass quadrupole radiation is usually far larger than the other multipoles. The current quadrupole field and mass octupole field are normally a factor $L/\lambda\!\!\!^{-}$ smaller. It is this fact which allows one to compute only the lowest few multipole moments when applying the slow-motion formalism.

In special cases (e.g., torsional oscillations of a neutron star) the mass quadrupole moment will vanish, or will be far smaller than its normal value. Then the current quadrupole or mass octupole radiation may dominate -- unless they, too, are abnormally small, allowing higher moments to make themselves felt.

Notice that, if one is interested in the linear momentum radiated by the source (eq. 3.4.7), one must compute not just the lowest significant multipole moment, but also the moments "adjacent" to it. Linear momentum is carried off only through the interference of adjacent multipole fields with each other.

4.3.1. A Sample Application

As a typical application of the slow-motion, multipole-moment formalism, consider a slowly rotating neutron star which is not quite axially symmetric. Ipser (1971) has formulated a general relativistic analysis of the interior of such a star, using the Regge-Wheeler (1957) formalism for small, strong-field deviations from spherical symmetry. Ipser's analysis shows how internal stresses, supported by the crystal structure of the star's mantle, maintain the star's deformation. It also gives formulas for the star's near-zone multipole moments in terms of the star's internal structure. By inserting those near-zone multipole moments into the slow-motion wave-generation formalism, one obtains the gravitational-wave field $\bar{h}^{TT}_{jk}$ produced by the star's rotation, and also the energy and angular momentum radiated. (Ipser computed the gravitational-wave field from his near-zone multipole moments by brute force, and discovered the then surprising result that it had the same form as in weak-field, slow-motion theory. That discovery was the original motivation for my constructing the above formalism.)

4.4 Weak-Field Limit of the Slow-Motion Formalism

For a slow-motion source ($L/\lambda\!\!\!^{-} << 1$) with weak internal fields ($\varepsilon << 1$) and weak internal stresses ($S^2 << 1$), one can use Newtonian theory to analyze the interior region; cf. §2.6. Using Newtonian theory, one can express the multipole moments

$$\mathcal{I}_{A_\ell} \text{ and } \mathcal{S}_{A_\ell}$$

in terms of volume integrals over the source [see, eg., TV for detailed proof].

The multipole volume integrals must be performed in a mass-centered Cartesian coordinate system--i.e., in Cartesian coordinates with

$$\mathcal{I}_j(t) \equiv \int \rho(\underset{\sim}{x},t)\, x_j\, d^3x = 0. \tag{4.4.1}$$

[Note that such a coordinate system is automatically the rest frame of the source, since the time-derivative of equation (4.4.1) can be put in the form

$$\begin{aligned} 0 = d\,\mathcal{I}_j(t)/dt &= \int [\partial\rho(\underset{\sim}{x},t)/\partial t]\, x_j\, d^3x \\ &= \int \rho(\underset{\sim}{x},t)\, v_j\, d^3x = (\text{momentum of source}). \end{aligned} \tag{4.4.2}$$

The third equality follows from the equation of mass conservation (2.6.3a) and an integration by parts.]

In a mass-centered Cartesian coordinate system, the volume integrals for $\mathcal{I}_{A_\ell}$ and $\mathcal{S}_{A_\ell}$ are

$$\mathcal{I}_{A_\ell} = \text{Symmetric trace-free part of } I_{A_\ell}, \tag{4.4.3a}$$

$$I_{A_\ell} = \begin{pmatrix}\ell\text{'th moment of} \\ \text{mass distribution}\end{pmatrix} = \int \rho x_{a_1} \cdots x_{a_\ell}\, d^3x; \tag{4.4.3b}$$

$$\mathcal{S}_{A_\ell} = \text{Symmetric trace-free part of } S_{A_\ell}, \tag{4.4.4b}$$

$$\begin{aligned} S_{A_\ell} &= \begin{pmatrix}\ell\text{-1 moment of angular} \\ \text{momentum distribution}\end{pmatrix} \\ &= \int (\varepsilon_{a_1 jk} x_j \rho\, v_k)\, x_{a_2} \cdots x_{a_\ell}\, d^3x. \end{aligned} \tag{4.4.4b}$$

The "symmetric, trace-free part" can be computed from equations (3.2.1) and (3.2.2).

The volume integrals (4.4.3) and (4.4.4) are not precisely equal to the exact near-zone multipole moments of the source, of course. They contain errors of post-Newtonian order (cf. eq. 2.6.1):

$$(\delta \mathcal{I}_{A_\ell}) \text{ due to weak-field assumption}$$

$$\sim ML^\ell[\varepsilon+(L/\lambda\!\!\!^-)^2+s^2], \tag{4.4.5a}$$

$$(\delta \mathcal{S}_{A_\ell}) \text{ due to weak-field assumption}$$

$$\sim M(L/\lambda\!\!\!^-)L^\ell[\varepsilon+(L/\lambda\!\!\!^-)^2+s^2]. \tag{4.4.5b}$$

These errors are in addition to the amounts (eqs. 4.3.2) by which the wave-zone multipole moments fail to agree with the near-zone multipole moments.

The weak-field limit of the slow-motion wave-generation formalism can be summarized by the following set of rules: (1) Use the Newtonian theory of gravity to analyze the structure and evolution of the source. (2) Calculate the lowest few multipole moments by evaluating expressions (4.4.3) and (4.4.4) in the mass-centered, Cartesian coordinates. (3) Insert those multipole moments into the local-wave-zone formulas of §3. Those formulas will then describe the lowest few multipoles of the radiation field. (4) Check that the radiation field is larger than the errors inherent (a) in the matching of near zone onto local wave zone (eqs. 4.3.2. with $R \simeq L$), and (b) in the weak-field assumption

$$(\delta h_{jk}^{TT})_{\text{mass } \ell\text{-pole, errors}} \sim (M/r)\ (L/\lambda\!\!\!^-)^\ell[\varepsilon+(L/\lambda\!\!\!^-)^2+s^2], \tag{4.4.6a}$$

$$(\delta h_{jk}^{TT})_{\text{current } \ell\text{-pole, errors}} \sim (M/r)(L/\lambda\!\!\!^-)^{\ell+1}[\varepsilon+(L/\lambda\!\!\!^-)^2+s^2]. \tag{4.4.6b}$$

This weak-field, slow-motion formalism dates back to Einstein (1918), for the mass quadrupole part. The mass octupole and current quadrupole parts were first derived (so far as I know) by Papapetrou (1962, 1971); and the full formalism (in different notations from this) was first derived by Mathews (1962); see also Campbell and Morgan (1971). The earliest versions of the formalism assumed gravity so weak that one had to use Linearized Theory rather than Newtonian Theory in analyzing the source ("no self-gravity"). However, it was soon realized that a modest amount of self-gravity causes no problems in the formalism.

4.5 Post-Newtonian Multipole Formalism

We now return to the Epstein-Wagoner (1975) Post-Newtonian Wave-Generation Formalism, which was described qualitatively in §2.7. In this formalism Epstein and Wagoner find it most convenient to analyze the structure and motion of their source using a different Post-Newtonian gauge from that of Chandrasekhar (1965). Like Chandrasekhar, they describe the matter of their source by perfect-fluid thermodynamic variables measured in its local rest frame

$$\rho_o = \text{mass density},\quad P = \text{pressure},\quad \Pi = \text{specific internal energy}; \tag{4.5.1a}$$

and by the coordinate velocity of the fluid

$$v_j \equiv \frac{dx^j}{dt}. \tag{4.5.1b}$$

The internal gravity of the source they describe by the potentials U, V_j, Ψ, and χ which satisfy

$$U_{,jj} = -4\pi\rho_o,\quad V_{j,kk} = -4\pi\rho_o v_j,$$

$$\Psi_{,jj} = -4\pi\rho_o(v^2+U+\tfrac{1}{2}\Pi + \tfrac{3}{2}P/\rho_o),\quad \chi_{,jj} = -2U, \tag{4.5.2}$$

and which are related to the post-Newtonian metric by

$$g_{00} = -1+2U-2U^2+4\Psi-\chi_{,00}+O(\varepsilon^6) \tag{4.5.3a}$$

$$g_{0j} = -4V_j+O(\varepsilon^5) \tag{4.5.3b}$$

$$g_{jk} = 2U\delta_{jk}+O(\varepsilon^4) \tag{4.5.3c}$$

[In the Post-Newtonian formalism one assumes $\varepsilon \sim S^2 \sim (L/\lambda\!\!\!^-)^2$.] The equations governing the evolution of the source in the Epstein-Wagoner gauge are the equation of mass conservation

$$[\rho_o(1+\tfrac{1}{2}\underset{\sim}{v}^2+3U)]_{,0}+[\rho_o(1+\tfrac{1}{2}\underset{\sim}{v}^2+3U)v_j]_{,j}=0, \tag{4.5.4a}$$

the equation of state

$$P=P(\rho_o,\Pi), \tag{4.5.4b}$$

the adiabatic equation of energy conservation

$$\rho_o\, d\Pi/dt + Pv_{j,j}=0,\quad d/dt \equiv \partial/\partial t + v_j\, \partial/\partial x_j, \tag{4.5.4c}$$

and the Euler equations of motion

$$\rho_o dv_i/dt - \rho_o U_{,i} + P_{,i} - (\Pi + \underset{\sim}{v}^2 + 4U + P/\rho_o)P_{,i}$$

$$+ (3\rho_o U_{,0} + P_{,0} + 4\rho_o v_j U_{,j})v_i + (4U - \underset{\sim}{v}^2)\rho_o U_{,i} - 4\rho_o V_{i,0}$$

$$+ 4\rho_o v_k (V_{k,i} - V_{i,k}) + \rho_o(\tfrac{1}{2}\chi_{,00i} - 2\Psi_{,i}) = 0 .$$

After one has computed the structure and evolution of the source using these equations, one can then compute the multipole moments which govern the radiation field. For typical sources -- to which Epstein and Wagoner restrict their attention -- the multipole components of the radiation field have the magnitudes (4.3.4). The mass quadrupole dominates with magnitude $(M/r)(L/\lambda\!\!\!^{-})^2$ and the post-Newtonian formalism is able to compute it with a fractional error $\sim \varepsilon^2 \sim (L/\lambda\!\!\!^{-})^4$. If one wishes to compute the other multipole contributions to similar absolute accuracy, then one must have fractional errors no larger than the following in the various multipole moments (cf. eq. 4.3.4):

$$\mathcal{I}_{A_2} : (L/\lambda\!\!\!^{-})^4, \quad \mathcal{I}_{A_3} : (L/\lambda\!\!\!^{-})^3, \quad \mathcal{I}_{A_4} : (L/\lambda\!\!\!^{-})^2, \quad \mathcal{I}_{A_5} : (L/\lambda\!\!\!^{-}),$$

$$\mathcal{S}_{A_2} : (L/\lambda\!\!\!^{-})^3 \quad \mathcal{S}_{A_3} : (L/\lambda\!\!\!^{-})^2 \quad \mathcal{S}_{A_4} : (L/\lambda\!\!\!^{-}). \qquad (4.5.5)$$

Up to this accuracy the multipole moments can be expressed in the following form:

$$(\mathcal{I}_{jk}, \mathcal{I}_{ijk}, \mathcal{S}_{jk}) = \text{Symmetric trace-free parts of } (I_{jk}, I_{ijk}, S_{jk}),$$

$$I_{jk} = \int\left[\tau_{00}x_j x_k + \frac{11}{21}\tau_{jk}r^2 + \frac{4}{21}\tau_{pp}x_j x_k - \frac{4}{7}x_p\tau_{pj}x_k\right]d^3x, \qquad (4.5.6)$$

$$I_{ijk} = \int\left[\tau_{00}x_i x_j x_k + x_i\tau_{jk}r^2 + \frac{1}{3}\tau_{pp}x_i x_j x_k - x_p\tau_{pi}x_j x_k\right]d^3x,$$

$$S_{ij} = \int\left[x_i\varepsilon_{jpq}x_p\tau^0{}_q - \frac{3}{28}\varepsilon_{ipq}x_p r^2\partial_t\tau_{qj} + \frac{1}{28}x_i\varepsilon_{jpq}x_p\partial_t\tau_{qs}x_s\right]d^3x,$$

$(\mathcal{S}_{ijk}, \mathcal{I}_{ijk\ell})$ equal the Newtonian expressions (4.4.3), (4.4.4).

Here, the "effective energy density". τ_{00}, "effective momentum density" $\tau^0{}_j$, and "effective stress" τ_{jk} are

$$\tau_{00} = \rho_o\,(1 + \Pi + \underset{\sim}{v}^2 + 4U) - \frac{3}{8\pi}U_{,j}U_{,j} \qquad (4.5.7a)$$

$$\tau^0{}_j = \rho_o(1+\Pi+\underset{\sim}{v}^2+4U)v_j+Pv_j + \frac{3}{4\pi}U_{,0}U_{,j} + \frac{1}{2\pi}U_{,k}V_{k,j}$$

$$- \frac{1}{2\pi}\dot{V}_k U_{,kj} - 2\rho_o V_j , \qquad (4.5.7b)$$

$$\tau_{jk} = \rho_o v_j v_k - \frac{1}{4\pi} U_{,j}U_{,k} - \frac{1}{2\pi} UU_{,jk}$$

$$+ \delta_{jk}(P + \frac{3}{8\pi} U_{,i}U_{,i} - 2\rho_o U). \qquad (4.5.7c)$$

Epstein and Wagoner use a different notation than ours for the multipole moments. The above formulas for the moments (eqs. 4.5.6) are derived in TV.

5. PERTURBATION FORMALISMS

Thus far I have described two large classes of wave-generation formalisms: weak-field formalisms (§2) and slow-motion formalisms (§4). Now I turn attention to a third large class: Perturbation formalisms.

The fundamental assumption underlying all perturbation formalisms is this: that the entire wave-generation region of space-time can be treated as a small perturbation which radiates, superimposed on a nonradiative but strongly curved "background" Examples are: (1) Small-amplitude pulsations of fully relativistic stars. Here the background is an unperturbed, equilibrium stellar model; and the perturbation is the pulsation. (2) Slow rotation of a slightly nonspherical neutron star (pulsar). Here the background is a nonrotating, spherical star; and the perturbation is both the deformation and the rotation. (3) Motion of a small object in the gravitational field of a black hole. Here the background is the Kerr metric of the black hole; and the small perturbation is the stress-energy tensor of the object, plus the gravitational field it produces.

There are a variety of different perturbation formalisms, each designed to handle a specific type of problem. Some applications make use of several formalisms combined together--and many applications combine a perturbation formalism for the wave-generation region with a multipole analysis of the radiation field (i.e., with a variant of the formalism described in §3).

In a recent review article (§II.C of Thorne 1977c) I have described most of the perturbation formalisms with which I am familiar. Rather than repeat that material here, I simply refer the reader to it.

6. WAVE GENERATION BY SOURCES WITH STRONG INTERNAL FIELDS AND FAST, LARGE-AMPLITUDE MOTIONS

The strongest sources of gravitational waves in the Universe should be sources with strong internal fields ($\varepsilon \sim 1$) and fast ($v \sim L/\lambdabar \sim 1$), large-amplitude internal motions. Examples are the highly nonspherical collapse of a star to form a black hole (or naked singularity, or whatever it does form), and a collision between two black holes. Unfortunately, such sources have eluded all efforts at analytic analysis.[5] There exists no formalism today by which one can calculate the waves they generate. Obviously, an accurate analysis of such systems is the most important and most difficult task lying ahead of us in the theory of gravitational-wave generation.

Fortunately great progress has been made on this task recently by Bryce DeWitt, Larry Smarr, Kenneth Eppley, and others (see Smarr 1977 for a review). Abandoning all hope of a truly analytic analysis, they turn to massive electronic computers as their key tool. Their method is elegant numerical solution of the full, nonlinear Einstein field equations. By now they have encountered and surmounted a number of serious numerical problems. The resulting numerical methods are nearly good enough to give reliable results for strong-field, high-speed, large-amplitude sources--and we can expect true reliability within another year or two. This, when it is achieved, will be very useful in planning gravitational-wave-detection efforts.

[5]An exception is the special situation of a collision between two black holes with relative velocity very nearly the speed of light [$\gamma=(1-v^2)^{-1/2} >>1$], for which D'Eath (1977) has formulated a remarkable "colliding-plane-wave" approximation that yields the dominant features of this radiation.

7. PROPAGATION OF WAVES TO EARTH

7.1. The Geometric-Optics Formalism

Once the gravitational-wave field h_{jk}^{TT} is known in the local wave zone, one can then propagate it out through the surrounding Universe to Earth, using the vacuum Einstein field equations. In nearly all situations one can use the geometric-optics approximation to the Einstein equations (e.g. Exercise 35.15 of MTW). In this section I describe the geometric optics formalism in a language different from, but equivalent to MTW.

This formalism, which is valid for $r >> r_I$ (wave zone), describes spacetime by a background metric $g_{\mu\nu}^{(B)}$ through which the waves propagate. The waves are described by a gravitational-wave field $\psi_{\mu\nu}$ which reduces to

$$\psi_{jk} = h_{jk}^{TT}, \quad \psi_{j0} = 0, \quad \psi_{00} = 0 \quad \text{in local wave zone and in asymptotic rest frame of source.} \tag{7.1.1}$$

In an appropriate coordinate system (gauge) the full metric of spacetime is

$$g_{\mu\nu} = g_{\mu\nu}^{(B)} + \psi_{\mu\nu} + 0\,(\psi^2). \tag{7.1.2}$$

The waves are distinguished from the background by the very small length scale $\lambda\!\!\!^-$ on which they vary

$$\lambda\!\!\!^- \equiv \begin{pmatrix}\text{length scale of} \\ \text{variations in } \psi_{\mu\nu}\end{pmatrix} << \mathcal{R}_B \equiv \begin{pmatrix}\text{radius of curvature} \\ \text{of background spacetime}\end{pmatrix}. \tag{7.1.3}$$

The geometric optics formalism remains valid so long as the propagating waves do not encounter regions of extremely strong curvature--i.e. regions where $\mathcal{R}_B \lesssim \lambda\!\!\!^-$.

In applying the geometric optics formalism to a specific problem, one proceeds as follows: (i) In the local wave zone and in the asymptotic rest frame of the source one describes the background metric by the line element

$$ds^2 = -dt^2 + dr^2 + r^2(d\theta^2 + \sin^2\theta d\phi^2) + 0(M/r)dx^\alpha dx^\beta, \tag{7.1.4}$$

with the source located at the origin. (ii) One constructs null geodesics of $g_{\mu\nu}^{(B)}$ extending radially away from the source. Near the source each geodesic has the form

$$t - r = \tau_e = \text{const.}, \quad \theta = \text{const.}, \quad \phi = \text{const.} \quad \text{in local wave zone.} \tag{7.1.5}$$

These geodesics are called "rays." The gravitational waves propagate along them. (iii) Each ray is labeled by the proper time ("emission time" or "retarded time") τ_e at which it intersects the source; and by this ray labeling τ_e becomes a scalar field extending through all spacetime. (iv) Each ray has an affine parameter ζ, a world line $\mathcal{P}(\zeta)$ with coordinates $x^\alpha(\zeta)$, and a tangent vector ("propagation vector")

$$\vec{k} = d/d\zeta, \qquad k^\alpha = dx^\alpha/d\zeta. \tag{7.1.6}$$

Of course, since $\mathcal{P}(\zeta)$ is a null geodesic, $\vec{k}$ must satisfy

$$\vec{k}^2 \equiv k_\alpha k^\alpha = 0, \qquad k_{\alpha|\beta}k^\beta = 0. \tag{7.1.7}$$

Here and below a slash denotes covariant derivative with respect to the background metric. (v) One normalizes the affine parameter of each ray so that near the source

$$\zeta = r = \left\{\begin{array}{l}\text{proper distance from source}\\ \text{as measured in asymptotic}\\ \text{rest frame; eq.(7.1.4)}\end{array}\right\} \quad \text{in local wave zone.} \tag{7.1.8a}$$

As a result

$$k^0 = 1, \qquad \underset{\sim}{k} = \underset{\sim}{n} = \left\{\begin{array}{l}\text{unit radial vector}\\ \text{pointing away from}\\ \text{source}\end{array}\right\} \quad \begin{array}{l}\text{in source's asymptotic}\\ \text{rest frame and local}\\ \text{wave zone.}\end{array} \tag{7.1.8b}$$

(vi) From the above definitions and constructions one can show that throughout spacetime the gradient of the retarded time is equal to the propagation vector, except for sign

$$\vec{k} = -\vec{\nabla}\tau_e; \qquad k_\alpha = -\tau_{e,\alpha}. \tag{7.1.9}$$

(vii) In the local wave zone one imposes the starting conditions (7.1.1) on the gravitational-wave field $\psi_{\alpha\beta}$. Note that because h^{TT}_{jk} is transverse and traceless, $\psi_{\alpha\beta}$ is initially trace-free and orthogonal to the propagation vector

$$\Psi \equiv \Psi^\alpha_\alpha \equiv \Psi_{\alpha\beta}\, g^{\alpha\beta}_{(B)} = 0, \; \Psi_{\alpha\beta}k^\beta = 0. \tag{7.1.10}$$

These properties are preserved as the waves propagate (cf. eq. 7.1.12 below). (viii) Initially, and after it has propagated, $\Psi_{\alpha\beta}$ is a rapidly varying function of retarded time τ_e and in addition is a slowly varying function of location along surfaces of constant τ_e:

$$\psi_{\alpha\beta} = \psi_{\alpha\beta}\,(\tau_e; x^0, x^1, x^2, x^3), \tag{7.1.11a}$$

$$\dot{\psi}_{\alpha\beta} \equiv \frac{\partial\psi_{\alpha\beta}}{\partial\tau_e} = 0\left[\frac{\psi_{\alpha\beta}}{\lambda\!\!\!^{-}}\right], \qquad \ddot{\psi}_{\alpha\beta} \equiv \frac{\partial^2\psi_{\alpha\beta}}{\partial\tau_e^2} = 0\left[\frac{\psi_{\alpha\beta}}{\lambda\!\!\!^{-2}}\right] \tag{7.1.11b}$$

$$\left(\frac{\partial\psi_{\alpha\beta}}{\partial x^\mu}\right)_{\tau_e} = 0\left[\frac{\psi_{\alpha\beta}}{\zeta} + \frac{\psi_{\alpha\beta}}{\mathcal{R}_B}\right], \tag{7.1.11c}$$

$$\psi_{\alpha\beta|\gamma} = \dot{\psi}_{\alpha\beta}k_\gamma + 0\left[\frac{\psi_{\alpha\beta}}{\zeta} + \frac{\psi_{\alpha\beta}}{\mathcal{R}_B}\right]. \tag{7.1.11d}$$

Here the slash ($\psi_{\alpha\beta|\gamma}$) denotes covariant derivative with respect to the background metric, and 0[X] means "of order X." (ix) $\psi_{\alpha\beta}$ propagates along the rays in accordance with the propagation equation

$$\psi_{\alpha\beta|\mu}k^\mu = -\frac{1}{2}k^\mu{}_{|\mu}\psi_{\alpha\beta}. \tag{7.1.12}$$

(x) One uses equation (7.1.12) to propagate the initial, local-wave-zone field out through the Universe to Earth.

Because, in an appropriate gauge, $\psi_{\alpha\beta}$ is the metric perturbation associated with the waves (eq. 7.1.2), one can use the usual formulas (Chapters 35 and 37 of MTW) to calculate from $\psi_{\alpha\beta}$ whatever properties of the waves one wishes. For example, one can use equations (35.62a,e)of MTW to compute the contribution of the waves to the Riemann curvature tensor of spacetime. By virtue of equations (7.1.11) that contribution is

$$R^{(GW)}_{\alpha\beta\gamma\delta} = \frac{1}{2}(\ddot{\psi}_{\alpha\delta}k_\beta k_\gamma + \ddot{\psi}_{\beta\gamma}k_\alpha k_\delta - \ddot{\psi}_{\beta\delta}k_\alpha k_\gamma - \ddot{\psi}_{\alpha\gamma}k_\beta k_\delta) \tag{7.1.13}$$

$$+ \text{ fractional errors of } 0[\lambda\!\!\!^{-}/\zeta + \lambda\!\!\!^{-}/\mathcal{R}_B].$$

Similarly, one can use equation (35.70) of MTW to compute the Isaacson (1968) stress-energy tensor associated with the waves. It reduces to

$$T^{(GW)}_{\alpha\beta} = \frac{1}{32\pi} \langle \dot{\psi}_{\mu\nu}\dot{\psi}^{\mu\nu} \rangle k_\alpha k_\beta, \tag{7.1.14}$$

where < > means "average over several wavelengths of the waves."

The gravitational-wave field is not fully measurable. Only that part which contributes to the Riemann curvature tensor is measurable; the remainder can be changed at will by gauge transformations (infinitesimal coordinate transformations). The form of the generic gauge transformation which preserves all the above equations is

$$\psi_{\mu\nu}^{NEW} = \psi_{\mu\nu}^{OLD} + \Phi_\mu k_\nu + \Phi_\nu k_\mu \,, \tag{7.1.15a}$$

where ϕ_μ is a vector field which has the same variability properties as $\psi_{\alpha\beta}$ [eqs. 7.1.11], which is orthogonal to the propagation vector

$$\Phi_\mu k^\mu = 0, \tag{7.1.15b}$$

and which satisfies the propagation equation

$$\Phi_{\mu|\alpha} k^\alpha = -\tfrac{1}{2} k^\alpha{}_{|\alpha} \phi_\mu \,. \tag{7.1.15c}$$

An arbitrary observer with 4-velocity $\vec{u}$, at an arbitrary event in spacetime, may find it convenient to make $\psi_{\mu\nu}$ "spatial, transverse, and traceless" in his own rest frame. He can accomplish this by a gauge transformation of the above form with

$$\Phi_\mu = -\frac{u^\alpha \psi_{\alpha\mu}}{u^\gamma k_\gamma} + \frac{u^\alpha \psi_{\alpha\beta} u^\beta}{2(u^\gamma k_\gamma)^2} \, k_\mu \,. \tag{7.1.16a}$$

The resulting ("NEW"; "Transverse Traceless") gravitational-wave field $\psi_{\mu\nu}^{TT}$ is related to the original ("OLD") one $\psi_{\mu\nu}$ by

$$\psi_{xx}^{TT} = \psi_{xx} = -\psi_{yy} = -\psi_{yy}^{TT}; \quad \psi_{xy}^{TT} = \psi_{xy} = \psi_{yx}^{TT}\,; \quad \text{all other } \psi_{\alpha\beta}^{TT} = 0$$

in local rest frame of observer, with Minkowskii coordinates so oriented that $\vec{k} = k^0(\vec{e}_0 + \vec{e}_z)$. (7.1.16b)

In other words, this gauge transformation simply throws away all parts of $\psi_{\alpha\beta}$ except those that are purely spatial and are transverse to the propagation direction; and in the process it preserves the tracelessness of $\psi_{\alpha\beta}$.

7.2 Example of Propagation

As an example of the above formalism, consider a gravitational wave emitted by the gravitational collapse of a 10^6 solar-mass star at a time when the universe was only a few million years old.

Idealize the background metric, through which the waves propagate toward Earth, as a perfectly smooth Friedmann metric (MTW chapter 29)

$$ds^2 = g_{\mu\nu}^{(B)} dx^\mu dx^\nu = a^2[-d\eta^2 + d\chi^2 + \Sigma^2(d\theta^2 + \sin^2\theta d\phi^2)] \quad (7.2.1a)$$

$$a = a(\eta), \ \Sigma = \begin{cases} \sin\chi & \text{if universe is "closed"} \\ \chi & \text{if universe is "flat"} \\ \sinh\chi & \text{if universe is "open"} \end{cases} ; \quad (7.2.1b)$$

and at each event in the universe introduce the orthonormal frame ("local proper rest frame of universe")

$$\vec{e}_{\hat\eta} = \frac{1}{a}\frac{\partial}{\partial\eta}, \ \vec{e}_{\hat\chi} = \frac{1}{a}\frac{\partial}{\partial\chi}, \ \vec{e}_{\hat\theta} = \frac{1}{a\Sigma}\frac{\partial}{\partial\theta}, \ e_{\hat\phi} = \frac{1}{a\Sigma\sin\theta}\frac{\partial}{\partial\phi}. \quad (7.2.1c)$$

Place the supermassive star (source) at the origin of the Friedmann spatial coordinate system, $\chi_s = 0$; and let it emit its gravitational waves during an interval of time $\eta_s \leq \eta \leq \eta_s + \Delta\eta$. The duration of the burst [$\Delta\tau_e = a(\eta_s)\Delta\eta$ = (a few seconds)] will be very short compared to the age of the universe at the time of emission; and hence "a" will change negligibly during the emission

$$a_s \equiv a(\eta_s) >>> (da/d\eta)_s \ \Delta\eta. \quad (7.2.2)$$

The rays, along which the waves propagate, are null geodesics emanating from the star's world line ($\chi_s = 0$, $\eta_s \leq \eta \leq \eta_s + \Delta\eta$). By solving the geodesic equation in the Friedmann metric one obtains the following equations for the ray originating at ($\chi = 0$, $\eta = \eta_s + \tau_e/a_s$) and propagating in the (θ_e, ϕ_e) direction:

$$\chi = \eta - (\eta_s + \tau_e/a_s), \quad \theta = \theta_e, \ \phi = \phi_e, \quad (7.2.3a)$$

$$k^\eta = k^\chi = \frac{d\eta}{d\zeta} = \frac{d\chi}{d\zeta} = \frac{a_s}{a^2}, \quad k^\theta = k^\phi = 0. \quad (7.2.3b)$$

Note that the "physical components" of the propagation vector (components in the local proper rest frame of the universe [7.2.1c]) are

$$k^{\hat\theta} = k^{\hat\phi} = 0, \quad k^{\hat\eta} = k^{\hat\chi} = (a_s/a) \ \text{at general event}$$

$$= 1 \ \text{at source.} \quad (7.2.4)$$

The source's retarded time, expressed as a function of Friedmann coordinates, is

$$\tau_e = a_s\,(\eta - \eta_s - \chi); \tag{7.2.5}$$

cf. equation (7.2.3a). The gradient of τ_e is the negative of the propagation vector, as required by equation (7.1.9).

In the asymptotic rest frame of the source ($\chi \ll 1$; $\eta_s \lesssim \eta \lesssim \eta_s + \Delta\eta$) the Minkowskii radial and time coordinates are related to Friedmann coordinates and to retarded time by

$$r = a_s\chi, \quad t = a_s(\eta-\eta_s); \quad \tau_e = t-r; \tag{7.2.6a}$$

and the line element is

$$ds^2 = -\,dt^2 + dr^2 + r^2\,(d\theta^2 + \sin^2\theta\, d\phi^2) \tag{7.2.6b}$$

$$+\,O(M/r)\,dx^\alpha\,dx^\beta \leftarrow \text{contributions from gravity of source}$$

$$+\,O(r^2/a^2 + t\,d\,\ln a/dt)\,dx^\alpha dx^\beta \leftarrow \text{cosmological corrections.}$$

The gravitational waves, expressed in a gauge that is "TT" with respect to this asymptotic rest frame, have as their only nonzero components

$$\left.\begin{aligned} \psi_{\hat\theta\hat\theta} &= -\psi_{\hat\phi\hat\phi} \equiv \frac{1}{r}\,A_+(\tau_e,\theta,\phi) \\ \psi_{\hat\theta\hat\phi} &= \psi_{\hat\phi\hat\theta} \equiv \frac{1}{r}\,A_\times(\tau_e,\theta,\phi) \end{aligned}\right\} \text{in local wave zone.} \tag{7.2.7}$$

The quantities A_+ and $A_\times$ are amplitudes for the two orthogonal polarization states "+" and "x". They are rapidly varying functions of retarded time $\tau_e = t-r$, and slowly varying functions of θ and ϕ :

$$\dot A \equiv \frac{\partial A}{\partial \tau_e} \sim \frac{A}{\lambda\!\!\!^{-}}, \quad \frac{1}{r}\frac{\partial A}{\partial\theta} \backsim \frac{A}{r}, \quad \frac{1}{r\sin\theta}\frac{\partial A}{\partial\phi} \backsim \frac{A}{r} \tag{7.2.8}$$

[cf. eq. (7.1.11) and note that in the asymptotic rest frame ζ = (affine parameter) = r].

The gravitational-wave field (7.2.7) propagates out into the Friedmann universe by means of the geometric-optics propagation equation (7.1.12). By solving this equation one discovers that

$$a\Sigma\psi_{\hat{\alpha}\hat{\beta}} = \text{constant along each ray,} \tag{7.2.9}$$

where Σ is the cosmological circumference function defined in equation (7.2.1b). Consequently (cf. eqs. [7.2.1b], [7.2.6a], [7.2.7], and [7.2.5]) the gravitational-wave field at an arbitrary event $(\eta, \chi, \theta, \phi)$ in spacetime is

$$\psi_{\hat{\theta}\hat{\theta}} = -\psi_{\hat{\phi}\hat{\phi}} = \frac{1}{a\Sigma} \quad A_+[a_s(\eta-\eta_s-\chi), \theta, \phi] \tag{7.2.10a}$$

$$\psi_{\hat{\theta}\hat{\phi}} = \psi_{\hat{\phi}\hat{\theta}} = \frac{1}{a\Sigma} \quad A_\times[a_s(\eta-\eta_s-\chi), \theta, \phi]. \tag{7.2.10b}$$

Because the source is far from Earth, when these waves reach Earth they look plane waves. An observer on Earth can interpret them in terms of a local Minkowskii coordinate system with basis vectors

$$\frac{\partial}{\partial t} \equiv \vec{e}_0 = \vec{e}_{\hat{\eta}}, \quad \frac{\partial}{\partial z} \equiv \vec{e}_z = e_{\hat{\chi}}, \quad \frac{\partial}{\partial x} \equiv \vec{e}_x = \vec{e}_{\hat{\theta}}, \quad \frac{\partial}{\partial y} \equiv \vec{e}_y = \vec{e}_{\hat{\phi}}. \tag{7.2.11a}$$

If the location of Earth today is $(\eta_o, \chi_o, \theta_o, \phi_o)$, then the observer's Minkowskii coordinates and the global Friedmann coordinates near Earth are related by

$$t = a_o(\eta-\eta_o), \quad z = a_o(\chi-\chi_o), \quad x = a_o\Sigma_o(\theta-\theta_o), \quad y = a_o\Sigma_o\sin\theta_o(\phi-\phi_o). \tag{7.2.11b}$$

The observer on Earth describes the universe's cosmological structure in terms of a Hubble expansion rate H_o and a deceleration parameter q_o; and he describes the source of the waves as having a cosmological redshift Z_s, which is related to the expansion factor by

$$1 + Z_s = a_o/a_s. \tag{7.2.12}$$

The equations of Friedmann cosmology permit one to express $a_o\Sigma_o \equiv$ (circumference of a circle that passes through Earth and is centered on the source at redshift Z_s)/2π in terms of H_o, q_o, and Z_s:

$$a_o \Sigma_o \equiv R = \frac{H_o^{-1}}{q_o^2(1+Z_s)} \left[1-q_o+q_o Z_s-(1-q_o)(2q_o Z_s+1)^{1/2}\right]; \tag{7.2.13}$$

cf. eq. (29.33) of MTW. By combining equations (7.2.10) -(7.2.13) one obtains the following expressions for the gravitational-wave field that sweeps past Earth:

$$\psi_{xx} = -\psi_{yy} = \frac{1}{R} A_+\left[\frac{t-z}{1+Z_s} + \text{const.},\ \theta_o, \phi_o\right], \tag{7.2.14a}$$

$$\psi_{xy} = \psi_{yx} = \frac{1}{R} A_x\left[\frac{t-z}{1+Z_s} + \text{const.}, \theta_o, \phi_o\right], \tag{7.2.14b}$$

all other $\psi_{\mu\nu}$ vanish.

Notice that the time dependence, $A[(t-z)/(1+Z_s)+\text{const.}, \theta_o, \phi_o]$, of the waves as they sweep past Earth is identical to the time dependence of the emitted waves as measured by the source, $A[\tau_e, \theta_o, \phi_o]$, except for a redshift of $1+Z_s$ -- the same redshift as one sees in electromagnetic spectral lines. Notice also that for very large redshifts, $Z_s >> 1$, the amplitude of the waves is independent of redshift:

$$\psi \sim \frac{A}{R} = \frac{A}{H_o q_o} \quad \text{for } Z_s >> 1 \text{ and } Z_s >> 1/q_o. \tag{7.2.15}$$

8. CONCLUDING REMARKS

Although it may appear from these lectures that the theory of gravitational-wave generation is a highly sophisticated and complex subject, one should not let this blind one to its gross inadequacies.

The strongest sources of gravitational waves in the universe-- and the most promising sources for ultimate detection -- are those with strong internal gravity and fast large-amplitude internal motions. For them the only reliable technique of analysis is massive computer calculations (§6). All the fancy analytic tools of these lectures are helpless in the face of such sources!

I am indebted to Joseph Weber for his patience, and to Alessandra Exposito and Jim Isenberg for valuable assistance in the preparation of the typed manuscript.

REFERENCES

Bonnor, W.B., and Rotenberg, M.A. 1966, Proc. Roy. Soc. A, 289, 247.

Campbell, W.B. and Morgan, T. 1971, Physica, 53, 264.

Chandrasekhar, S. 1964, Astrophys. J., 140, 417.

Chandrasekhar, S. 1965, Astrophys. J., 142, 1488.

Clarke, C.J.S. and Sciama, D.W. 1971, Gen. Rel. Grav., 2, 331.

Crowley, R.J. and Thorne, K.S., 1977, Astrophys. J., in press; also available as Caltech Orange-Aid Preprint 450.

D'Eath, P. 1977, paper submitted for publication in Phys. Rev.

DeWitt, B.S. 1971, Lectures on General Relativity-Stanford, Fall 1971 (unpublished manuscript).

Eardley, D.M., Lee, D.L., and Lightman, A.P. 1974, Phys. Rev. D, 10, 3308.

Eddington, A.S. 1922, Proc. Roy. Soc. A, 102, 268.

Einstein, A. 1918, Berlin Sitzungsberichte, p. 154.

Epstein, R. 1976, Ph.D. Thesis, Stanford University.

Epstein, R. and Wagoner, R.V., 1975, Astrophys.J., 197, 717.

Feynman, R.P., 1963, Acta Physica Polonica, 24, 697.

Geroch, R. 1970, J. Math. Phys., 11, 2580.

Havas, P. and Goldberg, J.N. 1962, Phys. Rev., 128, 398.

Ipser, J.R. 1971, Astrophys. J., 166, 175.

Isaacson, R.A. 1968, Phys. Rev., 166, 1272.

Janis, A.I. and Newman, E.T. 1965, J. Math. Phys., 6, 902.

Kovács, S.J., Jr. and Thorne, K.S. 1977a, paper submitted to Astrophys. J: also available as Caltech Orange Aid Preprint

Kovács, S.J., Jr. and Thorne, K.S. 1977b, paper in preparation.

Landau, L.D. and Lifshitz, E.M. 1962, The Classical Theory of Fields (Oxford: Pergamon Press).

Mashoon, B. 1973, Astrophys. J., 181, L65.

Mathews, J. 1962, J. Soc. Indust. Appl. Math, 10, 768.

Matzner, R.A. and Nutku, Y. 1974, Proc. Roy. Soc. A, 336, 285.

Misner, C.W., Thorne, K.S., and Wheeler, J.A. 1973, Gravitation (San Francisco, W.H. Freeman and Co.); cited in text as MTW.

Papapetrou, A. 1962, Comptes Rendues, 255, 1578.

Papapetrou, A. 1971, Ann. Inst. Henri Poincare, 14, 79.

Peters, P.C. 1970, Phys. Rev. D, 1, 1559.

Peters, P.C. and Mathews, J. 1963, Phys. Rev., 131, 435.

Pirani, F.A.E. 1965 in A. Trautman, F.A.E. Pirani, and H. Bondi, Lectures on General Relativity (Englewood Cliffs, N.J.: Prentice-Hall).

Press, W.H., 1977, paper in press.

Regge T. and Wheeler, J.A. 1957, Phys. Rev., 108, 1063.

Ruffini, R. and Wheeler, J.A. 1971, in Proceedings of the Conference on Space Physics (Paris: ESRO), p. 45.

Sachs, R. 1961, Proc. Roy. Soc. A, 264, 309, esp. Appendix B.

Shapiro, I.I. 1964, Phys. Rev. Lett., 13, 789.

Smarr, L. 1977, in Proceedings of the Eighth Texas Symposium on Relativistic Astrophysics, to be published in Proceedings of the New York Academy of Sciences.

Thorne, K.S. 1977a, paper to be submitted for publication. Cited in text as TV.

Thorne,, K.S. 1977b, paper to be submitted for publication. Cited in text as TVI.

Thorne, K.S., 1977c , in N. Lebovitz ed., Proceedings of the (Chandrasekhar) Symposium on Theoretical Principles in Astrophysics and Relativity (Chicago: University of Chicago Press).

Thorne, K.S. 1977d, paper submitted to Astrophys. J.

Thorne, K.S. and Kovács, S.J., Jr. 1975, Astrophys. J., 200, 245. Cited in text as TK.

Turner, M. 1977, paper submitted for publication; preprint, Stanford University.

van der Burg, M.G.J. 1968, Proc. Roy. Soc. A, 303, 37.

Wagoner R.V. and Will, C.M. 1976, Astrophys. J., 210, 764.

Wagoner, R.V. 1977, in Proceedings of the Academia Nazionale dei Lincei International Symposium on Experimental Gravitation, Pavia, Italy, September 1976; B. Bertotti ed., in press.

Wheeler, J.A. 1966, Ann. Rev. Astron. Astrophys., 4, 423.

EFFECTS OF GRAVITATIONAL RADIATION UPON ELECTROMAGNETIC WAVES IN A DISPERSIVE MEDIUM

B. Bertotti and R. Catenacci

Istituto Fisica Teorica

Università Pavia, Via Bassi, 4 (Italy)

INTRODUCTION

We summarize in this talk the main content of a paper published in G.R.G. (1), in which we studied the propagation of gravitational radiation in a dispersive medium. This problem arises because one hopes to detect gravitational waves through the photons they generate or scatter when traveling through interstellar or intergalactic space; in this case, in fact, a distant electromagnetic source would exhibit an extraterrestrial scintillation. This hope, however, is frustrated at least when one considers linearized gravitational disturbances in vacuo. This is essentially due to the circumstance that electromagnetic and gravitational waves are assumed to propagate with the same velocity and are both transversal.

We want now to study whether an effect arises when one of these two conditions is waived. The main result we find consists in the Čerenkov effect, occurring when a graviton propagates through a dispersive medium in which the index of refraction is larger than unity. Our calculations are carried out only for the case of an isotropic and random spectrum of gravitational waves.

CALCULATIONS

If we neglect the gravitational field produced by the medium, a spectrum of linearized gravitational waves in a flat space time is described by (see, e.g. (2)):

$$h_{\mu\nu} = \int d^3K \left\{ A(\vec{K})(n_\mu n_\nu + n^*_\mu n^*_\nu)\, e^{iK_\mu x^\mu} + c.c. \right\} , \qquad 1)$$

where K_μ is the propagation vector and n_μ is a unit null complex vector which can be chosen so as to satisfy the usual transverse traceless gauge:

$$h^\mu_\mu = 0 \quad , \quad h_{\mu 0} = 0 \quad , \quad h_\mu{}^\rho{}_{,\rho} = 0 \; . \qquad 2)$$

In the random phase approximation we have:

$$\langle A(\vec{K})\, A(\vec{K}') \rangle = \langle A^*(\vec{K})\, A^*(\vec{K}') \rangle = 0$$
$$\langle A(\vec{K})\, A^*(\vec{K}') \rangle = S(\vec{K})\, \delta(\vec{K} - \vec{K}') \quad , \qquad 3)$$

where $S(\vec{K})$ is the spectral density, determined by the spectral energy density

$$W(\omega) = 2\pi S(\omega)\, \omega^4 \qquad 4)$$

The medium is assumed homogeneous and is characterized by its four velocity v^μ: we take it to be a geodesic field, neglecting all nongravitational forces. At the lowest order we take

$$v^\mu = (1, 0, 0, 0) \qquad 5)$$

in the chosen gauge this is also true in the linearized approximation.

The general Hamiltonian formalism for the ray propagation in a dispersive medium is described in (3). The propagation vector p_μ and the position of the wave packet x^μ are solution of:

$$\frac{dx^\mu}{d\lambda} = \frac{\partial H}{\partial p_\mu} \qquad 6)$$

$$\frac{dp^\mu}{d\lambda} = -\frac{\partial H}{\partial x^\mu}, \qquad 7)$$

where the Hamiltonian

$$H \equiv \frac{1}{2}\left[p_\mu p^\mu + (1-n^2)(p_\mu v^\mu)^2\right] \qquad 8)$$

is set equal to zero after differentiation and n is the index of refraction, function of the local frequency $\Omega = -p_\mu v^\mu$.

Equation 7) gives:

$$\frac{dp^\mu}{d\lambda} + \Gamma^\mu_{\rho\sigma} p^\rho p^\sigma = -\chi p_\rho \frac{\partial v^\rho}{\partial x^\nu} g^{\mu\nu}, \qquad 9)$$

where

$$\chi = \frac{d\xi}{d\Omega} \qquad \xi \equiv \frac{1}{2}(n^2-1)\Omega^2 . \qquad 10)$$

With 6) one obtains for the time-like ray vector $V^\mu \equiv \frac{dx^\mu}{ds}$

$$\frac{ds}{d\lambda} V^\mu = p^\mu + \chi v^\mu , \qquad 11)$$

The energy density u of the electromagnetic waves is conserved along the ray:

$$\frac{du}{ds} + u\,\Theta = 0 \qquad 12)$$

and the intensity fluctuations are determined by the linearized Raychaudhuri's equation for a source at infinity and a small change $\delta\Theta$

$$\frac{d\delta\Theta}{ds} + R_{\mu\nu}V^{\mu}V^{\nu} = \dot{V}^{\mu}{}_{;\mu}\,. \qquad 13)$$

The effect of the term $R_{\mu\nu}V^{\mu}V^{\nu}$ can be neglected, as explained in (1).

When $\chi = 0$ $\dot{V}^{\mu} = 0$ from 9) and 11) and there is no change in intensity. In this case one can show that $n < 1$ and, from 10)

$$n^2 = 1 - \frac{\omega_p^2}{\Omega^2} \qquad 14)$$

moreover the "dancing" (position fluctuations) is not secular (see (1)); this negative result prompts us to investigate the general case.

Variation of the equations 9), 10), and 11) gives:

$$\frac{d\delta\Theta}{ds} = \frac{d}{ds}\left[\chi\left(\frac{d\chi}{d\Omega} + 1\right) h_{\varrho\sigma,0} V_0^{\varrho} V_0^{\sigma}\right] - \frac{1}{2}\frac{d\chi}{d\Omega} h_{\varrho\sigma,00} V_0^{\varrho} V_0^{\sigma} \;; \qquad 15)$$

the first term gives a nonsecular contribution and can be neglected. Introducing the ordinary velocity of the wave packet V, we set

$$V^{\varrho} = \left((1-V^2)^{-1/2}, 0, 0, (1-V^2)^{1/2}\right) \qquad 16)$$

the ordinary distance is:

$$D = sV(1-V^2)^{-1/2} \;; \qquad 17)$$

V is determined by the index of refraction. From equations 12), 13), we have:

$$\frac{d^2}{dD^2}\left(\frac{\delta n}{n}\right) = \frac{1}{2}\frac{d\chi}{d\Omega} h_{\varrho\sigma,00} \hat{V}_0^{\varrho} \hat{V}_0^{\sigma}, \qquad 18$$

where $\hat{V}^{\mu} = V(1-V^2)^{-1/2} V^{\mu}$. Note that there is an effect only if $\frac{d\chi}{d\Omega} \neq 0$.

We can compute the correlation function $\langle \delta w(0)\, \delta w(t)\rangle$ and its Fourier transform

$$Q(\omega) = \left(\frac{2}{\pi}\right)^{1/2} \int_0^\infty dt\, \cos\omega t \left\langle \frac{\delta w(0)\, \delta w(t)}{n^2} \right\rangle \qquad 19)$$

when $n < 1$ we have (see (1) for more details):

$$Q(\omega) \propto \left(\frac{d\chi}{d\Omega}\right)^2 S(\omega)\, \omega^4 D_0^2 \qquad 20)$$

and when $n > 1$

$$Q(\omega) \propto V^4 \left(\frac{d\chi}{d\Omega}\right)^2 \omega^5 S(\omega) \left(1 - \frac{1}{n^2}\right)^2 D_0^3 \quad . \qquad 21)$$

For the dancing we are interested in the change of the unit 3-vector

$$\hat{V}^m = V^m (V_k V^k)^{1/2} \qquad (n, m = 1, 2, 3).$$

Variation of equations 9) and 10) gives

$$\frac{d\delta V^m}{ds} = \frac{1}{2} h_{\rho\sigma}{}^{,m} V_0^\rho V_0^\sigma - \frac{\chi}{\sqrt{2\xi + \chi^2 + 2\chi\Omega}} h^m_{\rho,0} V_0^\rho$$
$$+ \chi \left(\frac{d\chi}{d\Omega} + 1\right) V^m h_{\rho\sigma,0} V_0^\rho V_0^\sigma \; ; \qquad 22)$$

again, introducing the ordinary distance D, we can compute the Fourier transform of the correlation tensor:

$$D_{mn}(\omega) = \sigma^2(\omega) \begin{vmatrix} 1 & 0 & 0 \\ 0 & 1 & 0 \\ 0 & 0 & 0 \end{vmatrix} = \left(\frac{2}{\pi}\right)^{1/2} \int_0^\infty dt\, \cos\omega t \langle \delta \hat{V}_m(0)\, \delta \hat{V}_n(t)\rangle . \qquad 23)$$

We conclude:

when $n < 1$ σ^2 is not secular,

when $n > 1$

$$\sigma^2(\omega) \approx S(\omega)\, \omega^3 D_0\, F(\Omega, n) \quad , \qquad 24)$$

(where $F(\Omega, n)$ is a complicated function of n and Ω).

CONCLUSION

The mathematical theory has shown that the main secular effect in scintillation and dancing appears when $n > 1$ and is proportional respectively to the powers 3/2 and 1/2 of the distance. There is also a weaker secular scintillation when $n < 1$ but

$$\frac{d^2}{d\Omega^2}\left\{(n^2-1)\Omega^2\right\} \neq 0 \ .$$

The effect is therefore absent in a purely electrostatic plasma, but arises when a magnetic field is present. Rough estimates, both in interstellar and in intergalactic space (see (1)) show that the effect is negligible unless perhaps, one considers regions of strong magnetic fields (like near pulsars). In this case however the situation is essentially inhomogeneous and requires separate considerations.

It is interesting to study the effect of a rest mass of the graviton, which waives both the transversality and the equality of propagation velocity between gravitons and photons. The covariant description of the graviton mass is contained in the vacuum field equations (see (4)).

$$R_{\mu\nu} + \Lambda g_{\mu\nu} = 0 \ .$$

Hence Λ cannot be larger than the square of Hubble constant. In any case in equation 13) the term is constant and therefore cannot contribute to fluctuations.

REFERENCES

1) B. Bertotti and R. Catenacci: GRG 6, 329 (1975).
2) D. M. Zipoy: Phys. Rev. 26, 1398 (1971).
3. J. L. Synge: Relativity: The General Theory (North Holland, Amsterdam) (1960), p. 375.
4) H. Amar: Phys. Rev. 89, 1298 (1953).

GRAVITATIONAL WAVES AND THEIR INTERACTION WITH ELECTROMAGNETIC FIELD

VENZO DE SABBATA

ISTITUTO DI FISICA DELLA UNIVERSITÀ

40126 BOLOGNA ITALY

We wish here to consider some problems connecting the gravitational waves in their interaction with electromagnetic field and precisely:

a) in a static electric and magnetic field
b) in a dipole magnetic field
c) more in general with photons

The aim of these considerations is to see if there are some possibilities to detect gravitational waves through the interactions with electromagnetic fields. Then in these lectures we shall be concerned with the production of gravitational waves not just by mechanical device or some massive astronomical objects, but by means of the interaction between e.m. waves and some electrostatic and magnetostatic fields.

Formerly some years ago Gertsenshtein ([1]) and Lupanov ([2]) have taken into account the former the possibility of creating gravitons in a magnetic field and the latter the possibility of detecting gravitational waves by means of a capacitor. In order to explore in more systematic way the perspectives offered by this kind of researches, Boccaletti, Fortini, Gualdi and myself ([3]) have investigate in more detail the possibility of producing and detecting gravitons in a static electromagnetic field.

1. An electromagnetic wave travelling through a static e.m. field produces a gravitational wave of the same frequency which, in its turn, travelling through another static e.m. field, creates an electromagnetic wave. We have calculated the ratio of the intensity of the produced e.m. wave to that of the incoming one. We shall see that in the production or detection of gravitational waves by means of a static magnetic field (or electric field), the

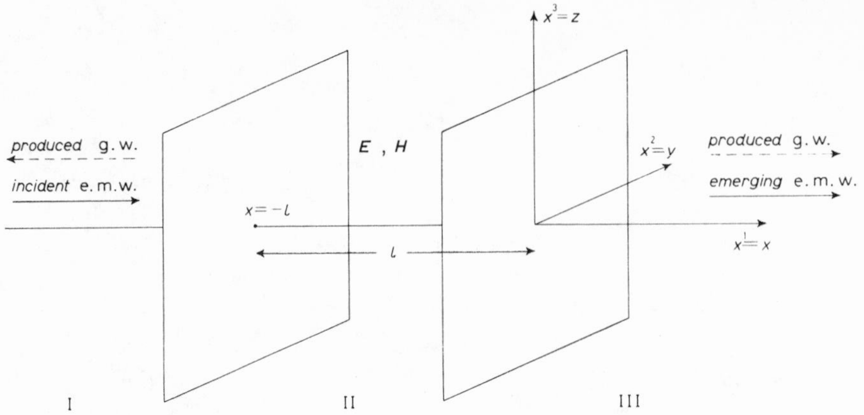

Fig. 1

efficiency of the process depends upon the square of the field as well as on the square of its linear dimensions.

We will consider some few details of this process: let us have a static e.m. field which is different from zero only in the region between two planes $x = -l$ and $x = 0$ and constant inside (see Fig.1). The plane e.m. wave comes from the negative x-axis and travels along this direction; the bordering plane surfaces are supposed to be transparent to the e.m. wave. We can study the gravitational effects of the e.m. wave crossing the static field using Einstein's equations for the weak fields:

$$\Box \psi^i_k = -\frac{16\pi G}{c^4} T^i_k \tag{1}$$

where

$$T^i_k = \frac{1}{4\pi}\left(F^{il}F_{kl} - \frac{1}{4}\delta^i_k F^{lm}F_{lm}\right) \tag{2}$$

and the F^{ik} contain both the static part of the field and that due to the electromagnetic wave.

The e.m. wave which propagate along the positive x-direction and which is assumed to be linearly polarized can be written as:

$$H_{z(\text{wave})} = a \exp[ik(x-ct)]$$

$$E_{y(\text{wave})} = a \exp[ik(x-ct)] \tag{3}$$

For the e.m. field in the region II we have:

$$F^{12} = H_z + a \exp[ik(x-ct)] \qquad F^{23} = H_x$$
(4) $$F^{13} = -H_y \qquad F^{20} = -E_y - a \exp[ik(x-ct)]$$
$$F^{10} = -E_x \qquad F^{30} = -E_z$$

Then it is easy to calculate the energy-momentum tensor: it can be shown that this tensor is formed by three parts: a quadratic term in the static field, a mixed term, and a quadratic term in the field of e.m. wave. The first and the third terms are not to be taken into account as they contribute nothing to the production of gravitational waves.The important term is therefore the mixed term. We have 10 components of the form (considering for instance the T_1^1 component):

(5) $$4\pi T_1^1 = a(H_z + E_y) \exp[ik(x-ct)]$$

and it is easy to verify that the condition $\frac{\partial T_i^k}{\partial x^k} = 0$ is always satisfied.
Then the equations (1) are:

$$\Box\, \psi_i^k = 0 \qquad \text{in regions I and III}$$
(6) $$\Box\, \psi_i^k = \lambda_i^k \exp[ik(x-ct)] \qquad \text{in region II}$$

where λ_i^k contain the constant part of (5).
Looking for those solutions of Einstein's equations which represent outgoing waves in region I and III, we can write the general solution in plane waves as

$$\psi_{I\,i}^{\;k} = \lambda_i^k C_{(ik)} \exp[-ik(x+ct)]$$
(7) $$\psi_{II\,i}^{\;k} = \lambda_i^k \left[\frac{x+l}{2ik} \exp[ik(x-ct)] + A_{(ik)} \exp[ik(x-ct)] + B_{(ik)} \exp[-ik(x+ct)]\right]$$
$$\psi_{III\,i}^{\;k} = \lambda_i^k D_{(ik)} \exp[ik(x-ct)]$$

where A, B, C, D are arbitrary constants which are determined by the continuity conditions.
Then the solution in region III may be written as:

(8) $$\psi_{III\,i}^{\;k} = \frac{2iGla}{kc^4}\, \alpha_i^k \exp[ik(x-ct)]$$

where α_i^k are just the static-field-dependent constants which appear in (5).

We are now able to calculate explicitly the metric tensor in the region III and then the energy flux of the gravitational wave that reduces to:

$$(9)\quad t^{01} = \frac{c^2}{16\pi G}\left[\dot{h}_{23}^2 + \frac{1}{4}\left(\dot{h}_{22} - \dot{h}_{33}\right)^2\right]$$

This formula (9) in our case gives (being $h_i^k = \psi_i^k$):

$$(10)\quad t^{01} = \frac{G}{4\pi c^4}\, l^2 a^2 \left[(H_y + E_z)^2 + (E_y + H_z)^2\right]$$

In the case of a magnetostatic field alone the formula (10) is identical (apart from a factor 1/4) with that given by Gertsenshtein ([1]) but it is more explicit as here are specified the field components. According to our calculations the energy flux is zero if the constant magnetic field is directed along the x-axis that is along the direction of propagation of the e.m. wave.

It should also be noticed that the quadratic dependence on l derives explicitly from the fact that the velocity of propagation of both the e.m. waves and gravitational waves is the same. The presence of a medium with an index of refraction different from 1, will modified this dependence.

2. On the same lines we may calculate the production of an e.m. wave by a gravitational wave travelling through a static e.m. field.

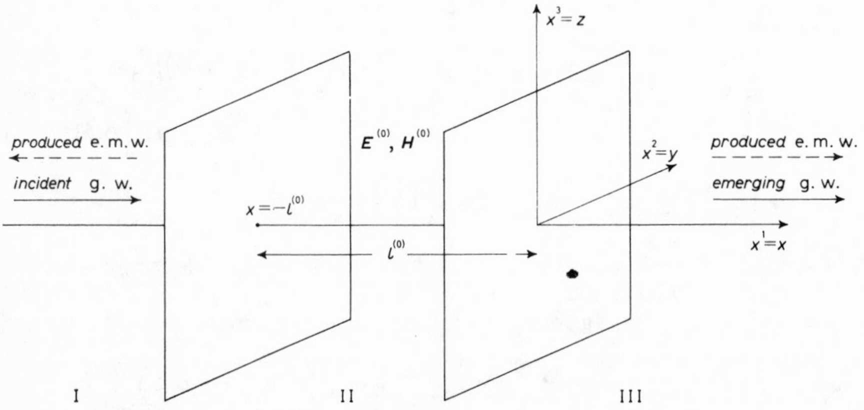

Fig. 2

We have the same situation of Fig.1 when now the e.m. and gravitational waves are exchanged and in region II we have only static $E^{(0)}$ and $H^{(0)}$ fields (see Fig.2). Without entering in the details of the calculations (see for this ref.3) we may emphasize that for instance in the region III we have an outgoing e.m. wave of the type:

$$
(11) \quad \begin{cases}
E_y = -\dfrac{A\ell^{(0)}}{2ik} \exp[ik(x-ct)] \\
H_z = -\dfrac{A\ell^{(0)}}{2ik} \exp[ik(x-ct)] \\
E_z = -\dfrac{B\ell^{(0)}}{2ik} \exp[ik(x-ct)] \\
H_y = -\dfrac{B\ell^{(0)}}{2ik} \exp[ik(x-ct)]
\end{cases}
$$

where

$$
(12) \quad \begin{aligned}
A &= -k^2\left\{[E_y^{(0)} - H_z^{(0)}]\alpha_{22} + [E_z^{(0)} + H_y^{(0)}]\alpha_{23}\right\} \\
B &= -k^2\left\{[E_y^{(0)} - H_z^{(0)}]\alpha_{23} + [E_z^{(0)} + H_y^{(0)}]\alpha_{33}\right\}
\end{aligned}
$$

(being α_i^k the coefficients that appear in (8)).
If in particular we have $E_y^{(0)} = H_z^{(0)}$, $H_y^{(0)} = -E_z^{(0)}$ we have an e.m. wave which propagates in the backward x-direction being in this case A = B = 0. If moreover we have different from zero only the $E_x^{(0)}$ component, we find the same solution as Lupanov ([2]) (a part from the amplitude: the difference is due to the fact that Lupanov considers a gravitational wave produced by mechanical oscillations while we consider gravitational waves produced by e.m. waves). If instead of $E_x^{(0)}$ we have different from zero only $H_x^{(0)}$ we have still the same results as before but the outgoing e.m. wave is in another state of linear polarization.

Combining the results corresponding to Fig.1 and Fig.2 together, that is considering an e.m. wave which produces a gravitational wave which in turn is reconverted into an e.m. wave (and therefore considering the device in Fig.1 as generator of gravitational waves and the device in Fig.2 as detector of gravitational waves (see Fig.3 where G is a generator of linearly polarized e.m. wave, D is a detector of e.m. waves, S is a screen opaque to the e.m. waves), we have for the energy density of emerging e.m. wave (if the incident wave has an energy density $W_i = a^2/4\pi$):

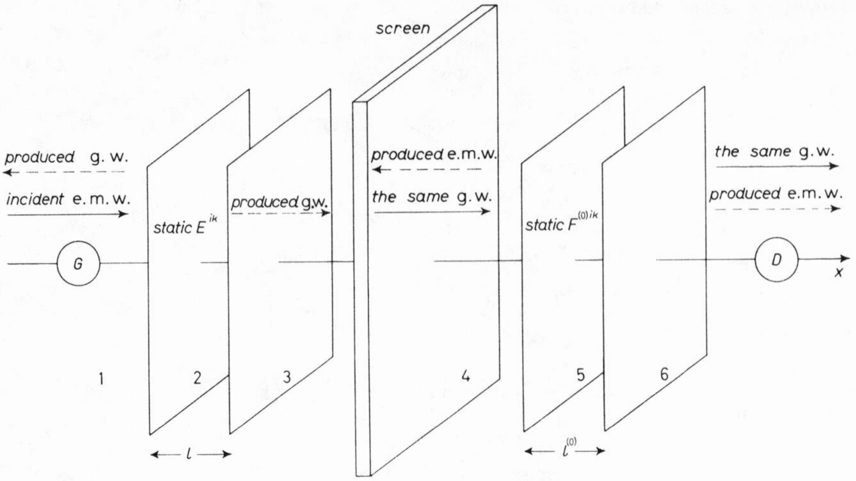

Fig. 3

$$W_p = \frac{a^2}{4\pi}\frac{G^2 \ell^2 \ell^{(0)2}}{c^8}\left\{\left[\left(E_y^{(0)} - H_z^{(0)}\right)\alpha_{22} + \left(E_z^{(0)} + H_y^{(0)}\right)\alpha_{23}\right]^2 \right.$$

(13)
$$\left. + \left[\left(E_y^{(0)} - H_z^{(0)}\right)\alpha_{23} + \left(E_z^{(0)} + H_y^{(0)}\right)\alpha_{33}\right]^2\right\}$$

where, as we know, the α depend on the static field in the generator. If in particular the detector contains only magnetic field we have

(14) $$W_p = W_i \frac{G^2 \ell^2 \ell^{(0)2}}{c^8}\left\{\left(H_z H_z^{(0)} + H_y H_y^{(0)}\right)^2 + \left(H_z^{(0)} H_y - H_y^{(0)} H_z\right)^2\right\}$$

where the factor G^2/c^8 makes the energy density very small. This energy density increase with ℓ^4 and this may be interesting in some astrophysical situations. It should also be noted that the produced power does not depend on the frequency of the e.m. wave and this allows for a more suitable choice of the frequency with respect to the characteristics of the detector. However we see that in order to have a high energy density, we must have very strong magnetic fields over long distances and for the moment this is outside the present technology.

Then in order to see if these processes are of some importance in astrophysical situation, we now will explore the possibility of emission of e.m. wave when a gravitational wave interacts with a dipole magnetic field, as we can find these fields in some peculiar astrophysical objects as neutron stars and black holes.

3. In this lecture we will give some simple calculations regarding the production of e.m. radiation by gravitational waves falling on a spherically symmetric magnetic field. The details of these calculations may be found in a work made with Fortini and al.([4]).

The method used there, is the usual treatment for which Maxwell equations are written in the field of a gravitational wave and linearized by means of the weak field approximation. These equations are then solved in the case of a uniformly magnetized sphere and the power emitted is calculated in the dipole-approximation. Some astrophysical considerations are made in the case of a gravitational radiation hitting a magnetic neutron star.

Maxwell equations in a gravitational field in vacuum are:

$$\text{(15)} \quad \begin{cases} \dfrac{\partial\left(\sqrt{-g}\, g^{\alpha\beta} g^{\gamma\delta} F_{\beta\delta}\right)}{\partial x^{\gamma}} = 0 \\ F_{[\alpha\beta,\gamma]} = 0 \end{cases}$$

where $g^{\alpha\beta}$ is the metric tensor with signature -2, which in the weak field approximation are:

$$\text{(16)} \quad g_{\alpha\beta} = \eta_{\alpha\beta} + h_{\alpha\beta}\,; \qquad g^{\alpha\beta} = \eta^{\alpha\beta} - h^{\alpha\beta}$$

and power of greater than 1 are neglected.
In this approximation the e.m. tensor is the sum of two terms: one which represents the unperturbed field $f^{(0)}_{\alpha\beta}$ and the other that represents the perturbation due to the gravitational field; we call these perturbations $f_{\alpha\beta}$ and being of the same order of $h_{\alpha\beta}$ we neglect second and higher order terms.
We consider the case of an unperturbed $f^{(0)}_{\alpha\beta}$ as a static magnetic field that is

$$\text{(17)} \quad f^{(0)}_{23} = H^{(0)}_{1}\,; \quad f^{(0)}_{31} = H^{(0)}_{2}\,; \quad f^{(0)}_{12} = H^{(0)}_{3}\,; \quad f^{(0)}_{0k} = 0$$

Now if a gravitational wave produced by a mass quadrupole oscillator falls on a uniformly magnetized sphere, an e.m. wave will be produced and we may calculate the perturbed terms $f^{(0)}_{\alpha\beta}$ and then the energy flux of e.m. wave. In the linear approximation the metric is given by

$$
(18)\qquad
\begin{aligned}
h_{11} &= \frac{\varphi}{2}\left(2\cos^2\vartheta - \mathrm{sen}^2\vartheta\right) & h_{13} &= \varphi\, \mathrm{sen}\,\vartheta \cos\vartheta \\
h_{22} &= -\frac{\varphi}{2}\,\mathrm{sen}^2\vartheta & h_{03} &= -\varphi\, \mathrm{sen}\,\vartheta \cos\vartheta \\
h_{33} &= \frac{\varphi}{2}\,\mathrm{sen}^2\vartheta & h_{01} &= -\varphi \cos^2\vartheta \\
h_{00} &= \frac{\varphi}{2}\left(1 + \cos^2\vartheta\right) & h_{02} &= h_{12} = h_{23} = 0
\end{aligned}
$$

with $g = -1 + \varphi\,\mathrm{sen}^2\vartheta$ and $\varphi = a\,\exp(ik(x_1 - x_0))$; ϑ is the angle between the oscillator direction of the quadrupol and the x_1-axis is taken as the direction of propagation of the gravitational wave.
We assume that the constant magnetic field $H^{(0)}$ (with components $H^{(0)}_1, H^{(0)}_2, H^{(0)}_3$) is different from zero only inside a sphere and zero outside where the radius of the sphere R_1 is subject to the condition $kR_1 \ll 1$. Then in the dipole approximation we have [4]:

$$
(19)\qquad H_x = \frac{e^{ikR - i\omega t}}{R^2} \cdot \frac{k^2 a R_1^3}{3}\left(\frac{\mathrm{sen}^2\vartheta}{2} H^{(0)}_z z - \frac{\mathrm{sen}^2\vartheta}{2} H^{(0)}_y y\right) \qquad (R \gg R_1)
$$

and similar expressions for the other magnetic and electric components. a is the amplitude of the incoming gravitational wave. The energy flux, when integrated on a spherical surface of radius R is:

$$
(20)\qquad W_{em} = \frac{c k^4 a^2 R_1^6}{27}\left\{\frac{\mathrm{sen}^4\vartheta}{4} H^{(0)2}_y + \mathrm{sen}^2\vartheta \cos^2\vartheta\, H^{(0)2}_y + \frac{\mathrm{sen}^4\vartheta}{4} H^{(0)2}_z\right\}
$$

and averaging over all possible directions of the mass

quadrupole oscillations (that is over ϑ) we find:

$$W_{em} = \frac{\omega^4 a^2 R_1^6}{27\, c^3} \left(\frac{4}{15} H_y^{(o)2} + \frac{2}{15} H_z^{(o)2} \right) \tag{21}$$

and finally averaging aver all possible orientations of the magnetic field

$$W_{em} = \frac{2}{405} \frac{\omega^4 a^2 R_1^6}{c^3} H^{(o)2} \tag{22}$$

where $H^{(o)}$ is the magnetic field strength. As the gravitational wave amplitude a is related to the energy flux F_g of the incident gravitational wave by:

$$a^2 = \frac{128\, \pi\, G\, F_g}{\omega^2 c^3} \tag{23}$$

we get

$$W_{em} = 1.8 \cdot 10^{-70} \omega^2 R_1^6 F_g H^{(o)2} \tag{24}$$

Now actually the external field is proportional to μ/R^3 where μ is the magnetic dipole moment of the sphere. We notice that in (24) a factor $H^2 R_1^6 \sim \mu^2$ appears. We therefore expect that the order of magnitude of the interaction between gravitational radiation and a true magnetic dipole is not very different from the one given by (24).

Then we can rewrite (24) in more expressive way as

$$W_{em} = 1.8 \cdot 10^{-70} \omega^2 F_g \mu^2 \tag{25}$$

These results may be applied to some astrophysical objects as for instance to magnetic neutron stars. If, in particular, we apply these results to a model of galactic center consisting of a cluster of neutron stars

we may consider the possible effects of the interaction of gravitational waves produced in the cluster itself with the magnetic dipoles of the neutron stars. The model [5],[6],[7], consists of a cluster of some 10^{11} neutron stars within a radius of $\sim 10^{17}$ cm; in analogy to the ratio of normal to magnetic stars in our galaxy we assume that in the cluster 10% of neutron stars have a magnetic moment $\mu \sim 10^{33}$ erg G^{-1} (see [8]).
The gravitational waves are produced in the cluster during the collisions of two neutron stars. These events occur with a frequency of 10 year^{-1} which is less than Weber's estimate [9]; the flux of gravitational energy per event in the cluster is about $10^{17} \div 10^{18}$ $\text{erg/cm}^2\text{sec}$ (if we assume that about 1/10 of the rest mass of neutron star converts in gravitational waves during the collision) and therefore from (25) we have for the whole cluster an energy of $10^{31} \div 10^{32}$ erg/sec in electromagnetic waves, with a frequency of $1.6 \cdot 10^3$Hz if we limit our considerations to Weber results.

As to the Earth, if we apply these results to the magnetic field of the Earth, in the assumption that on the Earth arrives a flux of gravitational waves of $10^6 \div 10^7 \text{erg/cm}^2\text{sec}$ (that is Weber[9] results), being $\mu \sim 8 \cdot 10^{25}$ erg G^{-1} we have an emission of e.m. waves $\sim 10^{-4} \div 10^{-3}$erg/sec per pulse (at a frequency of $1.6 \cdot 10^3$Hz).

4. To conclude this brief survey on the interaction between gravitational and e.m. waves we like to mention the direct interaction between photons and gravitons treated with the methods of the quantum theory [10], [11], as for instance the processes:

$$\gamma + \gamma \rightleftarrows g + g \qquad (26)$$
$$\gamma + g \rightleftarrows \gamma + g$$

In particular if these reactions connected with the creation and absorption of gravitons by photons occur in the presence of a massive body, the cross section is much higher as the large mass compensates for the smallness of gravitational coupling constant. In a certain sense one can said that the massive body acts as a catalyst [12]. For instance in the case of the reaction

$$\gamma + M \longrightarrow M + \gamma + g \qquad (27)$$

taking a laser beam with a flux of the order of $\sim 10^{28}$ photons/cm^2sec and using as a catalyst the mass of the Earth, we have a cross section

(28) $$\sigma \simeq 10^{-37} cm^2$$

To obtain these results, we start from the usual interaction Lagrangian density ([13]) between the gravitational and e.m. fields:

(29) $$\mathcal{L} = -\lambda \tilde{h}_{\mu\nu} T_{\mu\nu}$$

where $\tilde{h}_{\mu\nu} = h_{\mu\nu} - \frac{1}{2}\delta_{\mu\nu} h_{\sigma\sigma}$
and $T_{\mu\nu} = F_{\mu\alpha} F_{\nu}^{\alpha} - \frac{1}{4}\delta_{\mu\nu} F_{\alpha\beta} F^{\alpha\beta}$, $\lambda = \sqrt{8\pi G}$
that is:

(30) $$\mathcal{L} = -\lambda \delta^{\mu\nu} (\tilde{h}^{\alpha\varsigma} - \frac{1}{4} \delta^{\alpha\varsigma} \tilde{h}) F_{\alpha\varsigma} F_{\mu\nu}$$

In order to calculate the emission of a graviton by a photon which interacts with an external gravitational field, we consider the Feynman diagrams of the type illustrated in Fig.4; then the calculation of the matrix element is performed in the usual way. We do not give here the calculations that are developed in detail in ref.([10]). The results are,in the case of the photon energies in the range of visible light, the following: the cross section for the process (27) (described by diagrams like that in Fig.4) when the photon is considered in a gravitational field of a galaxy, is of the order of $10^{24} \div 10^{25}$ cm^2.

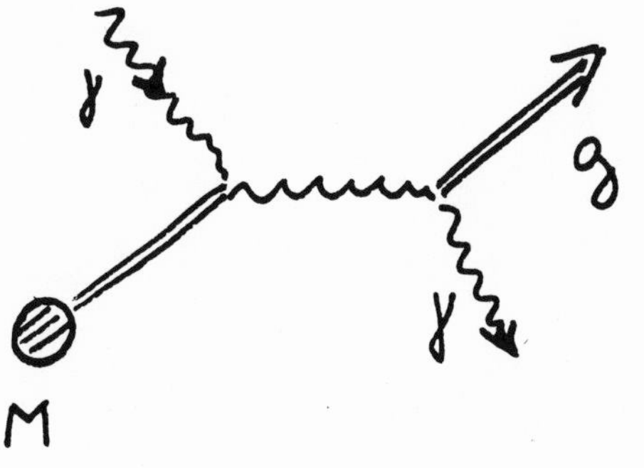

Fig. 4

If we consider that the target is a galaxy, the cross--section is very small and then it does not give any contribution to the red-shift of the light coming from the distant stars. In fact if, for example, we evaluate the energy loss of a photon which emits a graviton, we find a completely negligible figure: considering a path of the order of the radius of the universe, the loss of energy is $\Delta k \sim 10^{-50}$ ev. Therefore the red-shift of light cannot be explined as a "tired-light" phenomenon.

To conclude these concise lectures on the interaction between e.m. and gravitational waves, we may said that in future one can hope to do some laboratory experiments with the use of a laser beam of great intensity in strong magnetic field of some length. For instance with a laser beam with an intensity of 10^3W in the magnetic field of 10^5G for a length of $\sim 10^4$cm we obtain for the ratio of the gravitational radiation to the intensity of e.m. radiation a value of

$$\left(\frac{a}{b}\right)^2 \sim 10^{-25}$$

which means about 10^{-3} graviton/cm^{2}sec with an energy, per graviton, of approximately 1 ev.
This process is apparently one of the most effective (see ref.([12])).

BIBLIOGRAPHY

(1) M.E.Gertsenshtein - Sov.Phys. JETP, 14, 84 (1962)

(2) G.A.Lupanov - Sov.Phys. JETP, 25, 76 (1967)

(3) D.Boccaletti, V.De Sabbata, P.Fortini and C.Gualdi - Nuovo Cimento, 70B, 129 (1970)

(4) V.De Sabbata, P.Fortini, C.Gualdi and L.Fortini Baroni - Acta Physica Polonica, 115, 13 (1974)

(5) V.De Sabbata, P.Fortini and C.Gualdi - Astrophys. Lett., 12, 87 (1972)

(6) D.Boccaletti, V.De Sabbata, P.Fortini and C.Gualdi - Nature, 232, 53 (1971)

(7) V.De Sabbata, P.Fortini, L.Fortini Baroni and C.Gualdi - Acta Physica Polonica, B7, 19 (1976)

(8) D.Boccaletti, V.De Sabbata and C.Gualdi - Nuovo Cimento, 36, 685 (1965)

(9) J.Weber - Phys.Rev.Lett., 22, 1320 (1969)

(10) D.Boccaletti, V.De Sabbata, C.Gualdi and P.Fortini - Nuovo Cimento, 48, 58 (1967)

(11) D.Boccaletti, V.De Sabbata, C.Gualdi and P.Fortini - Nuovo Cimento 60B, 320 (1969)

(12) V.De Sabbata, D.Boccaletti and C.Gualdi - Soviet Journal of Nuclear Phys., 8, 537 (1969) (Yad. Fiz. 8, 924 (1968))

(13) T.W.B.Kibble - The quantum theory of gravitation in "High-Energy Physics and Elementary Particles (Vienna, 1965).

SHOCK WAVES IN GENERAL RELATIVITY

A. Papapetrou

Laboratoire de Physique Théorique
Institut Henri Poincare
Paris, France

1. Introduction

We start with an intuitive example from the Maxwell theory in special Relativity. Consider a charged particle which is initially at rest and starts being accelerated e.g. at $t = 0$. The retarded electromagnetic field produced by this particle will be initially time-independent, changing later to a variable field. The two regions of the space-time, in which the field is time-independent or variable, will be separated by a hypersurface Σ which is seen at once to be the future light cone of the point of space-time representing the position of the particle at $t = 0$. An accelerated charge emits electromagnetic radiation. Therefore the surface Σ will be the front of the wave emitted by the accelerated particle.

From the mathematical point of view the surface Σ is characterized by the fact that certain derivatives of the electromagnetic potential A_λ are discontinuous across Σ. This follows at once from the remark that all the derivatives of A_λ with respect to the time are zero below Σ, but this is certainly not true above Σ. These discontinuities are propagated along Σ in a way determined by the Maxwell equations. We say that there is an electromagnetic shock wave on the surface Σ.

Shock waves exist in any field theory having hyperbolic field equations. Therefore there are shock waves also in general relativity, since the Einstein equations for the gravitational field are hyperbolic. The detailed study of the gravitational shock waves is the main aim of these lectures. It turns out that this discussion is completely analogous to that of electromagnetic shock waves. This suggests to discuss first electromagnetic shock waves in special relativity, since this is somehow simpler and in any case interesting. Gravitational shock waves will then be treated in a similar way. Finally we shall discuss briefly shock waves in the Einstein-Maxwell theory of the combined gravitational and electromagnetic field.

I. ELECTROMAGNETIC SHOCK WAVES IN SPECIAL RELATIVITY

2. The Shock Wave of Order n = 2

We start from the Maxwell equations for vacuum :

(1) $$F^{\lambda\mu}{}_{,\mu} = 0.$$

The electromagnetic field tensor $F^{\lambda\mu}$ is expressed in terms of the electromagnetic potential as follows :

(2) $$F_{\lambda\mu} = A_{\mu,\lambda} - A_{\lambda,\mu} , \qquad F^{\lambda\mu} = \eta^{\lambda\alpha}\eta^{\mu\beta}F_{\alpha\beta} ,$$

$\eta^{\lambda\mu} = \eta_{\lambda\mu}$ being the Minkowski metric in inertial coordinates :

$$ds^2 = \eta_{\lambda\mu}dx^\lambda dx^\mu = (dx^0)^2 - (dx^1)^2 - (dx^2)^2 - (dx^3)^2 .$$

Considered as an equation for A_λ the field equation (1) is of the second order. Indeed, introducing (2) in (1) we obtain, after lowering the index λ :

(3) $$F_\lambda{}^\mu{}_{,\mu} = \eta^{\mu\nu}(A_{\mu,\lambda\nu} - A_{\lambda,\mu\nu}) = 0$$

It follows that the mathematically simplest case will be a shock wave of order n = 2 : The functions A_λ and their first derivatives $A_{\lambda,\mu}$ are continuous across Σ ; but at least some second derivatives $A_{\lambda,\mu\nu}$ are discontinuous. This is the case which will be discussed in detail.

Let the hypersurface Σ be defined by the equation

(4) $$z(x^\alpha) = 0 .$$

The space-time is divided by Σ in the half-spaces U^- $(z<0)$ and U^+ $(z>0)$. The discontinuity of a function $f(x^\alpha)$ across Σ is defined as follows. Consider points P^- in U^- and P^+ in U^+ , both tending to a point P on Σ . The discontinuity $[f]$ of f across Σ is :

(5) $$[f] = \lim_{P^- \text{ and } P^+ \to P} \{f(P^+) - f(P^-)\} .$$

With this notation the assumptions characterizing an electromagnetic shock wave of order n = 2 are :

(6) $$[A_\lambda] = 0 \ , \quad [A_{\lambda,\mu}] = 0$$

but $[A_{\lambda,\mu\nu}] \neq 0$ for at least some $A_{\lambda,\mu\nu}$. Derivatives of order $n > 2$ will in general be also discontinuous.

Let us consider some auxiliary variables $\tilde{x}^\lambda$ defined in the neighbourhood of Σ as follows :

$$\tilde{x}^0 = z(x^\alpha) , \quad \tilde{x}^i = f^i(x^\alpha) \quad \text{for } i = 1, 2, 3.$$

We shall demand that the functions $z(x^\alpha)$ and $f^i(x^\alpha)$ have continuous first and second derivatives and also that $\det(\partial\tilde{x}/\partial x) \neq 0$. Since $\tilde{x}^i$ are coordinates on Σ , it follows that if a function $F(\tilde{x}^\alpha)$ is continuous on Σ and has derivatives $F_{,\tilde{i}} \equiv \partial F/\partial \tilde{x}^i$, $F_{,\tilde{i}\tilde{k}}$ etc, then these derivatives are continuous on Σ :

$$[F] = 0 \implies [F_{,\tilde{i}}] = 0, \ [F_{,\tilde{i}\tilde{k}}] = 0 \qquad \text{etc.}$$

We therefore obtain from (6) :

$$[A_{\lambda,\tilde{i}}] = [A_{\lambda,\tilde{i}\tilde{k}}] = \cdots = 0 \ ; \ [A_{\lambda,\tilde{0}}] = [A_{\lambda,\tilde{0}\tilde{i}}] = \cdots = 0.$$

But according to the assumption $[A_{\lambda,\mu\nu}] \neq 0$ we shall have :

(7) $$[A_{\lambda,\tilde{0}\tilde{0}}] \equiv \left[\frac{\partial^2 A_\lambda}{\partial z^2}\right] \equiv \varphi_\lambda \neq 0 \qquad \text{for at least some } \lambda.$$

Now from the elementary formula

$$A_{\lambda,\mu\nu} = A_{\lambda,\tilde{\alpha}\tilde{\beta}} \frac{\partial \tilde{x}^\alpha}{\partial x^\mu}\frac{\partial \tilde{x}^\beta}{\partial x^\nu} + A_{\lambda,\tilde{\alpha}} \frac{\partial^2 \tilde{x}^\alpha}{\partial x^\mu \partial x^\nu}$$

we obtain the result :

(8) $$[A_{\lambda,\mu\nu}] = \varphi_\lambda P_\mu P_\nu \ ,$$

P_μ being the normal to the hypersurface Σ :

(9) $$P_\mu \equiv \frac{\partial \tilde{x}^0}{\partial x^\mu} = \frac{\partial z}{\partial x^\mu} .$$

3. The Local Condition. Trivial and Essential Discontinuities

A first condition imposed on the <u>discontinuity vector</u> φ_λ by the field equation (3) is derived as follows. This field equation has to be valid on both sides of Σ . Therefore :

(10') $$[F_\lambda{}^\mu{}_{,\mu}] = \eta^{\mu\nu}([A_{\mu,\lambda\nu}] - [A_{\lambda,\mu\nu}]) = 0.$$

Taking into account (8) we obtain :

(10) $$\eta^{\mu\nu}(\varphi_\mu p_\lambda p_\nu - \varphi_\lambda p_\mu p_\nu) = p_\lambda \varphi_\mu p^\mu - \varphi_\lambda p_\mu p^\mu = 0.$$

In discussing this condition we have to consider separately the two cases :

(11) (i) $p_\mu p^\mu \neq 0$ or (ii) $p_\mu p^\mu = 0$.

We start with the case (i), in which we get from (10):

(12) $$\varphi_\lambda = a p_\lambda, \qquad a = \varphi_\mu p^\mu \cdot (p_\nu p^\nu)^{-1}.$$

We shall prove that discontinuities of the form (12) are <u>trivial</u>, in the sense that they can be eliminated or produced by suitable gauge transformations. Indeed, consider the transformation

(13) $$A^*_\lambda = A_\lambda + \Lambda_{,\lambda}$$

and assume that the function Λ and the derivatives $\Lambda_{,\lambda}$ and $\Lambda_{,\lambda\mu}$ are continuous on Σ but the third derivatives $\Lambda_{,\lambda\mu\nu}$ are discontinuous. Then according to the argument leading to (8) :

(14) $$[\Lambda_{,\lambda\mu\nu}] = \left[\frac{\partial^3\Lambda}{\partial z^3}\right] p_\lambda p_\mu p_\nu$$

and consequently :

$$[A^*_{\lambda,\mu\nu}] = [A_{\lambda,\mu\nu}] + \left[\frac{\partial^3\Lambda}{\partial z^3}\right] p_\lambda p_\mu p_\nu .$$

Since we have in the present case

$$[A_{\lambda,\mu\nu}] = a p_\lambda p_\mu p_\nu ,$$

we see that it is sufficient to take

$$\left[\frac{\partial^3\Lambda}{\partial z^3}\right] = -a$$

in order to obtain

(15) $$[A^*_{\lambda,\mu\nu}] = 0 .$$

Consequently the case (11 i) is uninteresting.

4. The Algebraic Structure of the Discontinuity Vector φ_λ

We now examine the case (11 ii), in which p_μ is a null vector. The relation (11 ii) must of course be satisfied at all points of Σ . The condition (10) reduces now to

(16) $$\varphi_\mu p^\mu = 0.$$

This is the local condition, determining the algebraic structure of the discontinuity vector φ_μ of an electromagnetic shock wave. The vector p_μ is the propagation vector of the shock wave.

Before deriving the consequences following from (16) we shall discuss the geometrical meaning of (11 ii). The vector p_μ being the normal to Σ , any vector a_μ satisfying the relation

$$a_\mu p^\mu = 0$$

lies on Σ . Therefore equation (11 ii) means that the normal p_μ to Σ lies on Σ , this being of course possible because of the indefinite character of the Minkowski metric. Surfaces Σ having this property are called null surfaces or characteristic surfaces of the Maxwell equations (from the discussion of the corresponding Cauchy problem).

Since we have assumed that (11 ii) is satisfied at every point of Σ we have on Σ the null vector field $p_\mu(x^\alpha)$. The trajectories of this field are null-geodesics of the space-time. This is proved at once if we calculate the quantity

$$p_{\mu;\nu} p^\nu = p_{\mu,\nu} p^\nu.$$

Since

$$p_{\mu,\nu} = \frac{\partial^2 z}{\partial x^\mu \partial x^\nu} = p_{\nu,\mu} ,$$

it follows :

$$p_{\mu;\nu} p^\nu = p_{\nu,\mu} p^\nu = \frac{1}{2}\left(p_\nu p^\nu\right)_{,\mu} = 0 ,$$

which is the geodesic equation in its normal form, valid for an affine parameter on the geodesic. These null geodesics are called the bicharacteristic curves of the null surface.

The surface Σ being 3-dimensional, there will be 3 independent directions on it. As the first of them we shall take the vector p_μ ; then there will be two other independent directions, given by e.g. the vectors a_μ and b_μ . We shall have :

(17) $$a_\mu p^\mu = b_\mu p^\mu = 0.$$

From these relations it is easily derived that a_μ and b_μ are space-like. It will be convenient to use vectors a_μ and b_μ which are orthonormal :

(18) $$a_\mu b^\mu = 0 \; ; \quad a_\mu a^\mu = b_\mu b^\mu = -1.$$

Moreover we can choose a_μ and b_μ arbitrarily, but in a sufficiently continuous manner, only at one point of each bicharacteristic and then determine them on the whole of Σ by parallel transport on each bicharacteristic. The corresponding mathematical relations are :

(19) $$a_{\mu;\nu}p^\nu = b_{\mu;\nu}p^\nu = 0.$$

We now discuss the equation (16). Since φ_μ is a vector on Σ , we can decompose it as follows :

(20') $$\varphi_\mu = \alpha a_\mu + \beta b_\mu + \gamma p_\mu .$$

The last term represents, according to (15), a trivial part of the discontinuity vector. The essential part is therefore :

(20) $$\varphi'_\mu = \alpha a_\mu + \beta b_\mu .$$

We say that the electromagnetic shock wave is transverse - φ'_μ being orthogonal to the propagation vector - and it has two degrees of freedom corresponding to the two arbitrary scalar coefficients α and β .

It will be useful to introduce a complex null vector m_μ by the relation :

(21) $$m_\mu = \frac{1}{\sqrt{2}}(a_\mu + i b_\mu) \implies m_\mu m^\mu = 0 , \quad m_\mu \bar{m}^\mu = -1 .$$

We shall then have instead of (20) :

(22) $$\varphi'_\mu = \sqrt{J}\left(e^{i\psi} m_\mu + e^{-i\psi}\bar{m}_\mu\right).$$

J is called the amplitude and ψ the phase of the shock wave. The comparison with (20) gives at once :

(23) $$J = \tfrac{1}{2}(\alpha^2 + \beta^2) , \quad tg\,\psi = -\beta/\alpha .$$

5. The Propagation of the Discontinuities

Besides the algebraic condition (16) the discontinuity vector φ_μ must obey a differential relation which we shall derive now. This new relation will be derived from the equation

(24) $$F_\lambda{}^\mu{}_{,\mu\nu} = 0 ,$$

which follows immediately from the equation (3) and which of course has to be valid on both sides of Σ .

We shall need now the discontinuities of the third order derivatives $A_{\lambda,\mu\nu\rho}$. In order to calculate these discontinuities we proceed as follows. Let A^-_λ and A^+_λ be the functions representing the potential A_λ in the regions U^- and U^+ . Let $A^-_{\lambda(+)}$ be some continuation of A^-_λ into the layer $0 < z < \varepsilon$ of U^+ such that the function defined by A^-_λ and $A^-_{\lambda(+)}$ has continuous derivatives up to the order 3 on Σ . We can then write in this interval of z :

$$A^+_\lambda = A^-_{\lambda(+)} + \frac{1}{2} z^2 \varphi_\lambda(x^\alpha) + \frac{1}{6} z^3 \chi_\lambda(x^\alpha) + \cdots . \tag{25}$$

The functions $A^-_{\lambda(+)}$ are not uniquely determined, but the ambiguity does not affect the values of φ_λ on Σ , in which we are interested here.

From (25) we find :

$$A^+_{\lambda,\mu\nu\rho} = A^-_{\lambda(+),\mu\nu\rho} + (\varphi_\lambda p_\mu p_\nu)_{,\rho} + \{(\varphi_\lambda p_\mu)_{,\nu} + p_\nu(\varphi_{\lambda,\mu} + \chi_\lambda p_\mu)\} p_\rho + \cdots, \tag{26}$$

the omitted terms having at least one factor z . Remembering that according to the construction of $A^-_{\lambda(+)}$ we have

$$A^-_{\lambda(+),\mu\nu\rho} = A^-_{\lambda,\mu\nu\rho} \qquad \text{at } z = 0,$$

we obtain from (26) :

$$[A_{\lambda,\mu\nu\rho}] = (\varphi_\lambda p_\mu p_\nu)_{,\rho} + \{(\varphi_\lambda p_\mu)_{,\nu} + (\varphi_{\lambda,\mu} + \chi_\lambda p_\mu) p_\nu\} p_\rho . \tag{27}$$

Introducing this result in the equation

$$[F_\lambda{}^\mu{}_{,\mu\nu}] = 0$$

we find :

$$\eta^{\mu\nu}(\varphi_\mu p_\lambda p_\nu - \varphi_\lambda p_\mu p_\nu)_{,\rho} + \eta^{\mu\nu}\{(\varphi_\mu p_\lambda - \varphi_\lambda p_\mu)_{,\nu} + (\varphi_{\mu,\lambda} + \chi_\mu p_\lambda - \varphi_{\lambda,\mu} - \chi_\lambda p_\mu) p_\nu\} p_\rho = 0. \tag{28}$$

We shall discuss this equation for the case (11 ii), for which we derived the condition (16). Note, however, that these two relations are valid only on the surface Σ and consequently the derivatives $(p_\mu p^\mu)_{,\rho}$ and $(\varphi_\mu p^\mu)_{,\rho}$ do not need to vanish.

Therefore equation (28) will reduce to :

$$(\varphi_\mu p^\mu)_{,\rho} p_\lambda + (\varphi_\mu p^\mu)_{,\lambda} p_\rho - (p_\mu p^\mu)_{,\rho} \varphi_\lambda + \{\varphi^\mu{}_{,\mu} p_\lambda - \varphi_\lambda p^\mu{}_{,\mu} - 2\varphi_{\lambda,\mu} p^\mu + p_\lambda \chi_\mu p^\mu\} p_\rho = 0. \tag{28'}$$

This equation can be simplified with the help of the following remark. The surface Σ will be represented also by the equation

$$z^*(x^\alpha) = z(x^\alpha) . f(x^\alpha) = 0 , \tag{29}$$

provided that $f(x^\alpha) \neq 0$ in some neighbourhood of Σ . We now have

$$p^*_\alpha = \frac{\partial z^*}{\partial x^\alpha} = f p_\alpha + z f_{,\alpha} \rightarrow f p_\alpha \quad \text{on } \Sigma , \tag{30}$$

and we can prove at once the following lemma : If $(p^\alpha p_\alpha)_{,\rho} \neq 0$ on Σ for some ρ , we can obtain :

$$(p^*_\alpha p^{*\alpha})_{,\rho} = 0 \quad \text{on } \Sigma \tag{31}$$

by an appropriate choice of the function f .

With the vector p^α replaced by $p^{*\alpha}$ equation (28') reduces to :

$$(\varphi_\mu p^{*\mu})_{,\rho} p^*_\lambda + (\varphi_\mu p^{*\mu})_{,\lambda} p^*_\rho + \{(\varphi^\mu{}_{,\mu} + \chi_\mu p^{*\mu}) p^*_\lambda - \varphi_\lambda p^{*\mu}{}_{,\mu} - 2\varphi_{\lambda,\mu} p^{*\mu}\} p^*_\rho = 0. \tag{32}$$

From this equation we can obtain two relations which do not contain the functions χ_μ . The first relation is obtained if we multiply (32) by φ^λ . Remembering that according to (16) φ^λ lies on Σ we find :

$$(\varphi_\mu p^{*\mu})_{,\lambda} \varphi^\lambda = 0$$

and the final result is :

$$\varphi^\lambda (2\varphi_{\lambda,\mu} p^{*\mu} + \varphi_\lambda p^{*\mu}{}_{,\mu}) = (\varphi^\lambda \varphi_\lambda p^{*\mu})_{,\mu} = 0. \tag{33}$$

Using (22) we can rewrite this equation in the form :

$$(J p^{*\mu})_{,\mu} = 0. \tag{33'}$$

The second relation is obtained if we multiply (32) by the quantity

$$\tilde{\varphi}^\lambda = \frac{1}{i} (e^{i\psi} m^\lambda - e^{-i\psi} \bar{m}^\lambda). \tag{34}$$

We have again

$$(\varphi_\mu p^{*\mu})_{,\lambda} \tilde{\varphi}^\lambda = 0 ,$$

since $\tilde{\varphi}^\lambda$ lies also on Σ . We also verify the relation :

(35) $$\varphi_\lambda \tilde{\varphi}^\lambda = 0 ,$$

so that the result of the multiplication of (32) by $\tilde{\varphi}^\lambda$ reads :

(36) $$\tilde{\varphi}^\lambda \varphi_{\lambda,\mu} p^{*\mu} = 0.$$

Taking into account the condition (19), or equivalently :

$$m_{\mu;\nu} p^{*\nu} = 0$$

we find finally that equation (36) reduces to :

(37) $$\psi_{,\mu} p^{*\mu} = 0.$$

What remains now from the equation (32) is an algebraic relation for χ_μ, which we shall not discuss further.

Equations (33) or (33') and (37) are the propagation relations. The first is a conservation law for the amplitude of the shock wave. The second relation expresses the fact that the direction of polarisation of the wave is undergoing parallel transport along each bicharacteristic. It follows that the quantities J and ψ can be chosen arbitrarily only on a 2-dimensional surface cutting all bicharacteristics. The propagation relations will then determine these quantities on the whole hypersurface Σ.

6. Shock Waves of Order n ≠ 2

Shock waves of order $n = 2 + m$, $m > 0$ are discussed in a similar way. One starts by remarking that the derivatives of A_λ of order n have discontinuities of the form :

(38) $$[A_{\lambda,\mu\nu\ldots}] = \varphi_\lambda p_\mu p_\nu \cdots ,$$

p_μ being again the normal to Σ . Evidently, in this case the field equation (3) cannot lead to a condition for φ_λ : It is the derivative of order m of the field equation which will lead now to the algebraic condition on φ_λ and this condition turns out to be identical with (16) . It follows then again that essential discontinuities, i.e. an electromagnetic shock wave can exist only on null surfaces. The propagation relations are now derived from the derivative of order $m + 1$ of the field equation and they are identical with the relations (33) and (37).

Electromagnetic shock waves of order $n = 1$ are also possible and they are described by exactly the same equations as for $n \geq 2$. However, the derivation of these equations is obtained now by a different reasoning. Indeed, in the presence of essential discontinuities of the first derivatives,

$$[A_{\lambda,\mu}] = \varphi_\lambda p_\mu \quad , \tag{39}$$

the left-hand side of equation (3) will in general contain a term proportional to the Dirac function $\delta(z)$. This represents a surface distribution of electromagnetic sources on Σ . In order to have the vacuum equations satisfied we must put the coefficient of $\delta(z)$ equal to zero. It is this equation which leads now to the algebraic condition (16). The propagation relations are derived in this case directly from the field equation (3).

Finally we mention that electromagnetic shock waves of order n = 0 are formally compatible with the field equation (3). In this case, however, the electromagnetic field tensor $F_{\lambda\mu}$ contains a factor $\delta(z)$. Consequently the electromagnetic energy density will be proportional to $\{\delta(z)\}^2$. This remark shows that electromagnetic shock waves of order n = 0 are physically meaningless.

II. GRAVITATIONAL SHOCK WAVES

7. The Shock Wave of Order n = 2

The vacuum field equation is in this case :

$$R_{\lambda\mu} = 0 . \tag{40}$$

This is an equation of the second order for the metric tensor $g_{\lambda\mu}$, which has to be considered as the gravitational potential. The terms containing derivatives of the second order are given in the following formula :

$$R_{\lambda\mu} = \frac{1}{2} g^{\alpha\beta}\left(g_{\lambda\mu,\alpha\beta} + g_{\alpha\beta,\lambda\mu} - g_{\alpha\lambda,\beta\mu} - g_{\alpha\mu,\beta\lambda}\right) + \cdots . \tag{41}$$

The omitted terms are quadratic in the first derivatives of $g_{\lambda\mu}$.

Comparing with the Maxwell equation (3) we remark that equation (41) is non-linear. It is of course hyperbolic, since $g_{\lambda\mu}$ has the same signature as the metric of the Minkowski space. Moreover it is linear in the second derivatives of $g_{\lambda\mu}$, though with non-constant coefficients. It is this linearity in the second derivatives which permits the exact discussion of the gravitational shock waves, as we shall see in the following. This is particularly important in view of the fact that all realistic problems involving gravitational radiation have to be treated by approximation methods. Compared with the discussion of electromagnetic shock waves there will be now some minor complications resulting from the fact that equation (41) is essentially more complicated than the Maxwell equation (3).

We shall consider in detail a shock wave of order $n = 2$. Let the hypersurface Σ be defined again by the equation

$$z(x^\alpha) = 0.$$

Then in the case of a shock wave of order $n = 2$ we shall have :

(42) $$[g_{\lambda\mu}] = 0 , \quad [g_{\lambda\mu,\nu}] = 0 ;$$

(43) $$[g_{\lambda\mu,\nu\rho}] = \gamma_{\lambda\mu} p_\nu p_\rho$$

with at least some $\gamma_{\lambda\mu} \neq 0$. The vector p_λ,

$$p_\lambda = \frac{\partial z}{\partial x^\lambda}$$

is again the normal to Σ.

As we shall show in detail now, the local condition on the discontinuity tensor $\gamma_{\lambda\mu}$ will follow from the demand that the field equation (41) be valid on both sides of Σ. The propagation relations will follow from the similar demand on the first derivatives of the equation (41), in complete analogy with what we have found in the electromagnetic case.

Demanding the validity of equation (41) on both sides of Σ leads to the relation :

$$[R_{\lambda\mu}] = 0$$

Since $g_{\lambda\mu}$ and $g_{\lambda\mu,\nu}$ are continuous on Σ, we find from (41):

(44') $$2[R_{\lambda\mu}] = g^{\alpha\beta}\{[g_{\lambda\mu,\alpha\beta}] + [g_{\alpha\beta,\lambda\mu}] - [g_{\alpha\lambda,\beta\mu}] - [g_{\alpha\mu,\beta\lambda}]\} = 0.$$

Therefore from (43) :

(44) $$g^{\alpha\beta}\left(\gamma_{\lambda\mu} p_\alpha p_\beta + \gamma_{\alpha\beta} p_\lambda p_\mu - \gamma_{\alpha\lambda} p_\beta p_\mu - \gamma_{\alpha\mu} p_\beta p_\lambda\right) = 0.$$

For the discussion of this equation we have to consider separately each one of the following two cases :

(45) (i) $g^{\alpha\beta} p_\alpha p_\beta \equiv p^\alpha p_\alpha \neq 0$; (ii) $p^\alpha p_\alpha = 0$,

which are the direct generalization of the two cases (11).

In the case (45 i) equation (44) shows that $\gamma_{\lambda\mu}$ will be of the general form :

(46) $$\gamma_{\lambda\mu} = h_\lambda p_\mu + h_\mu p_\lambda .$$

We shall now prove that discontinuities of the form (46) are trivial : They can be eliminated or produced by a transformation to new coordinates $\tilde{x}^\mu$,

$$x^\mu = f^\mu(\tilde{x}^\nu),$$

which are determined by functions f^μ having discontinuous derivatives of the third order. Indeed, from the transformation formula

$$\tilde{g}_{\lambda\mu} = \frac{\partial x^\alpha}{\partial \tilde{x}^\lambda}\frac{\partial x^\beta}{\partial \tilde{x}^\mu} g_{\alpha\beta} \equiv f^\alpha_{,\lambda} f^\beta_{,\mu} g_{\alpha\beta}$$

we find after two derivations :

$$\tilde{g}_{\lambda\mu,\nu\rho} = (f^\alpha_{,\lambda\nu\rho} f^\beta_{,\mu} + f^\alpha_{,\lambda} f^\beta_{,\mu\nu\rho}) g_{\alpha\beta} + f^\alpha_{,\lambda} f^\beta_{,\mu} g_{\alpha\beta,\gamma\delta} f^\gamma_{,\nu} f^\delta_{,\rho} + \dots,$$

the omitted terms being continuous on Σ . It follows :

(47) $$[\tilde{g}_{\lambda\mu,\nu\rho}] = \{[f^\alpha_{,\lambda\nu\rho}] f^\beta_{,\mu} + f^\alpha_{,\lambda}[f^\beta_{,\mu\nu\rho}]\} g_{\alpha\beta} + [g_{\alpha\beta,\gamma\delta}] f^\alpha_{,\lambda} f^\beta_{,\mu} f^\gamma_{,\nu} f^\delta_{,\rho} .$$

Now we have :

(48) $$[f^\alpha_{,\lambda\nu\rho}] = F^\alpha p_\lambda p_\nu p_\rho , \qquad F^\alpha = \left[\frac{\partial^3 f^\alpha}{\partial z^3}\right],$$

and according to (43) and (46) :

$$[g_{\alpha\beta,\gamma\delta}] = \gamma_{\alpha\beta} p_\gamma p_\delta = (h_\alpha p_\beta + h_\beta p_\alpha) p_\gamma p_\delta .$$

Remembering that

(49) $$\tilde{p}_\lambda = f^\alpha_{,\lambda} p_\alpha$$

we obtain finally :

(50) $$[\tilde{g}_{\lambda\mu,\nu\rho}] = (F^\alpha g_{\alpha\beta} + h_\beta)(f^\beta_{,\mu}\tilde{p}_\lambda + f^\beta_{,\lambda}\tilde{p}_\mu)\tilde{p}_\nu \tilde{p}_\rho .$$

Consequently it is sufficient to use functions $f^\mu(\tilde{x}^\alpha)$ with discontinuities of the third derivatives satisfying the relation

$$F^\mu = -g^{\mu\nu} h_\nu$$

in order to obtain

$$[\tilde{g}_{\lambda\mu,\nu\rho}] = 0.$$

The same conclusion can be reached by the following reasoning. Consider the Riemann tensor :

(51) $$R_{\lambda\mu\nu\rho} = \frac{1}{2}(g_{\lambda\rho,\mu\nu} + g_{\mu\nu,\lambda\rho} - g_{\lambda\nu,\mu\rho} - g_{\mu\rho,\lambda\nu}) + \dots ;$$

the omitted terms being quadratic in the first derivatives. For the discontinuities of the components of the Riemann tensor we find with the help of (43) the general formula :

(52) $$[R_{\lambda\mu\nu\rho}] = \frac{1}{2}(\gamma_{\lambda\rho}p_\mu p_\nu + \gamma_{\mu\nu}p_\lambda p_\rho - \gamma_{\lambda\nu}p_\mu p_\rho - \gamma_{\mu\rho}p_\lambda p_\nu).$$

When $\gamma_{\lambda\mu}$ has the form (46), one verifies at once that

$$[R_{\lambda\mu\nu\rho}] = 0,$$

a result which shows again that these discontinuities of $g_{\lambda\mu,\nu\rho}$ are trivial.

8. The Algebraic Structure of the Tensor $\gamma_{\lambda\mu}$

From the preceding result it follows that essential discontinuities can exist only in the case (45 ii) : A gravitational shock wave can be propagated only on a null or characteristic surface of the riemannian space with the metric $g_{\lambda\mu}$. Here also we can prove without difficulty that the trajectories of the null-vector field p_λ defined on Σ are null-geodesics of the riemannian space. They are called the bicharacteristics of the surface Σ. Besides the vector p_λ we have on Σ two other independent directions : We introduce again two vectors a_λ and b_λ satisfying the relations (17) , (18) and (19).

With (45 ii) equation (44) is reduced to :

(53) $$\gamma p_\lambda p_\mu - \gamma_{\lambda\alpha}p^\alpha p_\mu - \gamma_{\mu\alpha}p^\alpha p_\lambda = 0 \ ; \qquad \gamma \equiv g^{\alpha\beta}\gamma_{\alpha\beta}$$

This equation can be written in the form :

(54) $$q_\lambda p_\mu + q_\mu p_\lambda = 0 \ ;$$

(55) $$q_\lambda \equiv -\gamma_{\lambda\alpha}p^\alpha + \frac{1}{2}\gamma p_\lambda \ ,$$

and is therefore equivalent to :

(56) $$q_\lambda = 0 \iff \gamma_{\lambda\alpha}p^\alpha = \frac{1}{2}\gamma p_\lambda .$$

These are 4 algebraic equations for the 10 components $\gamma_{\lambda\mu}$. Consequently the general solution $\gamma_{\lambda\mu}$ of (56) will contain 6 arbitrary quantities.

One can verify at once that a $\gamma_{\lambda\mu}$ of the form (46), which contains 4 arbitrary quantities h_λ , satisfies the equation (56). It follows that the general solution of (56) can be written in the form :

(57)
$$\gamma_{\lambda\mu} = h_\lambda p_\mu + h_\mu p_\lambda + \gamma'_{\lambda\mu} ,$$

the last term $\gamma'_{\lambda\mu}$ containing two arbitrary quantities. Note that $\gamma'_{\lambda\mu}$ is the essential part of the discontinuity tensor $\gamma_{\lambda\mu}$.

The detailed structure of $\gamma'_{\lambda\mu}$ is obtained easily if we introduce a coordinate system such that at the point we are considering we have :

$$g_{\lambda\mu} = \eta_{\lambda\mu} \quad \text{and} \quad p_\lambda = (1,0,0,1).$$

The final result is the following : The essential discontinuity tensor $\gamma'_{\lambda\mu}$ can be written in the form :

(58)
$$\gamma'_{\lambda\mu} = A_\lambda B_\mu + A_\mu B_\lambda$$

with :

(58')
$$A_\lambda = \alpha a_\lambda + \beta b_\lambda \quad , \quad B_\lambda = -\beta a_\lambda + \alpha b_\lambda .$$

The coefficients α and β are the two arbitrary quantities contained in $\gamma'_{\lambda\mu}$. In what follows we shall omit the trivial part of $\gamma_{\lambda\mu}$,i.e. we shall write :

(59)
$$\gamma_{\lambda\mu} = \gamma'_{\lambda\mu} .$$

It is interesting to compare this result with what we found in the Maxwell case. Here we have the discontinuity tensor $\gamma'_{\lambda\mu}$, while in the Maxwell case we had the vector φ'_λ. What we have in both cases in common is that the shock wave is transverse and has two degrees of freedom, corresponding to two possible states of polarization of the wave.

As in the Maxwell case, here also we can introduce the complex null-vector m_λ defined by (21). The result (58') takes then the form :

(60)
$$A_\lambda = E^{1/4}\left(e^{i\delta} m_\lambda + e^{-i\delta}\bar{m}_\lambda\right) , \quad B_\lambda = \frac{1}{i} E^{1/4}\left(e^{i\delta} m_\lambda - e^{-i\delta}\bar{m}_\lambda\right).$$

E is the amplitude and δ the phase of the gravitational wave.

9. The Propagation Relations

These relations are obtained from the field equation (41) derived with respect to x^ν. We find :

(61)
$$R_{\lambda\mu,\nu} = \frac{1}{2} g^{\alpha\beta}\left(g_{\lambda\mu,\alpha\beta\nu} + g_{\alpha\beta,\lambda\mu\nu} - g_{\alpha\lambda,\beta\mu\nu} - g_{\alpha\mu,\beta\lambda\nu}\right) + \cdots .$$

The omitted terms are bilinear in the first and second derivatives of $g_{\lambda\mu}$. It follows that in order to obtain the detailed form of the equation

$$[R_{\lambda\mu,\nu}] = 0 \tag{62}$$

we have to calculate the discontinuities of the third order derivatives of $g_{\lambda\mu}$.

Let us denote by $\bar{g}_{\lambda\mu}$ and $\overset{+}{g}_{\lambda\mu}$ respectively the $g_{\lambda\mu}$ in the regions U^- and U^+. We now write a relation of the form (25) :

$$\overset{+}{g}_{\lambda\mu} = \bar{g}_{\lambda\mu(+)} + \frac{1}{2} z^2 \gamma_{\lambda\mu} + \frac{1}{6} z^3 \varepsilon_{\lambda\mu} + \cdots ; \tag{63}$$

$\bar{g}_{\lambda\mu(+)}$ is a continuation of $\bar{g}_{\lambda\mu}$ in the region $0 < z < \varepsilon$ of U^+, such that the derivatives up to the order 3 of the function defined by $\bar{g}_{\lambda\mu}$ and $\bar{g}_{\lambda\mu(+)}$ are continuous on Σ. We then find at once :

$$\overset{+}{g}_{\lambda\mu,\nu\rho\sigma} = \bar{g}_{\lambda\mu(+),\nu\rho\sigma} + (\gamma_{\lambda\mu} P_\nu P_\rho)_{,\sigma} + \{(\gamma_{\lambda\mu} P_\nu)_{,\rho} + (\gamma_{\lambda\mu,\nu} + \varepsilon_{\lambda\mu} P_\nu) P_\rho\} P_\sigma + \cdots , \tag{64}$$

the omitted terms having at least one factor z. Therefore :

$$[g_{\lambda\mu,\nu\rho\sigma}] = (\gamma_{\lambda\mu} P_\nu P_\rho)_{,\sigma} + \{(\gamma_{\lambda\mu} P_\nu)_{,\rho} + (\gamma_{\lambda\mu,\nu} + \varepsilon_{\lambda\mu} P_\nu) P_\rho\} P_\sigma . \tag{65}$$

In (61) it will be sufficient to consider only the first 4 terms : All remaining terms vanish in a coordinate system which is geodesic at the point we are considering. Introducing in (62) the expressions (61) and (65) we find the following lengthly relation :

$$\left\{ \begin{aligned} & g^{\alpha\beta} \Big\{ (\gamma_{\lambda\mu} P_\alpha P_\beta + \gamma_{\alpha\beta} P_\lambda P_\mu - \gamma_{\alpha\lambda} P_\beta P_\mu - \gamma_{\alpha\mu} P_\lambda P_\beta)_{,\nu} + \big((\gamma_{\lambda\mu} P_\alpha)_{,\beta} + (\gamma_{\alpha\beta} P_\lambda)_{,\mu} \\ & \qquad - (\gamma_{\alpha\lambda} P_\mu + \gamma_{\alpha\mu} P_\lambda)_{,\beta} + \gamma_{\lambda\mu,\alpha} P_\beta + (\gamma_{\alpha\beta,\lambda} + \varepsilon_{\alpha\beta} P_\lambda) P_\mu \\ & \qquad - (\gamma_{\alpha\lambda,\mu} + \gamma_{\alpha\mu,\lambda} + \varepsilon_{\alpha\lambda} P_\mu + \varepsilon_{\alpha\mu} P_\lambda) P_\beta \big) P_\nu \Big\} = 0. \end{aligned} \right. \tag{66}$$

Since we are using a geodesic coordinate system, we can move the factor $g^{\alpha\beta}$ across the differential operator. The first 4 terms of (66) can then be written in the form :

$$(\gamma_{\lambda\mu} P_\alpha P^\alpha + q_\lambda P_\mu + q_\mu P_\lambda)_{,\nu} = \gamma_{\lambda\mu} (P^\alpha P_\alpha)_{,\nu} + q_{\lambda,\nu} P_\mu + q_{\mu,\nu} P_\lambda ,$$

the vector q_λ being defined by (55). We have used the fact that $P_\alpha P^\alpha$ and q_λ vanish on Σ. The remaining terms of (66) are treated similarly and, if we remember that

$$P_{\lambda,\mu} = \frac{\partial^2 z}{\partial x^\lambda \partial x^\mu} = P_{\mu,\lambda}$$

and that, according to (58) and (58') :

$$\gamma = g^{\alpha\beta}\gamma_{\alpha\beta} = g^{\alpha\beta}\gamma'_{\alpha\beta} = 0,$$

we find finally from equation (66) :

$$\gamma_{\lambda\mu}(p_\alpha p^\alpha)_{,\nu} + q_{\lambda,\nu}p_\mu + q_{\mu,\nu}p_\lambda + p_\nu\{q_{\lambda,\mu} + q_{\mu,\lambda} + 2\gamma_{\lambda\mu,\alpha}p^\alpha \tag{67}$$
$$+ \gamma_{\lambda\mu}p^\alpha_{,\alpha} - \gamma_{\lambda}{}^{\alpha}{}_{,\alpha}p_\mu - \gamma_{\mu}{}^{\alpha}{}_{,\alpha}p_\lambda + \varepsilon p_\lambda p_\mu - \varepsilon_{\lambda\alpha}p^\alpha p_\mu - \varepsilon_{\mu\alpha}p^\alpha p_\lambda\} = 0.$$

In this equation we eliminate the first term by passing from the vector p_λ to p^*_λ satisfying the condition (31). We then derive two relations which do not contain $\varepsilon_{\lambda\mu}$. The first relation is obtained if we multiply (67) by $\gamma^{\lambda\mu} = \gamma'^{\lambda\mu}$. Since $\gamma^{\lambda\mu}$ does lie on Σ , according to (58) and (58'), the terms containing a factor p_λ or p_μ and also the terms with $q_{\lambda,\mu}$ or $q_{\mu,\lambda}$ disappear. The final result is :

$$\gamma^{\lambda\mu}(2\gamma_{\lambda\mu,\alpha}p^{*\alpha} + \gamma_{\lambda\mu}p^{*\alpha}_{,\alpha}) = 0. \tag{68'}$$

This equation can be written in the simpler form :

$$(\gamma^{\lambda\mu}\gamma_{\lambda\mu}p^{*\alpha})_{,\alpha} = 0. \tag{68}$$

This is the first propagation relation. In an arbitrary coordinate system this relation has to be written in the general form :

$$(\gamma^{\lambda\mu}\gamma_{\lambda\mu}p^{*\alpha})_{;\alpha} = 0 \iff (\gamma^{\lambda\mu}\gamma_{\lambda\mu}p^{*\alpha}\sqrt{-g})_{,\alpha} = 0. \tag{69}$$

The scalar $\gamma^{\lambda\mu}\gamma_{\lambda\mu}$ can be calculated from (58) and (60) and is :

$$\gamma^{\lambda\mu}\gamma_{\lambda\mu} = 8E. \tag{70}$$

Therefore equation (69) can be written also in the form :

$$(Ep^{*\alpha}\sqrt{-g})_{,\alpha} = 0, \tag{71}$$

which expresses the conservation of the amplitude of the gravitational shock wave. Note that this equation is of exactly the same form as the electromagnetic equation (33').

The second propagation relation will be obtained if we multiply equation (67) by

$$\tilde{\gamma}^{\lambda\mu} = A^\lambda A^\mu - B^\lambda B^\mu.$$

The tensor $\tilde{\gamma}^{\lambda\mu}$ lies again on Σ :

$$\tilde{\gamma}^{\lambda\mu}p_\mu = 0,$$

and consequently the result of this multiplication is :

(72) $$\tilde{\gamma}^{\lambda\mu}\left(2\gamma_{\lambda\mu,\alpha}\overset{*}{p}{}^{\alpha}+\gamma_{\lambda\mu}\overset{*}{p}{}^{\alpha}{}_{,\alpha}\right)=0.$$

One verifies at once that

(73) $$\tilde{\gamma}^{\lambda\mu}\gamma_{\lambda\mu}=0,$$

so that equation (72) reduces to its first term. Remembering the relation imposed on m_λ :

$$m_{\lambda;\mu}p^\mu=0,$$

which follows immediately from (21) and (19), we find after some elementary calculations the following simple result :

(74) $$\delta_{,\alpha}p^\alpha=0.$$

This equation also has the same form as the electromagnetic equation (37).

What remains now from equation (67) is an algebraic condition for the functions $\varepsilon_{\lambda\mu}$, which we shall not discuss further.

10. Shock Waves of order n ≠ 2

Gravitational shock waves of order $n = 2 + m$, $m > 0$ are treated in the same way as the corresponding electromagnetic shock waves. The discontinuity tensor $\gamma_{\lambda\mu}$ is now defined from the derivatives of $g_{\lambda\mu}$ of order $2 + m$:

(75) $$[g_{\lambda\mu,\nu\rho\cdots}]=\gamma_{\lambda\mu}p_\nu p_\rho\cdots.$$

The algebraic relation for $\gamma_{\lambda\mu}$ is obtained from the field equation (41) differentiated m times with respect to the coordinates x^ν .One finds again that non-trivial discontinuities can exist only on surfaces Σ which are null-surfaces :

(76) $$p_\lambda p^\lambda=0.$$

The propagation relations are derived from the field equation differentiated $m + 1$ times. The final algebraic as well as propagation relations satisfied by $\gamma_{\lambda\mu}$ are identical with those found in the case n = 2 .

Gravitational shock waves of order n = 1 are compatible with the field equation (41). They are treated in the same way as the electromagnetic shock waves of order n = 1 . The final relations obtained for $\gamma_{\lambda\mu}$, which is now defined by the derivatives of first order:

(77) $$[g_{\lambda\mu,\nu}] = \gamma_{\lambda\mu} p_\nu ,$$

are again identical with those derived for the case $n = 2$.

We only mention that gravitational shock waves of order $n = 0$ have been proved to be incompatible with the field equation (41). This had to be expected, if we remember that electromagnetic shock waves of order $n = 0$ carry an infinite amount of electromagnetic energy : We know that there are no solutions of the Einstein equation (41) corresponding to an infinite total energy.

III. SHOCK WAVES IN THE EINSTEIN-MAXWELL THEORY

11. Local Conditions and Propagation Relations

In this case, we have to start from the " electrovacuum " field equations. There will be no electromagnetic sources and consequently we have the Maxwell equation for vacuum, the only difference being that we have now a curved riemannian space with the metric $g_{\lambda\mu}$:

(78) $$\begin{cases} F^{\lambda\mu}{}_{;\mu} = 0 \quad ; \\ F_{\lambda\mu} = A_{\mu,\lambda} - A_{\lambda,\mu} , \qquad F^{\lambda\mu} = g^{\lambda\alpha} g^{\mu\beta} F_{\alpha\beta} . \end{cases}$$

In the gravitational field equation we shall have as a source the Maxwell tensor of the electromagnetic field :

(79) $$R_{\lambda\mu} = - \varkappa S_{\lambda\mu} ,$$

(80) $$S_\lambda{}^\mu = - F_{\lambda\alpha} F^{\mu\alpha} + \frac{1}{4} \delta_\lambda^\mu F_{\alpha\beta} F^{\alpha\beta}$$

We have put in (79)

$$R \equiv R^\alpha{}_\alpha = 0$$

because of the vanishing of the trace of the Maxwell tensor :

$$S^\alpha{}_\alpha = 0.$$

For the discussion of a shock wave in this theory we have to consider the Maxwell as well as the Einstein equation. The calculations for the Maxwell equation will be now slightly more general than those in special relativity, because of the use of the riemannian metric $g_{\lambda\mu}$ instead of $\eta_{\lambda\mu}$. Similarly the calculations for the Einstein equation will be a little more complicated because of the presence of the Maxwell tensor $S_{\lambda\mu}$. We shall not give details of the calculations, but only the final results, which are obtained on the assumption that we have a shock wave of the same order for both, the gravitational and the electromagnetic field. The results are the same for any $n \geq 1$.

The local conditions, determining the algebraic structure of the discontinuities, are obtained in the case $n = 2$ from the field equations (78) and (79). It is easy to see that these relations are identical with the equations (22), (58) and (60) derived previously for each one of the two fields separately :

$$\varphi_\lambda = \sqrt{J}\left(e^{i\psi} m_\lambda + e^{-i\psi}\bar{m}_\lambda\right), \tag{81}$$

$$\left\{ \begin{aligned} &\gamma_{\lambda\mu} = A_\lambda B_\mu + A_\mu B_\lambda , \\ &A_\lambda = E^{1/4}\left(e^{i\delta} m_\lambda + e^{-i\delta}\bar{m}_\lambda\right), \quad B_\lambda = \frac{1}{i} E^{1/4}\left(e^{i\delta} m_\lambda - e^{-i\delta}\bar{m}_\lambda\right). \end{aligned} \right. \tag{82}$$

In the propagation relations the two fields are no more separated. The explanation is straightforward : In order to derive the propagation relations in the case $n = 2$ we have to use for the electromagnetic field the first derivative of equation (78), which contains second derivatives of $g_{\lambda\mu}$. Similarly for the gravitational field we have to use the first derivative of equation (79) , which contains second derivatives of A_λ . There are two propagation relations for the electromagnetic and two for the gravitational field. They are the following :

$$\left\{ \begin{aligned} &\left(2 J p^\alpha\right)_{;\alpha} = F_{\alpha\beta}\gamma^{\alpha\mu}\varphi_\mu p^\beta , \\ &\sqrt{J}\, \psi_{,\alpha} p^\alpha = \frac{1}{2}\sqrt{E}\, F_{\alpha\beta}\left\{e^{i(2\delta-\psi)} m^\alpha + e^{-i(2\delta-\psi)}\bar{m}^\alpha\right\} p^\beta ; \end{aligned} \right. \tag{83}$$

$$\left\{ \begin{aligned} &\left(2 E p^\alpha\right)_{;\alpha} = -\varkappa F_{\alpha\beta}\gamma^{\alpha\mu}\varphi_\mu p^\beta , \\ &\sqrt{E}\, \delta_{,\alpha} p^\alpha = \frac{\varkappa}{4}\sqrt{J}\, F_{\alpha\beta}\left\{e^{i(2\delta-\psi)} m^\alpha + e^{-i(2\delta-\psi)}\bar{m}^\alpha\right\} p^\beta . \end{aligned} \right. \tag{84}$$

These formulae show that, if there is an initial electromagnetic field on Σ :

$$F_{\alpha\beta} \neq 0 ,$$

the amplitudes and phase angles of each field are not conserved. There is, however, a conservation law for a combined amplitude of the two fields expressed by the equation :

$$\left\{\left(E + \varkappa J\right) p^\alpha\right\}_{;\alpha} = 0 , \tag{85}$$

which follows immediately from the first equations (83) and (84). Similarly we get from the two last equations (83) and (84) the relation :

$$(\delta_{,\alpha} p^{\alpha}) / (\psi_{,\beta} p^{\beta}) \equiv \frac{d\delta}{d\psi} = \frac{\varkappa}{2} \frac{J}{E} . \tag{86}$$

References

For shock waves in the Einstein and the Einstein-Maxwell theory:

F.K. Stellmacher, Math. Ann. 115 (1938), p. 740.

H. Treder , Ann. Phys. (7) 2 (1958), p. 225.

A. Papapetrou and H. Treder , Math. Nachr. 20 (1959), p. 53 and 23 (1961), p. 371.

A. Lichnerowicz, Comptes Rendus Acad. Sc. Paris A273 (1971), p. 528 and A276 (1973), p. 1385.

EDITORIAL COMMENTS

GRAVITATIONAL RADIATION ANTENNAS AND DATA ANALYSES

By 1975 a number of gravitational radiation antenna installations were operating, and research was being carried on to develop lower temperature, more sensitive antennas. The sensitivity and data analyses continue to be controversial. Some, but not all, points of view are discussed here.

THE DETECTION OF GRAVITATIONAL WAVES AND QUANTUM NONDISTURBTIVE MEASUREMENTS*

V.B. BRAGINSKY
MOSCOW STATE UNIVERSITY
Moscow, U.S.S.R.

These lectures describe the different problems which have to be solved by the experimentalists working in one of the most interesting areas of astrophysics - the detection of nonterrestrial origin gravitational waves.

It is possible to divide this area into two parts: the detection of high frequency gravitational waves and the detection of low frequency gravitational waves. The first part of the lectures contains the discussions about the conditions for the high frequency gravitational antennae including the conditions for e.m. quantum detector. The second part of the lectures concerns the prospects for the low frequency gravitational antennae.

HIGH FREQUENCY GRAVITATIONAL DETECTORS

The high frequency gravitational waves in our Universe may be produced by such a very rare event as a collision or very close fly-by of two neutron stars or two black holes or nonsymmetric collapse. The theoretical-astrophysicists (K.S. Thorne, I.D. Novikov, J. Ostriker) predict that the total radiated energy in one such event is

$$\varepsilon_g \simeq 10^{52} \to 10^{54} \text{ erg}$$

and that the duration of a gravitational pulse must be

$$\hat{\tau} \simeq 10^{-3} \to 10^{-4} \text{ sec.}$$

*EDITORS NOTE: Discussions of Braginsky's most recent proposals will appear in the Journal of General Relativity and Gravitation, with title "The Quantum Particularity of a Ponderomotive Electromagnetic Energy Measuring Device" by Braginsky, Vorontsov, and Halili.

Unfortunately, at the present time there is no definite prediction as how frequent these events are. Planning the construction of a sensitive gravitational antenna , an experimentalist can make some primitive estimates which are based on the necessity to detect approximately 10 events per year. The number 10 is important regarding the duration of the experimentalist's life and the attitude of the University's Government towards long, not too promising experiments. Table I shows these estimates.

In the first column of Table I, there is the distance R between the terrestrial laboratory and the source of radiation; in the second - the number of galaxies in a sphere $4/3\ \pi R^3$; in the third - the frequency Ω of the collisions we must ask nature to give 10 events per year; the fourth column gives the level $\tilde{I}$ of the density of energy in erg/cm^2 for one pulse of radiation; in the fifth column - the amplitude of vibrations of Weber's type antenna which eigen frequency $\omega_M \simeq \omega_g \simeq 1/\hat{\tau}_g$ and length 50 cm. It is reasonable to assume that the first line of table I represents the optimistic point of view and the second - the pessimistic one.

The well-known fundamental Weber's equation for a quadrupole mass-detector gives the classical condition for the detection of a gravitational wave:

$$m\ddot{x}^\mu + H\dot{x}^\mu + Kx^\mu = -mc^2 \ell^\alpha R^\mu_{o\alpha o} + F_N. \tag{1}$$

The spectral density of Nyquist force is

$$(F_N^2)_f = 4kT_M H, \tag{2}$$

where k is Boltzmann's constant, T_M - is the temperature of the mechanical heat bath. The classical condition is very simple:

$$mc^2 \ell^\alpha R^\mu_{o\alpha o} \geq F_{min} \simeq \sqrt{4kT_M H\Delta f}, \tag{3}$$

where Δf - is equal to the main part of the gravitational wave bandwidth.

It is more convenient to rewrite the condition (3) using the density of gravitational energy $\tilde{I}$ in the pulse instead of $R^\mu_{o\alpha o}$:

$$[\tilde{I}]_{min} \gtrsim \frac{c^3 kT_M}{2\pi G\ m\ Q_M \omega_M \ell^2}, \tag{4}$$

TABLE I

R distance from the laboratory	N the number of the galaxies in a sphere $4/3\pi\ R^3$	Ω the frequency of events in one galaxy to detect 10 events per year	$\tilde{I}$ the density grav. wave energy near the antennae	δx_g the response of 50 cm length Weber's type antennae
1×10^{25} cm=3Mpc	3.10^{2}	3.10^{-2} year^{-1}	10^{+3} $\frac{erg}{cm^2}$	10^{-17}cm
3×10^{27}cm horizon of events	10^{10}	10^{-9}year^{-1}	10^{-3} $\frac{erg}{cm^2}$	10^{-20}cm

where G - is the gravitational constant and Q_M - the mechanical quality factor. The factor $\frac{T_M}{mQ_M}$ is in the hands of the experimentalists. In a series of Weber's type of antennae this factor was close to

$$\frac{300}{10^6 . 10^5} \simeq 3.10^{-9},$$

which corresponds to $[\tilde{I}]_{min} \simeq 10^{+6} \frac{erg}{cm^2}$. In the programme of Stanford University - Louisiana University - Istituto Marconi this factor is

$$\frac{3.10^{-3}}{5.10^6 . 10^5} \simeq 10^{-14} \quad \text{and} \quad [\tilde{I}]_{min} \simeq 1 \frac{erg}{cm^2} .$$

In the programme of Moscow University - Physical Institute of Earth this factor is expected to be close to

$$\frac{2}{10^4 \ . 10^{10}} \simeq 2.10^{-14}$$

if it is possible to use a sufficiently large monocrystal of Al_2O_3 (sapphire). It is reasonable to expect to obtain high Q_M factor using monocrystal of dielectric with high Debye temperature. The following Landau equation gives the limit for the product $Q_M\omega_M$ of a mechanical resonator whose heat conductivity is β, the linear heat expansion coefficient is α, the density is ρ and C_p - heat capacity:

$$Q_M\omega_M \simeq \frac{4\rho C_p^2}{\beta T_M \alpha^2} \tag{5}$$

Substituting in equation (5) the experimental data for β, α and C_p which were received for sapphire from the National Bureau of Standards one can obtain the first line in Table II. The second line in Table II is derived from the first line and the equation (4) in which m = 4.10^4 gr and ℓ = 40 cm:

Equation (5) does not take into account the losses for Q_M in the suspension and in the surface. The possibility to eliminate or reduce these types of losses depends only on the level of the skill of an experimentalist. Table II shows the results which were obtained by Bagdasarov, Mitrofanov and me.

TABLE II

T_M	$=300°K$	$4.2°K$	$2°K$	$0.01°K$
$Q_M \omega_M$	$=2.2x10^{16}$	$2.8x10^{17}$	$2x10^{18}$	$3.5x10^{22}$
$[\tilde{I}]_{min}$	$5 \frac{erg}{cm^2}$	$5x10^{-3} \frac{erg}{cm^2}$	$1.5x10^{-7} \frac{erg}{cm^2}$	$10^{-10} \frac{erg}{cm^2}$

TABLE III

$T_M =$	$300°K$	$80°K$	$7°K$	
$Q_M =$	$8x10^7$	$2x10^8$	$7x10^8$	$m=1x10^3 gr$
$\frac{2Q_M}{\omega_M} = \tau^*_M =$	$8x10^2$ sec	$2x10^3$ sec	$7x10^3$ sec	$\omega_M=2.10^5 \frac{rad}{sec}$ Exp. Res.
$\omega_M Q_M =$	1.6×10^{13}	$4x10^{13}$	$1.4x10^{14}$	$\ell = 15$ cm
$\omega_M Q_M =$	$2.2x10^{16}$	$8.4x10^{15}$	$3x10^{16}$	Theor. Pred.

Table III shows that the gaps between the Landau prediction limit and our results at the present time is approximately two orders but nevertheless these results are promising.

Returning to the last, the very optimistic line in table II, it is important to note that these estimates do not take into account two circumstances:

1. It is necessary to have a very sensitive system to detect a small change of amplitude (for example for

$$\tilde{I} = 10^{+3}\ \frac{\text{erg}}{\text{cm}^2} \quad \text{the} \quad \Delta x \simeq 1.10^{-17}\ \text{cm when the bandwidth is}$$

$$\Delta\omega \simeq 2\pi 10^3\ \frac{\text{rad}}{\text{sec}}\ )$$

2. The backward fluctuational influence of the detection system is to be taken into account.

The first circumstance is a serious one. If an experimentalist uses a capacity pickoff system (Figure 1), the smallest change of amplitude $[\Delta x]_{min}$ which is possible to be observed within the bandwidth Δf is equal to

$$[\Delta x]_{min} \simeq \frac{4d}{U_{\sim}} \sqrt{\frac{kT\ \Delta f}{\omega_e Q_e C_e}} \tag{6}$$

where d is the distance between the plates of the capacity, $U_{\sim}$ - is the amplitude of RF or VHF voltage between the plates, ω_e is the eigen frequency of the electrical resonator, Q_e is the quality factor of the resonator and C_e is the capacity. Equation (6) is based on the assumption that the unique source of noise is the electrical heat bath with the temperature T_e. If we substitute in equation (6) $C_e = 10^{-12}$F, $\omega_e = 6 \times 10^{10}$ rad/sec, $T_e = 2°$K, $Q_e = 10^{+11}$, $U/d \simeq 10^6$ v/cm we obtain $[\Delta x]_{min} \simeq 10^{-21}\ \sqrt{\Delta f}$ cm. This estimate shows that the present state of art for Q_e permits to reach the sensitivity

$$10^{+3}\ \frac{\text{erg}}{\text{cm}^2}$$

which corresponds to the "optimistic point of view" (see the first line in table I). To reach the level of sensitivity which corresponds to the "pessimistic point of view" (the second line in Table II) it is necessary to use higher Q_e or lower T_e. It is important to note here that if the averaging time is greater than

$$1/\Delta f_g \simeq 10^{-3} \rightarrow 10^{-4} \quad \text{sec}$$

then to maintain the level of sensitivity the experimentalist has to have a higher factor Q_M.

To analyze the second circumstance which was mentioned before, let us suppose that the Nyquist force is reduced almost to zero (decreasing T_M and increasing Q_M). In this case, it is necessary to take into account only the backward influence of the detection system. The calculations for the two cases give two very simple equations which substitute for equation (3):

$$F_{min} \simeq \frac{4}{\hat{\tau}} \sqrt{kT_e m \frac{\omega_M}{\omega_e}} \tag{7}$$

if $\hat{\tau} >> \tau_e^*$ and

$$F_{min} \simeq \frac{4}{\hat{\tau}} \sqrt{kT_e m \frac{\omega_M}{\omega_e}} \sqrt{\frac{2\hat{\tau}}{\tau_e^*}} \tag{8}$$

if $\hat{\tau} << \tau_e^*$.

In the equation (7) and (8) $\hat{\tau}$ is the averaging time and

$$\tau_e^* = \frac{2Q_e}{\omega_e} .$$

The equations (7) and (8) are classical and they are valid if $kT_e > \hbar\omega_e$. Comparing these two equations one can see that it is promising to decrease the ratio $\hat{\tau}/\tau_e^*$. At a definite level of the ratio $\hat{\tau}/\tau_e^*$ the minimum force F_{min} from equation (8) becomes less than

$$F_{quantum} \simeq \frac{4}{\hat{\tau}} \sqrt{\frac{\hbar\omega_M m}{n}} \tag{9}$$

$F_{quantum}$ - is the amplitude of sinusoidal force which adds or substracts one quantum of energy in the mechanical resonator whose mass

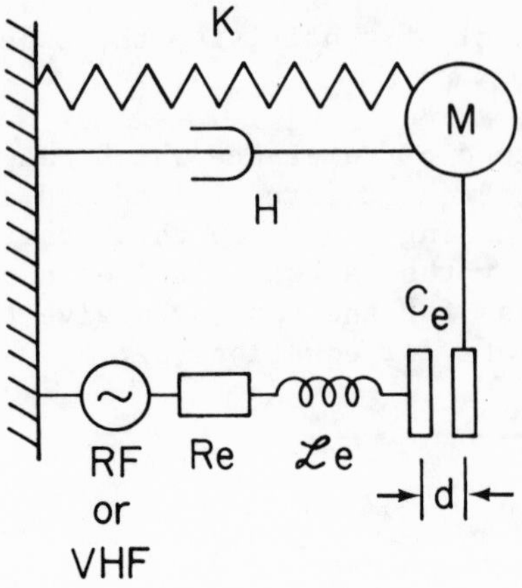

Figure 1

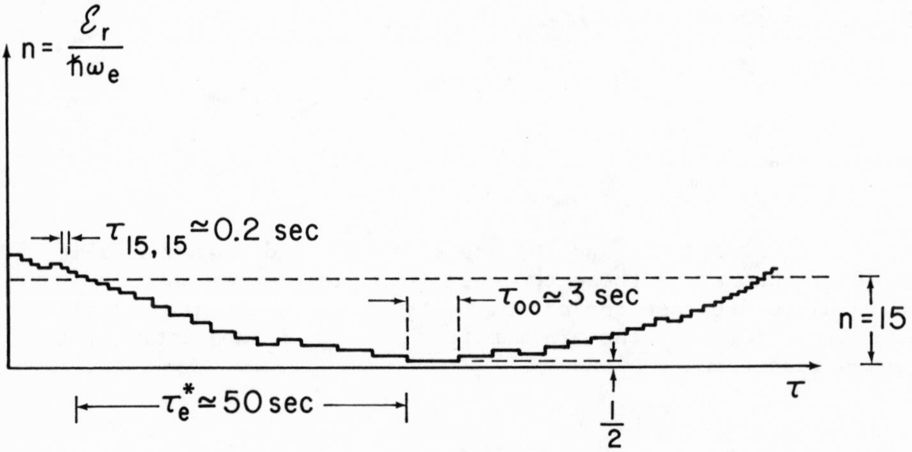

Figure 2

is m, the eigen frequency ω_m and the initial quantum level is n. In other words equation (9) is the limit for the validity of equation (8) when

$$\hat{\tau}/\tau_e^* \ll 1.$$

If we substitute in equation (8) some very optimistic data

$(m = 4 \times 10^{+4}$ gr, $\ell = 50$ cm, $\omega_m = 10^{+4}\ \frac{\text{rad}}{\text{sec}}$,

$T_e = 1°K$, $Q_e = 5.\ 10^{+12}$, $\omega_e = 2\ .\ 10^{10}\ \frac{\text{rad}}{\text{sec}}$,

$\hat{\tau} = 10^{-3}$ sec) and if we combine equation (8) with equation (1) we shall obtain the estimate for $[\tilde{I}]_{min} \simeq 1\ \frac{\text{erg}}{\text{cm}^2}$.

This sensitivity is much less than the estimates which we obtained, taking into account only the Nyquist force for the high Q_m of sapphire (see the last line in table II). This fact permits to make the conclusion that the backward fluctuational force gives the most severe limit for high sensitivity.

The estimates that we obtained ($[\tilde{I}]_{min} \simeq 1\ \frac{\text{erg}}{\text{cm}^2}$) are based on the assumption that we can use a nonexisting electronical voltmeter which permits us not to disturb a long relaxation time τ_e^* in the electrical resonator and at the same time to measure the small electrical signal ΔU_{B2} in the resonator which is very close to the random Brownian changes

$$\Delta U_{B2} \simeq \sqrt{\frac{2kT_e}{C_e}}\ \sqrt{\frac{2\hat{\tau}}{\tau_e^*}}\ , \tag{10}$$

If $T_e = 1°K$, $C_e = 10^{-12}$ F, $\tau_e^* = 50$ sec, $\hat{\tau} = 0.5$ sec, the value $\Delta U_{B2} \simeq 0.7 \times 10^{-6}$ volt. These estimates are very close to the change $\Delta U_{n+1,n}$ of the mean square amplitude of the electrical voltage due to the change of energy which is equal to one quantum:

$$\Delta U_{n+1,n} \simeq \sqrt{\frac{2\hbar\omega_e}{C_e}}\ (\sqrt{n+1} - \sqrt{n}) \tag{11}$$

For $C_e = 10^{-12}$F, $\omega_e = 2\ .\ 10^{+10}\ \frac{\text{rad}}{\text{sec}}$, $n = 7 = \frac{kT_e}{\hbar\omega_e}$

we obtain $\Delta U_{8,7} \simeq 0.4 \times 10^{-6}$ volt. These two estimates show that the electronical voltmeter which we need must measure exactly the number of quanta in the electrical resonator and must not disturb the number (if at all). This is the last serious obstacle in the hard way of the experimentalists which are eager to reach the level of sensitivity

$$\tilde{I} \simeq 1 \frac{\text{erg}}{\text{cm}^2} .$$

THE NONDISTURBTIVE QUANTUM MEASUREMENT

The superconductive microwave resonator have relatively long relaxation time

$$\tau_e^* = \frac{2Q_e}{\omega_e} .$$

John Turneaure from Stanford University reported about the $Q_e \simeq 5 \times 10^{11}$ for a niobium cavity. This Q_e and $\omega_e = 2 \times 10^{10} \frac{\text{rad}}{\text{sec}}$ corresponds to $\tau_e^* \simeq 50$ sec. If we suppose definite T_e of the heat bath it is easy to calculate the mean expected time τ which the eigen mode ω_e has the quantized energy $\varepsilon_e = (1/2 + n) \hbar\omega_e$. For example, if $T_e = 2°K$, then the

$$n_e = \frac{kT_e}{\hbar\omega_e} = 15 \quad \text{and}$$

$$\tau_{15,15} \simeq \frac{\hbar\omega_e}{kT_e} \cdot \frac{\tau_e^*}{n_e} \simeq 0.2 \text{ sec.}$$

If due to the Brownian motion, the eigen mode ω_e will reach the quantum level $\hbar\omega_e/2$ then

$$\tau_{o,o} \simeq \frac{\hbar\omega_e}{kT_e} \tau_e^* \simeq 3 \text{ sec.}$$

These two estimates show that for such cavity during the time $\hat{\tau} \simeq 10^{-3} \rightarrow 10^{-1}$ sec we can measure the energy of the eigen mode with an error $\Delta\varepsilon \simeq \frac{\hbar}{\hat{\tau}} \simeq 10^{-24} \rightarrow 10^{-26}$ erg which is much less than the distance between the quantum levels

$\Delta\varepsilon_{n,\,n-1} = \hbar\omega_e = 2 \times 10^{-17}$ erg (see Fig. 2). In other words, the principle of uncertainty permits to measure the number of quanta and not disturb this number.

It is important to emphasize here that the discreteness of the quantum levels of energy appears when $k\,T_e > \hbar\omega_e$ if the relaxation time τ_e^* is long. A simple calculation shows that this quantum effect appears when

$$\frac{n}{Q_e}\, kT_e \gtrsim \hbar\omega_e, \tag{12}$$

where n is the number of quantum level.

Now let us discuss one of the possible methods which gives the possibility to measure the number of quanta in an eigen mode of e.m. resonator and which also permits to be sure that the number of quanta after the measurement is the same as before the measurement. Suppose that we have a klystron type e.m. resonator which capacity is $C_e = 0.3$ cm and eigen frequency $\omega_e = 2 \times 10^{10}$ rad/sec. If the resonator is at the ground quantum state the mean square electrical voltage between the plates is equal to

$$\sqrt{\langle U_o^2\rangle} = \sqrt{\frac{\hbar\omega_e}{2C_e}} \simeq 5.7 \times 10^{-9}\ \text{CGS} = 1.7 \times 10^{-6}\,\text{V} \tag{13}$$

A beam of electrons which passes between the plates will be disturbed by this electrical field. This disturbance we can detect if we put electrodes (see Figure 2) in the vicinity of the focus of a lense A_1 using the methods from the scanning microscope.

The mean square displacement $\sqrt{\langle\delta y^2\rangle}$ for one random electron will be equal to

$$\sqrt{\langle\delta y^2\rangle} \simeq \frac{2eL}{m_e\omega_e Y v_x}\sqrt{\frac{\hbar\omega_e}{2C_e}} \tag{14}$$

where e - the electrical charge of electron, m_e - the mass of electron, Y - the distance between the plates of the capacity, v_x - the speed of electron. In equation (14) the flying time for electron is supposed to be close to π/ω_e.

If we substitute in equation (14) $Y = 5 \times 10^{-2}$ cm, $C_e = 0.3$ cm, $\omega_e = 2 \times 10^{10} \frac{\text{rad}}{\text{sec}}$, $v_x = 1 \times 10^{10} \frac{\text{cm}}{\text{sec}}$; $L = 10^2$ cm (or $L = 10^3$ cm) we obtain $\sqrt{\langle \delta y^2 \rangle} = 6 \times 10^{-8}$ cm (or 60×10^{-8} cm). These estimates for $\sqrt{\langle \delta y^2 \rangle}$ are not very small if we compare them with the distance D between the two minima of the electrons diffraction patterns near the focus

$$D \simeq \frac{\hbar L}{\pi Y m_e v_x} \tag{15}$$

The value of D is equal 260×10^{-8} cm for $L = 10^2$ cm (and 2600×10^{-8} cm for $L = 10^3$ cm). Using some thousand of electrons it is possible to detect the change of D due to $\sqrt{\langle U_o^2 \rangle}$. The most important question now is how big is the probability of quantum transition due to flight of the electrons through the capacity C_e.

One electron produces the classical force F (t) which excites the resonator

$$F(\tau) = \begin{cases} e\,\dfrac{y_o + v_y \tau}{Y} & \text{when the electron is between the plates} \\ 0 & \text{when the electron is out of the plates.} \end{cases} \tag{16}$$

y_o is the distance between the initial vertical coordinate of the electron and the middle plane (See Figure 3). In the equation (16) the boundary effects and the influence of the field $\sqrt{\langle U_o^2 \rangle}$ on the motion of the electron are neglected. It is well known that if a classical force $F(\tau)$ ($F(\tau) = 0$ when $\tau = \pm \infty$) is exiting a quantum oscillator then the probability of the transition from the ground state into any other states is equal to

$$\sum_{n=1}^{\infty} P_{on} = 1 - \exp(\xi),$$

where

$$\xi = \frac{\omega_e}{2\hbar C_e} \left| \int_{-\infty}^{+\infty} F(\tau) \exp[i\omega_e \tau]\, d\tau \right|^2 .$$

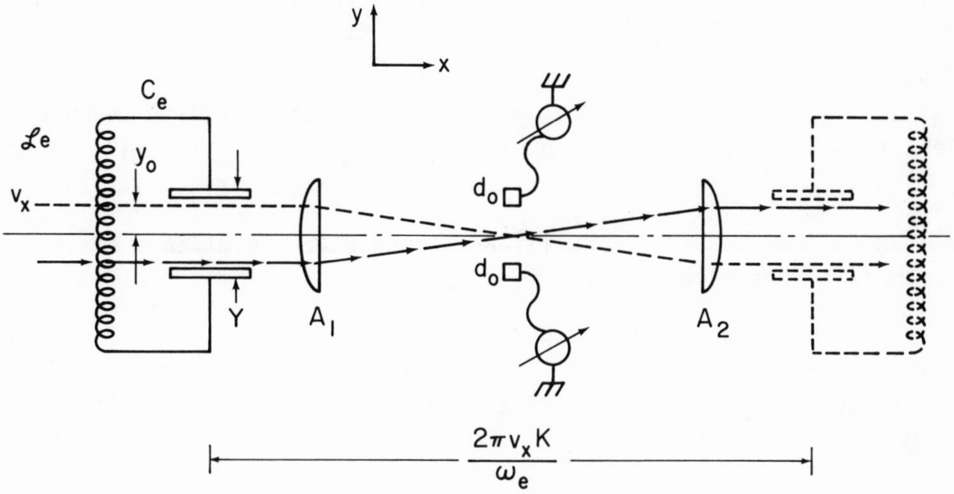

Figure 3

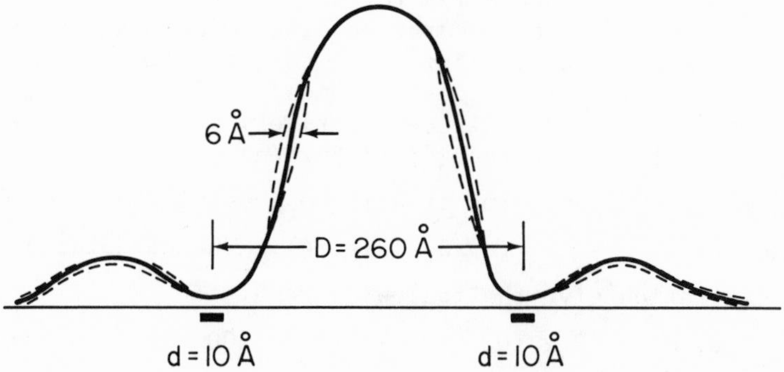

Figure 4

The values y_o and v_y are random. Supposing that $F(\tau) \neq 0$ during $\hat{\tau}_{FL} = \pi/\omega_e$, we obtain

$$\xi = \frac{e^2}{2\hbar\omega_e C_e} \left\{ \frac{1}{3} + \frac{(4+\pi^2) \langle v_y^2 \rangle}{Y^2 \omega_e^2} \right\} \tag{18}$$

The second member in the right part of the equation (18) is much less than the first member. Substituting in the equation (18) $\omega_e = 2 \times 10^{+10}\ \frac{\text{rad}}{\text{sec}}$ and $C_e = 0.3$ cm we obtain $\xi = 6.6 \times 10^{-3}$. This estimates show that the flight of one electron gives a small probability for transitions.

Now let us suppose that due to a system of mirrors and a symmetrically situated lense A_2 one electron flies twice in the resonator (see Fig. 3). If the installation is completely symmetrical and the distance between the lenses is exactly equal to $\frac{2\pi v_x}{\omega_e} \cdot K$ (where K = 1,2,3,..), then ξ is equal to zero. Thus the information before the measurements about v_x and ω_e permit to compensate the force $F(\tau)$ and reduce the value $\sum_{n=1}^{\infty} P_{on}$ close to zero. But to obtain the information about the field $\sqrt{\langle U_o^2 \rangle}$ produced by the energy $\frac{\hbar\omega_e}{2}$ in the resonator it is necessary to use some absorbtive (or semiabsorbtive) electrodes. If we shall use two narrow absorbtive electrodes having width d_o and if we shall situate them close to two minima in the diffractional patterns close to the common focus then the fraction of the absorbed electrons will be relatively small and it will be possible to resolve the effect of disturbance due to $\sqrt{\langle U_o^2 \rangle}$ and at the same time it will be possible to have relatively small value for $\sum_{n=1}^{\infty} P_{on}$. (See Fig. 4). The calculations for a real installation give the following results: if $L = 10^2$ cm, $d_o = 10 \times 10^{-8}$ cm, (or $d_o = 100 \times 10^{-8}$ cm for $L = 10^3$ cm) the total number of electrons $N = 2 \times 10^4$, then two electrodes absorb 1 + 2 electrons without $\sqrt{\langle U_o^2 \rangle}$ and 1 + 6 electrons at the presence of the field

$\sqrt{\langle U_o^2 \rangle} = 1.7 \times 10^{-6}$ volts. For 8 absorbed electrons

$\sum_{n=1}^{\infty} P_{on} \simeq 5 \times 10^{-2}$.

This result was obtained under the assumption that the electrodes are absorbtive, but the technology in scanning microscopy permit us to use very thin and very transparent electrodes. This circumstance increases the level of compensation and for transparent electrodes $\sum_{n=1}^{\infty} P_{on}$ will be less that 5×10^{-2} for the same N, L and d_o.

We have described here the idea of a nondisturbtive procedure for measuring the ground quantum state of the eigen mode of an electromagnetic resonator. The same procedure is possible to use when the energy in the resonator is higher. The reader is referred for details to the original paper. [3,4]

Concluding this part of the lectures, it is important to note that such relatively sophisticated installations will be necessary to reach the sensitivity of gravitational antennae

$$\tilde{I} \simeq I \frac{\text{erg}}{\text{cm}^2} \quad \text{or less.}$$

LOW FREQUENCY GRAVITATIONAL ANTENNAE

If a cluster of dense stars exists in the Center of our Galaxy it is possible to expect an effective conversion of the masses into the gravitational radiation due to frequent fly-by. Ya. B. Zeldovich and S. Polnarev predict that experimentalists located in the Solar System can expect approximately 20 per year gravitational pulses whose duration $\hat{\tau}_g \simeq 1 \rightarrow 100.$ sec and density of energy flux is $\tilde{I} = 10 \rightarrow 1 \frac{\text{erg}}{\text{cm}^2}$. The most reasonable gravitational antennae for such source of radiation is two drag-free satellites with a sensitive Doppler ranging system which measures small variation of the relative speed Δv between them. A pulse of gravitational radiation produces a variation of the relative speed equal to

$$\Delta v_g \simeq \ell \sqrt{\frac{8\pi G}{c^3} \frac{\tilde{I}}{\hat{\tau}_g}}\,, \tag{19}$$

where ℓ - is the distance between the satellites, G - the gravitational constant, c - the speed of light. If $\ell = 10^{12}$ cm, $\hat{\tau}_g = 30$ sec, $\tilde{I} = 3 \frac{\text{erg}}{\text{cm}^2}$ then $\Delta v_g = 1 \times 10^{-7} \frac{\text{cm}}{\text{sec}}$ and the relative acceleration $\Delta a_g \simeq 3 \times 10^{-9} \frac{\text{cm}}{\text{sec}}$

The random accelerations of the frag-free satellite "TRIAD-I" does not exceed 10^{-9} cm/sec^2 if averaging time is $\tau \simeq 10$ sec. (Dr. R. A. Van Patten, private communication). This fact shows that the unique real problem to be solved is the problem of Doppler-ranging system with convenient level of sensitivity. The conditions for the e.m. generator stability is the main problem because the relative variations the frequency $\Delta f/f$ in Doppler systems are close to the relative error for the speed $\Delta v/c$. If we shall use a Sapphire cavity which surface is covered by niobium, we can expect high level of quasi-statistical stability $\Delta f/f$ of the VHF generator due to high value of the Debye temperature for sapphire ($T_d \simeq 1040°K$).

$$\frac{\Delta v_g}{c} \lessapprox \frac{\Delta f_T}{f} \gtrsim K\alpha(T)\Delta T. \tag{20}$$

In the equation (20) K - is the dimentionless factor of compensation which depends on the shape of the cavity $\alpha(T)$ - the coefficient of thermal linear expansion, ΔT - the variation of the temperature. Substituting in equation (20) $K = 10^{-2}$, $\alpha(T) = 5 \times 10^{-12}$ degr^{-1} (for $T = 2°K$) and $T = 1\text{x}10^{-6}°K$ we obtain

$$\frac{\Delta f}{f} \simeq 5\text{x}10^{-20} \quad \text{and} \quad \Delta v \lessapprox c\frac{\Delta f_T}{f} \simeq 1.5 \times 10^{-9} \text{ cm/sec} .$$

Thus the quasi-statical stability with such resonator permits to detect the change of the speed Δv_g which we estimated before.

The second important condition is connected with random deviation of the e.m. frequency:

$$\frac{\Delta v_g}{c} \lessapprox \frac{\Delta f_{random}}{f} \gtrsim \frac{\sqrt{\frac{W(f)}{\hat{\tau}_g}}}{f} = \sqrt{\frac{k T_e}{2PQ_e^2\hat{\tau}_g}} \tag{21}$$

In the equation (21) W (f) - is the spectral density of the fluctuation of frequency, P - the power of the generator.

If we substitute $T_e = 2$ °K, $P = 10^3$ erg/sec, $Q_e = 5 \times 10^{11}$, $\hat{\tau}_g \simeq 30$ sec then we obtain $\Delta v_g \gtrsim 2\text{x}10^{-11}$ cm/sec. Equation (21) is valid for maser type e.m. generator. If one uses instead of maser a tunnel diode then it is necessary to substitute in equation (21) eU instead of kT_e. For a usual tunnel diode and the same P and Q_e the resolving value $\Delta v_g \gtrsim 3\text{x}10^{-10}$ cm/sec. These estimates show that it is important to maintain the high level of Q_e of the cavity. A real tunnel diode decrease the quality factor up to Q_e 10 → 20. For to escape from this difficulty it is possible to use three coupled cavities: one with low $(Q_e)_1$ and a tunnel diode as a negative resistance, another also with low $(Q_e)_2$ and a third cavity with a high $(Q_e)_3$. If the cavities are coupled optimally, the variations of the frequency of a first original generator is stabilized with a factor which is approximately equal to $(Q_e)_3/(Q_e)_2$. A usual tunnel diode permits to reach the quasi-statical stability $\Delta f/f \simeq 1\text{x}10^{-9}$ if the power supply stability is at the level $\Delta U/U \simeq 10^{-6}$. If $(Q_e)_3/(Q_e)_2 = 10^{10}/10 \simeq 10^9$ the estimates for the resolved Δv_g will be approximately $\Delta v_g \gtrsim 10^{-9}$ cm/sec. There are two additional conditions for the e.m. generator to reach the level $\Delta v_g \simeq 1\text{x}10^{-7}$ cm/sec. The dependence of the surface impedance on the temperature gives the frequency instability. For to reach $\Delta v_g \simeq 1 \times 10^{-7}$ cm/sec it is necessary at $T_e = 2°K$ to have $\Delta T_e \simeq 1\text{x}10^{-7}$ °K. A nonstable power supply due to the pondermotive effect also produces the frequency instability. For to reach $\Delta v_g \simeq 1 \times 10^{-7}$ cm/sec it is necessary to have $\Delta P/P \simeq 10^{-7}$ if $P \lesssim 10^{+3}$ erg/sec.

Concluding these remarks about the conditions and difficulties for low frequency gravitational antennae it is possible to expect that this problem may be solved in the next decade.

REFERENCES

1. V.B. Braginsky, The Prospects for High Sensitivity Gravitational Antennae, "Gravitational Radiation and Gravitational Collapse", ed. C. De Witt, Morette, 1974.

2. Ch. Bagdasarov, V.B. Braginsky, V.P. Mitrofanov, Krystallographia, 19, 883, 1974.

3. V.B. Braginsky, Yu.I. Vorontsov,Soviet Physics "Uspechi", 114, 41, 1974.

4. V.B. Braginsky, Yu. I. Vorontsov, V.D. Krivchenkov, JETP, 68, 55, 1975.

5. V.B. Braginsky, I.I. Minakova, V.I. Panov, - to be published.

6. Ya. B. Zeldovich, S. Polnarev, The Radiation of Gravitational Waves from Clusters of High Density Stars. Preprint I.A.M. 1973.

ONEDIMENSIONAL ANALYSIS AND COMPUTER SIMULATION OF A WEBER ANTENNA

G.V.PALLOTTINO

ISTITUTO DI FISICA, UNIVERSITÀ DI ROMA
LABORATORIO PLASMA SPAZIO, CNR, ROMA, ITALY

1. INTRODUCTION TO THE ONEDIMENSIONAL ANALYSIS

The analysis of the Weber type antenna systems for the detection of gravitational waves (GW) is based in general on the representation of the cylinder as a lumped system described by a second order ordinary differential equation, which provides a very simple and powerful tool for a first order analysis of the antenna system, but neglects a number of important characteristics of the real antenna that are related to its distributed nature.

The real antenna, in fact, even if considering only its longitudinal modes, exhibits an infinite number of resonances and can therefore be used as a spectrum analyzer of the GW(1).

Furtherly, the distributed or local nature of the gravitational pseudoforce acting on the single elements of the antenna, that is coherent in space, and of the random thermal pseudoforce, that is non coherent both in space and time, can be exploited for purposes of SNR improvement (2).

The aim of this work is to discuss the response of the gravitational antenna in terms of the onedimensional approximation by means of a time domain approach and by computer simulation.

2. TIME DOMAIN APPROACH TO THE ANALYSIS OF THE ANTENNA

The antenna cylinder, when considered as a distributed onedimensional system of length L can be described by the following partial differential equation (3)

$$\rho \frac{\partial^2 \xi}{\partial t^2} - E \frac{\partial^2 \xi}{\partial z^2} - D \frac{\partial}{\partial t} \frac{\partial^2 \xi}{\partial z^2} = \rho \frac{z}{2} \frac{\partial^2 h_{zz}}{\partial t^2} + f_B(z,t) \quad (1)$$

where ξ is the displacement of each mass element from its rest position, E is the modulus of Young of the material, ρ its density, D the dissipation, when excited by a GW $h_{zz}(t)$, that is the variation of the component of the metric tensor along the axis z, and by the Brownian pseudoforce for unit volume $f_B(z,t)$.

Since the piezoelectric crystals connected to the antenna are strain detectors, the solution must be found in terms of the strain, $s = \partial\xi/\partial z$, to be considered as response or output variable of the system, and eq.(1) can be written as

$$\frac{\partial^2 s}{\partial t^2} - \frac{E}{\rho} \frac{\partial^2 s}{\partial z^2} - \frac{D}{\rho} \frac{\partial}{\partial t} \frac{\partial^2 s}{\partial z^2} = \frac{1}{2} \frac{d^2 h_{zz}}{dt^2} + \frac{1}{\rho} \frac{\partial f(z,t)}{\partial z} \quad (2)$$

with boundary conditions in the form of $s(\pm L/2)=0$ because at the ends of the cylinder the strain is zero in absence of mechanical constraints.

The solution of eq.(2) for arbitrary gravitational excitation can by obtained by several techniques such as analytic direct methods, integral transforms, or superposition of responses to elementary canonical excitations.

Frequency domain techniques such as Fourier and Laplace transforms are based of the isomorphism property of linear differential systems for sinusoidal signals.

In our case a time domain technique such as the Green function or impulsive response method is more suited since the system described by eq.(2) exhibits isomorphism properties for intrisecally simpler signals such as delta functions, provided the dissipative term is sufficiently small.

The above property holds for algebraic system, for algebraic systems with delays and also for distributed nondispersive systems such as the gravitational antenna in our approximation. This means that the response to a GW impulse will always be given by an impulse, albeit delayed or reduced in amplitude, or more exactly by a sequence of such impulses.

If at t=0 we apply a unit impulsive force to the mass element of abscissa z_1 a portion of the force is applied as compression to the right section of the cylinder and the remainder as fraction to the left section.

The two fractions are, in effect, equal to 1/2 because the mechanical impedance is the same in the two directions as seen by the force generator.

The same considerations can be made also for the electric equivalent circuit, as shown in fig.1. We have in fact a floating voltage generator that sees the input impedances of two transmission lines in series: the floating generator is thus equivalent to a pair of grounded generators of half amplitude and opposite polarity. The above equivalence holds only as far as the excitation is concerned since in subsequent times the grounded generators must be eliminated to allow the computation of the free response along the cylinder.

The essence of the method lies in transforming the excitation into suitable boundary conditions for the system in autonomous form and is based on the assumption

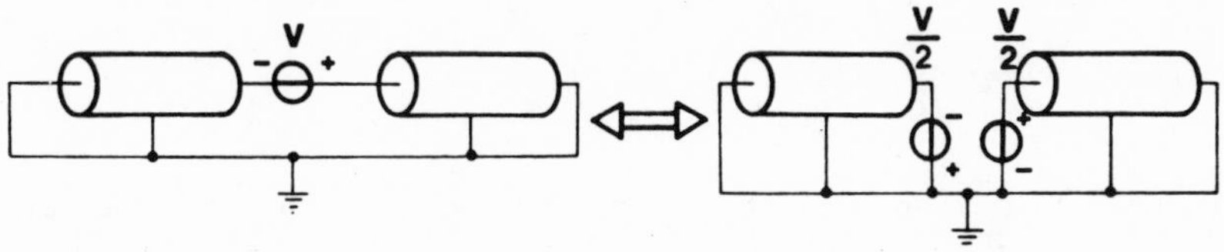

FIG. 1 Equivalence between a floating voltage generator and two grounded voltage generators.

that the impulsive force excitation can be assumed as being equivalent to impulsive boundary condition for the strain of the system.

The response of the system can thus be derived from eq. (2) written in homogeneous form, that is without distributed driving force.

In this case we have the classical wave equation, plus a dissipative term, whose response to an impulsive boundary condition is well known and consists of an impulsive wave that propagates with the speed of sound $u = (E/\rho)^{\frac{1}{2}}$ with reflection at the ends. We have both a compressive wave

$$\frac{1}{2E} \delta\left[t - \left(\frac{z-z_1}{u}\right)\right] \tag{3}$$

and a dilatation wave that propagates toward left

$$-\frac{1}{2E} \delta\left[t - \left(\frac{z_1-z}{u}\right)\right] \tag{4}$$

The effect of the dissipative term of eq.(2) can be approximately taken into account by introducing an attenuation factor for the strain waves as described by the law

$$e^{-\alpha t} = e^{-\frac{\omega_0 t}{2Q}} \tag{5}$$

where Q is the quality factor at the fundamental radian frequency of the cylinder

$$\omega_o = \frac{2\pi}{T_o} = \frac{\pi u}{L} ; \tag{6}$$

at that frequency it can be shown (3) that the quality factor can be expressed as

$$Q_o = \frac{E}{\omega_o D} \tag{7}$$

in terms of the parameter of the material and represents exactly the decay of the energy stored in the cylinder in the fundamental mode.

The impulse response $w_{12}(t)$ at an arbitrary coordinate z_2 can be obtained by summing the infinite contributions due to the reflections at the boundaries of the cylinder of both the right and the left waves, where each reflection provides a sign inversion, which involves easy but tedious calculations and is represented by a sequence of exponentially decaying delta functions. The effective strain response, as measured by a detector placed at abscissa z_2, to an arbitrary distributed force can be obtained by a double integration in space and time because the linearity and time invariance of the system allow the principle of superposition of effects as well as the principle of shift in time of cause with effect to be applied.

The space integral corresponds to a simple sum of the contributions to the strain at z_2 of the forces applied from $-L/2$ to $+L/2$.

The time integral is a convolution integral that provides at time t the contribution of the forces applied at times $t-\tau$ between $-\infty$ and t along the forward time abscissa t or between 0 and $+\infty$ along the backward abscissa τ, weighted according the pertinent values of the impulse response function:

$$s(t,z_2) = \int_{+\frac{L}{2}}^{+\frac{L}{2}} dz_1 \int_0^\infty d\tau \; f\left[(t-\tau),z_1\right] w_{12}(\tau) \qquad (8)$$

The computation of the strain response requires a single integration if the excitation force is impulsive in space or in time; if it is impulsive in both space and time the response coincides by definition with the impulsive response itself.

The strain response to a time impulsive force with the same space dependence of the gravitational pseudoforce

$$f(z,t) = -\frac{2z}{L}\,\delta(t) \qquad (9)$$

is given by the sum of two triangular waves with the phase shifts of $\pm \frac{\pi z_2}{L}$ as shown in fig. 2. The impulsive re-

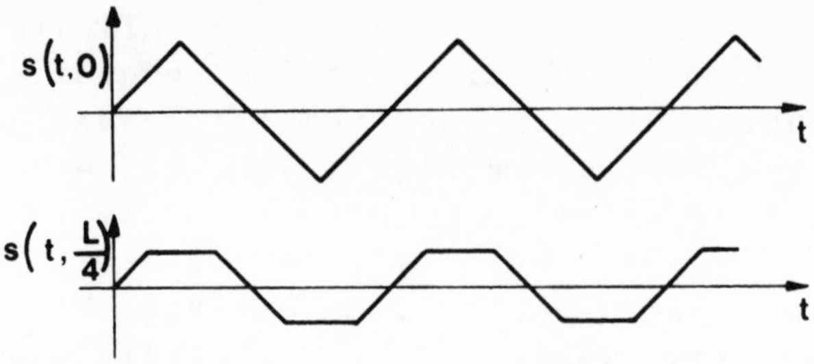

FIG. 2 Strain response to an impulse of gravitational pseudoforce.

sponse is an exponentially decaying triangular wave at $z_2=0$; for $z_2 \neq 0$ the wave is symmetrically clipped at a level that decreases linearly with the distance from the center of the cylinder providing a response identically zero at the ends.

It is to be emphasized here the different harmonic content of the response at strain detectors located in different position on the cylinder, that allows a direct implementation of a Fourier analyzer of the signal.

By Fourier series expansion of the triangular response wave we have infact

$$s(x,t) = \frac{e^{-\alpha t}}{E} \frac{4}{\pi^2} \sum_{1}^{n} {}_{i} \frac{1}{i} \cos\left(\frac{i\pi}{2} x\right) \sin\left(\frac{2\pi i t}{T}\right) \tag{10}$$

where the normalized abscissa $x=2z/L$ has been introduced. When using a number of detectors located at suitable abscissas x_i we can obtain the terms of the Fourier expansion of the input signal by a simple linear combination of their outputs.

3. DISCRETE SIMULATION OF THE ANTENNA

We can now divide the antenna into a number N (even) of sections in order to perform the computer simulation, as in fig. 3 where a) and b) show the equivalent mechanical and electrical lines and c) gives the model of the discrete simulation, where E denotes excitation and R response. For a N-section model of the antenna there are N driving points, with index I even, that is masses where the excitation can be applied, and N-1 measurement

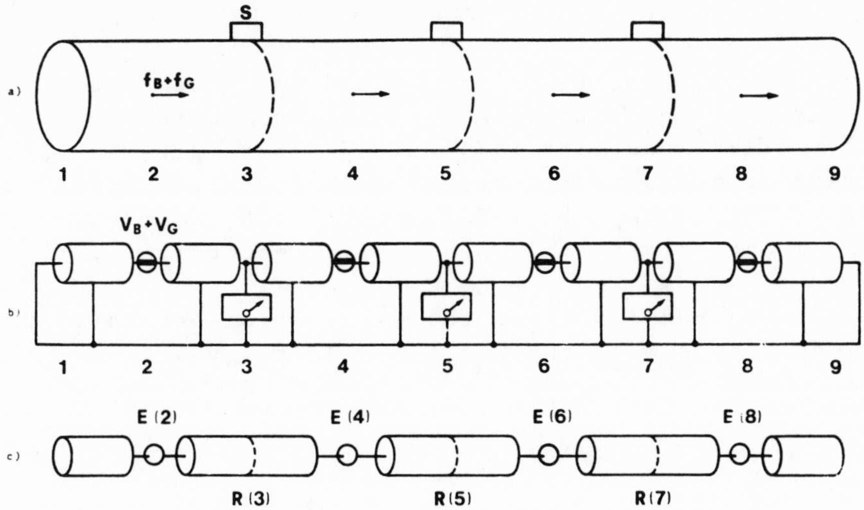

FIG. 3 Discrete models of the antenna.

points, with index I odd, that is strain sensors connected between two masses where the response is observed; the response at the extremes is identically zero because of the boundary conditions of the system.

Within each elementary time step of the simulation takes place the excitation of the N driving points, that is followed by the computation of the response at the N-1 measurement points.

The excitation is given by

$$E(I) = G(I) + T(I), \text{ I even} \qquad (11)$$

where G(I) represents the gravitational force e T(I) the Brownian force.

G(I) is obtained from the expression

$$G(I) = G. \left(\frac{N+1-I}{N}\right) \qquad (12)$$

in order to represent the desired space dependence of the gravitational force, and G is defined by the value at that time step obtained from the prescribed temporal behaviour of the gravitational signal.

T(I) is a white and quasi-Gaussian noise signal obtained very simply by summing three random numbers with uniform distribution and average value equal to zero.

The effect of the excitation is to give rise to two waves of opposite sign propagating toward right D(I) = E(I)/2 and left S(I)=-E(I)/2 that, after a delay corresponding to half section of the model, attenuated by a factor, A= exp $[-\pi/4QN]$, will reach the adjacent measurement points (I+1) and (I-1), respectively.

The response R(I) at the observation points is then evaluated as sum of the right and left waves that reach them

$$R(I) = D(I) + S(I) \qquad , I_{odd} \tag{13}$$

where the two waves are obtained by summing the contribution of the excitations during the same nominal time step and of the waves P and Q that were present at the adjacent observation points, (I-2) and (I+2) respectively, at the end of the precedent time step. At the ends of the line the situation is different since no response is to be observed but the waves are to be evaluated in order to be available for the next computations; one must take into account the reflection effects of the waves.

Before going to the next time interval the waves present at the points with odd index are memorized as "precedent" waves

$$P(I) = D(I) , \quad Q(I) = S(I) \tag{14}$$

The input signal, that is the gravitational force G, can be chosen as a doublet, as a sinusoidal wave packet, that is a burst, or as background, that is white noise, obtained by computing G at each time step as sum of three random numbers with uniform distribution and average value equal to zero.

In what follow some results of the computer simulation of the antenna are given (4).

TABLE 1

1	1200	124	124	124	124	124	124	124
2	0	249	374	374	374	374	374	249
3	0	249	499	624	624	624	499	249
4	0	249	499	749	874	749	499	249
5	0	249	499	749	874	749	499	249
6	0	249	499	624	624	624	499	249
7	0	249	374	374	374	374	374	249
8	0	124	124	124	124	124	124	124
9	0	-125	-125	-125	-125	-125	-125	-125
10	0	-250	-375	-375	-375	-375	-375	-250
11	0	-250	-499	-624	-624	-624	-499	-250
12	0	-250	-499	-749	-874	-749	-499	-250
13	0	-250	-499	-749	-873	-749	-499	-250
14	0	-250	-499	-624	-624	-624	-499	-250
15	0	-250	-374	-374	-374	-374	-374	-250
16	0	-125	-125	-125	-125	-125	-125	-125
17	0	124	124	124	124	124	124	124
18	0	249	373	373	373	373	373	249
19	0	249	498	622	622	622	498	249
20	0	249	498	747	871	747	498	249

In Table 1 is given the response of an 8 section antenna to a single pulse. The first colum represents the time, the second the value of the gravitational excitation G and the following columns give the response at the seven observation points: there is excellent agreement with the theoretical predictions. When considering the response to Brownian excitation the build up of stationary wave can be observed very clarly.

Fig.4 shows the response R(N+1) at central section of the bar for a 4 mass model with Q = 2000.

The system is driven by Brownian excitation and the time 1520 a gravitational doublet is applied. The cycle to cycle variations of the response are relative small both before and after the application of the doublet but there is a marked difference between the shape of the response before and after the doublet.

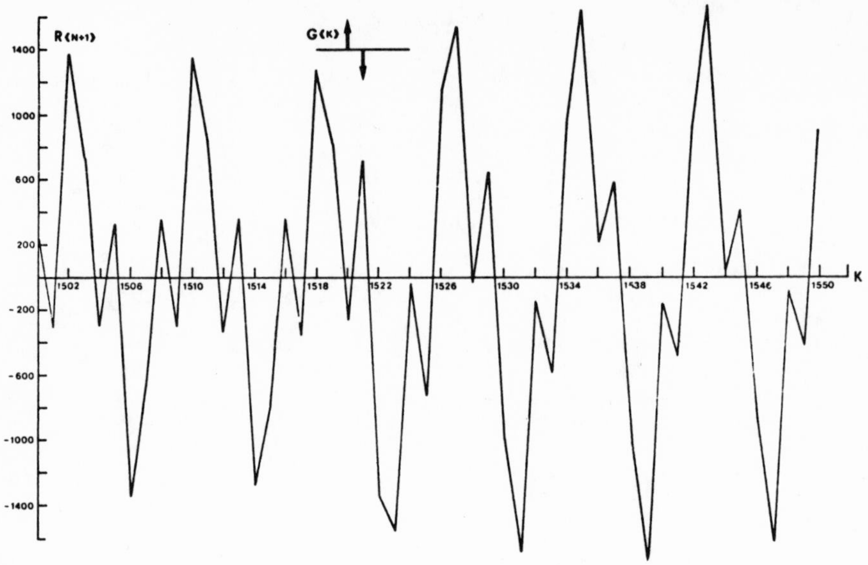

FIG. 4 Response of the simulation before and after a doublet of gravitational pseudoforce.

4. TWO SENSOR CORRELATION TECHNIQUE FOR SNR IMPROVEMENT

The signal to noise ratio (SNR) of the system can be improved by means of correlation techniques. The simplest approach consists of using two sensors that are symmetrically located respect to the center of the antenna at the abscissas $-z_1$ and $+z_2$ (Fig.5) so that they see exactly the same signal of gravitational origin s(t) plus a noise term $n_i(t)$ that is given by the sum of the Brownian response of the antenna and of the noise of the electronic amplifier.

If there is no correlation between the noises n_1 and n_2 then the time integral of the product of the two channel outputs will provide the power of the signal in

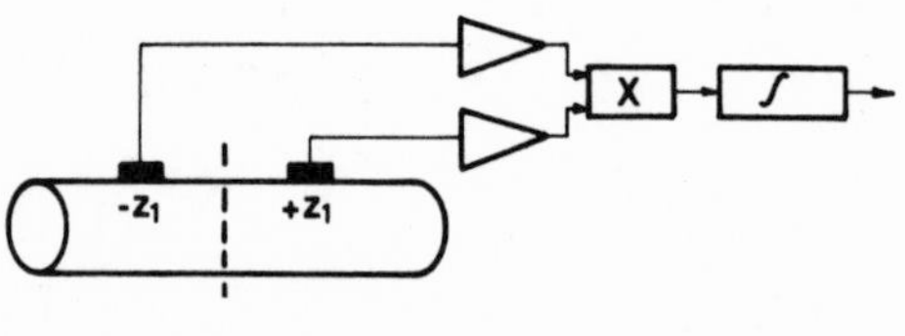

FIG. 5 Block diagram of the two sensor technique

the frequency range defined by the bandwidth of the two channels.

It is to be pointed out that this detection scheme provides the average power of the GW during the integration time and is more apt to the detection of continuous wave or random GW, rather than of bursts of radiation of larger short term power, but of short duration (5). It requires in fact rather long integration times, that increase with increasing values of the Q factor of the antenna. When considering the problem of a cryogenic antenna for which the Brownian noise has been greatly reduced due to the low temperature of operation, the two sensor technique can be applied from the viewpoint of reducing the noise of the electronics.

Most of the simulation results have been obtained with very low values of Q in order to increase the quality of the statistics without requiring a very long computation time; the computation time is in fact of the order of 1 s per time step of the simulation and therefore the correlation over 36000 time steps would require about $\sim$10 hours of computer (HP 2116-B). The results show improvements of the SNR in good agreement with the theory of the two sensor correlation technique for input SNR values ranging from .015 to 1: for example operating on a data sequence corresponding to 3850 independent samples with SNR of .021 the output relative deviation is equal to .55, while the theoretical relative variance is .79.

5. CONCLUSIONS

The first result of the work described in this note consists in the availability of a simulation method, well grounded into the physics of the problem, that is able to provide simulated data for any purpose of analysis. Another result of the work has consisted in demostrating the feasibility of the two sensor correlation technique for the detection of low power GW with time duration much larger than the relaxation time of the antenna, such as continuous waves (of any frequency) and random waves (gravitational background).

Other results have been obtained in the field of the detection of gravitational transients of very short duration by exploiting the concept of distance in the signal space that appears to be more powerful than the distance in the two-dimensional normal mode space commonly considered by the experimenters(6).

Further extension of the model will be done by adding white noise to the output of the various sensors to represent the noise of the electronics, and by taking into account the effect of dispersion other than dissipation in the bar.

Other computer experiments on the model will be performed to assess the performances of other algorithms of signal detection and also as far as the use of the gravitational antenna as spectrum analyzer of GW is concerned.

ACKNOWLEDGEMENTS

The author wishes to acknowledge the useful discussions with profs. A.De Gasperis and G.Pizzella.

REFERENCES

1) D.Maeder, Zeitschrift Ang.Mat.Phys., vol.22, pp.1146-1154, 1971.

2) G.V.Pallottino, Alta Frequenza, vol. 43, n.12, dec. 1974, pp. 1043-1044.

3) G.Pizzella, Rivista del Nuovo Cimento, vol.5, pp.369-397, July 1975.

4) G.V.Pallottino, Internal Report, LPS-75-29, sept.1975.

5) J.Hough, J.R.Pugh, R.Bland, R.W.P.Drever, Nature, vol. 254, 10 apr.1975, pp.498-501.

6) G.V.Pallottino, Internal Report, LPS-75-30, sept.1975.

GRAVITATIONAL RADIATION EXPERIMENTS*

J. WEBER

UNIVERSITY OF MARYLAND

COLLEGE PARK, MARYLAND 20742

In 1969 and 1970 I published results of observations of coincident changes in output of gravitational radiation antennas at the University of Maryland and the Argonne National Laboratory. A number of other physicists repeated these experiments, after making changes in the instrumentation and data analysis, which were expected to give improved sensitivity. Other groups did not confirm the Maryland results, and a controversy developed.

Human observer bias can be a factor in evaluation of data. For this reason I removed myself from direct participation in the data analyses in 1972. A "blind" experiment was set up making use of an analogue on line computer to record coincidences on pen and ink records. These were studied by a gifted young woman, Mrs. Alessandra Exposito, who was not a physicist. These were two sets of coincidence data, one with and the other without a time delay, for one detector. Coincidences caused by external signals will vanish with a sufficiently long delay in one channel. Therefore a measurement of the coincidence rate with the time delay measures the purely chance coincidence rate (accidentals). These charts were marked with a code which was changed at certain times. A magnetic tape recorder system with computer analyses was set up with programming performed by Mr. Michael Lee.

A major effort has been made to discover all computing errors. Copies of our magnetic tapes and lists of coincidences were sent to other laboratories. Significant errors were found and the programs

*Supported in part by the National Science Foundation

were carefully corrected. "Double Blind" experiments were carried out by Mr. William Davis and Mr. Bruce Webster. Davis inserted pulses in both detectors at certain periods unknown to the programmer. The computer did accurately locate the periods when pulses were present and when they were absent. Webster made copies of a number of tapes, some identical with originals, others with interchanged channels so that the channel corresponding to a given detector was replaced by a corresponding one for a different 4 day period. The content of these tapes was not known to Michael Lee. Lee identified the tapes correctly. As a result of these checks it can be concluded that recently published data[1] has neither computing errors nor human observer bias.

Another issue is data selection. We have published analyses of all data which can be read by the computer.

Data from the antenna at the Argonne National Laboratory was transmitted to a recorder at Maryland via a telephone circuit in coded digital form, and this appeared to be a possible cause for the coincidences.To check this possibility the telephone circuit was disconnected and separate clock-, synchronized recorders employed at the two antenna sites. An excess of coincidences over measured accidental ones was again obtained.

Other Experiments

I visited a number of other laboratories which have developed gravitational radiation antennas, and was impressed with many novel and clever features.

A beautiful design may have poor sensitivity in consequence of excess noise in electronics and data processing, and temperature control problems. At present two methods of data analysis[1] are employed at Maryland. All other groups have employed primarily one method which is responsive both to changes in phase and changes in amplitude of the antenna. This "preferred" algorithm observes such changes relative to the amplitude and phase of a quartz reference oscillator. If the cylinder is not temperature controlled, there will be a noise background associated with a temperature dependent phase difference of the cylinder mode and quartz oscillator, and excess noise associated with a cylinder not in thermal equilibrium.

For example my visits and measurements in 1973 indicated that the Bell Telephone Laboratories pre amplifiers had considerably more noise than the Maryland design, that the Rochester antenna had no effective temperature control and insufficient acoustic isolation. The Bell Laboratories Rochester data analysis employed computer programs which differentiated and squared data written on tape with 6 bit quantization, thus introducing additional, quantization noise.

The computer employed for their data analyses had insufficient memory and as a result their accidental coincidence rates were computed with delays smaller than 2 seconds. Maryland data processing differentiated and squared prior to quantization and employed delays exceeding 100 seconds.

A visit to Munich revealed no significant temperature control.

A recent study by Dr. Gustav Rydbeck[2] has again confirmed the great importance of temperature control with extremely small tolerances. A number of laboratories completely overlooked this problem and attempted to compensate the omission by computer program corrections and automatic tracking of the quartz reference oscillator. Such compensation cannot be complete because an error signal is required and because the excess noise in the cylinder itself cannot be removed.

Some experiments employed a single antenna and searched for departures of the noise distribution from thermal equilibrium values. This cannot discover external signals which have the same statistics as noise unless these external signals are very large.

Consideration of all of these problems led to the coincidence experiments and use of two algorithms at Maryland, one not requiring rigid temperature control. Nonetheless a major effort was devoted to achieving good temperature control, with uncompensated drifts usually less than one part in 60,000 per day.

For these reasons the published data[3] on the relative sensitivity of different installations do not in fact describe the operating installations.

[1] Lee, Gretz, Steppel, Weber, Phys. Rev. D., 14,4,893-906 Aug. 15,1976

[2] G. Rydbeck, Thesis, University of Maryland

[3] Douglass, Gram, Tyson, Lee, Phys. Rev. Lett.35, 8, p. 480, 25 August 1975

ESTIMATED SENSITIVITY OF THE LOW TEMPERATURE GRAVITATIONAL WAVE ANTENNA IN ROME

G. PIZZELLA

Istituto di Fisica, Università di Roma

Roma, Italy

INTRODUCTION

The University of Rome and the "Consiglio Nazionale delle Ricerche" have started an experimental activity for the search of gravitational waves emitted by extraterrestrial sources. It consists in the construction of a gravitational wave antenna cooled to low temperatures as suggested by Professors W.M. Fairbank and W.O. Hamilton. The design of the dewar is very similar to those of the Stanford and Louisiana Universities and, on a long range plan, it will make it possible to cool a 5 ton aluminum bar down to a temperature of the order of 10^{-2}K. The antenna located in Rome, when in operation, will work in coincidence with the two antennas at Stanford and Louisiana searching for gravitational waves and eventually measuring their velocity.

In this paper, we shall examine the sensitivity which we can expect from low temperature gravitational wave antennas as compared to those of the Weber-type operating at room temperature.

EXPERIMENTAL APPARATUS

The experimental work in Rome is developing following three lines. The line which will lead to the final stage consists in the construction of a dewar able to maintain a 5 ton aluminum bar at a temperature of the order of 10^{-2}K. The bar will be levitated with a magnetic field due to superconducting currents in order to be able to reach such very low temperatures and for mechanically isolating the antenna from the laboratory as much as possible.

For this project we had technical cryogenic problems and therefore, at present, some part of the dewar is being modified in order to overcome these difficulties.

A second line consists in realizing a dewar for a 300 kg antenna at 10^{-2}K. The dewar has been designed and is, at present, under construction. With this project, we think we can avoid many of the problems arising from a very large cryogenic system (including the running costs) without suffering, as it will be shown later, a large loss of sensitivity. In addition, in this system we shall use both the magnetic and wire suspension with a steel wire in a magnetically levitated container, in order to have at the same time a large Q and the magnetic suspensions.

The third line is the development of a small antenna (25 kg) at 1 K . The purpose is to test the various apparatus which will be employed for the larger antennas. This system has been used for the design of the magnetic suspensions(1), for determining the Q value(2) at liquid helium temperatures and for testing the electronics. One of the most interesting results was that Q increases by about a factor of 10 from room temperature to 1.1 K reaching for the small aluminum bar a maximum value of the order of 3×10^6.

As far as the vibration transducer is concerned, we are developing a transducer based on the use of a SQUID magnetometer(3) and at the same time we are testing the behavior of the piezoelectric ceramics at liquid helium temperatures. In what follows we shall be concerned only with the ceramics which, from our measurements, appear to be properly working at least down to 1.1 K.

RESPONSE OF THE ANTENNA TO A GRAVITATIONAL WAVE PULSE

Given two points along the z axis spaced by the distance ℓ and a gravitational wave propagating along the x axis, it is well known that the distance between the two points varies by the quantity $\xi(t)$ which satisfies the equation

$$\ddot{\xi} = - c^2 R^3_{030} \ell = \frac{\ell}{2} \ddot{h} \tag{1}$$

where R^3_{030} is a component of the Riemann tensor and h(t) is the tensor component for a linearly polarized gravitational awave.

For the response of a cylindrical gravitational wave antenna one could use the simple model of two point masses connected with a spring(4) and use equation (1) with the proper restoring and dissipation forces.

Here we shall solve the problem for a continuous antenna of length L on a more quantitative basis in the following way[5]. Let us consider the cylindrical antenna in the thin rod approximation with the axis in the z direction and center of the mass in the origin of the coordinate system. Taking two mass elements dm situated respectively in -z and +z we see, from equation (1) that the gravitational wave acts like a force of value $z\ddot{h}dm$ between them. Since the center of mass position is left unchanged by the wave, we can refer the forces to the origin and thus imagine that the single mass element be subjected to the force per unit volume F_G as due to the point origin,

$$F_G = \frac{\rho z}{2}\ddot{h} \tag{2}$$

having divided by two, for simmetry, the force between the two masses. We can then write the following equation for the thin rod

$$\rho \frac{\partial^2 \xi}{\partial t^2} = \frac{\partial \sigma}{\partial z} + F_G \tag{3}$$

where $\frac{\partial \sigma}{\partial z}$ is the force per unit volume along the z axis due to the surrounding material acting on the mass element at z which vibrates with displacement $\xi(z,t)$ and σ is the zz component of the stress sensor. This tensor obeys Hooke's law

$$\sigma = E\frac{\partial \xi}{\partial z} + D\frac{\partial}{\partial t}\frac{\partial \xi}{\partial z} \tag{4}$$

where E is the Young modulus and D a dissipation parameter. Introducing (2) and (4) in (3) we get an equation which can be solved with the temporal Fourier transforms. Indicating with $X(\omega,z)$ and $H(\omega)$ respectively, the Fourier transforms of $\xi(z,t)$ and $h(t)$ we get the solution

$$X(\omega,z) = X_1(\omega)e^{-i\alpha z} + X_2(\omega)e^{i\alpha z} + \frac{H(\omega)z}{2} \tag{5}$$

with

$$\alpha = \sqrt{\frac{\omega^2\rho}{E+i\omega D}} \tag{6}$$

We impose now the boundary conditions: that the center of mass does not move, $\xi(0,t) = 0$ for all values of t, and that the stress vanishes at the rod ends, $\left(\frac{\partial \xi}{\partial z}\right)_{z=\pm\frac{L}{2}} = 0$. We obtain the solution

$$\xi(z,t) = \frac{1}{2\pi} \int_{-\infty}^{+\infty} T(\omega,z)\, H(\omega) e^{i\omega t} d\omega \tag{7}$$

with the transfer function

$$T(\omega,z) = \frac{1}{2} \left(z - \frac{\sin \alpha z}{\alpha \cos \frac{\alpha L}{2}} \right) \tag{8}$$

These expressions can be simplified in the gravitational wave case because the merit factor

$$Q = \frac{E}{\omega D} \tag{9}$$

is generally much larger than unity. In such a case (6) becomes

$$\alpha = \frac{\omega}{v} \tag{10}$$

with $v = \sqrt{\frac{E}{\rho}}$ the sound velocity in the antenna and from (8) we see that there are resonances for $\cos \frac{\omega L}{2v} = 0$, that is at pulsations

$$\omega_K = (2K+1) \frac{\pi v}{L} = (2K+1)\omega_o \quad (K = 0,1,2,\ldots) \tag{11}$$

In order to actually compute the response, we have now to feed in the quantity $H(\omega)$ which depends on the particular gravitational wave. We consider interesting the case of a wave defined by

$$\begin{aligned} h(t) &= h_o \cos \Omega t & \text{for } |t| &< \tau/2 \\ h(t) &= 0 & \text{for } |t| &> \tau/2 \end{aligned} \tag{12}$$

whose Fourier transform is

$$H(\omega) = h_o \frac{\tau}{2} \left[\frac{\sin(\Omega-\omega)\tau/2}{(\Omega-\omega)\tau/2} + \frac{\sin(\Omega+\omega)\tau/2}{(\Omega+\omega)\tau/2} \right] \tag{13}$$

which represents a pulse of duration τ oscillating with pulsation Ω. The case of a continuous monochromatic wave is obtained from (12) and (13) by putting $\tau \to \infty$

For the sake of simplicity, we shall limit ourselves here only to the case $\Omega = \omega_o$ and $\tau \ll Q/\omega_o$. Substituting (13) and (8) into (7), with this approximation, we get for $t \gg \tau$

$$\xi(z,t) = \xi_o \sin \frac{\pi z}{L} \sin (\omega_o t+\pi) \tag{14}$$

with the amplitude

$$\xi_o = \frac{\omega_o h_o \tau L}{\pi^3} = \frac{h_o v\tau}{\pi^2} \tag{15}$$

This solution is valid at times $t<<\tau$ and it represents a stationary wave at the resonance pulsation ω_o. For $t\gtrsim\tau, \xi_o$ decreases exponentially with the time constant $Q/2\omega_o$. It can be shown that the solution valid for the two points mass oscillator is obtained multiplying ξ_o by the factor $\pi^2/4$.

In order to justify the approximations made in the derivation of our solution, we give some numberical values for the Weber[6] apparatus: Q = 190.000 and Q/ω_o = 19 sec. This means that (14) is valid for $\tau<<19$ sec which is very reasonable considering possible values of τ for gravitational waves of cosmic origin.

RESPONSE OF THE PIEZOELECTRIC CERAMIC

We consider the piezoelectric ceramic mounted on the center of mass section of the antenna with the voltage signal $V(t)$ sent to a high impedance amplifier, It can be shown[4] that

$$V(t) = \alpha \; \xi(\pm \frac{L}{2}, t) \equiv \alpha \; \xi(t) \tag{16}$$

with

$$\alpha = \frac{d\zeta\pi}{\varepsilon s L} \tag{17}$$

where d is the piezoelectric constant, ε the piezoelectric dielectric constant, s the piezoelectric elastic compliance and ζ the ceramic thickness between the conducting layers. $\xi(t)$ should include the effect of the ceramic reaction on the antenna. Weber[4] has treated this problem in detail and it has showed that, for a small ceramic, $\xi(t)$ is essentially the undisturbed value which we have given with (14) and (15) for $z = L/2$. Therefore, the response of the piezoelectric ceramic to a gravitational wave pulse will be

$$V(t) = \alpha\xi_o \sin (\omega_o t + \pi) \tag{18}$$

with ξ_o given by (15).

The value of α depends also on the particular mounting. In the Weber case $\alpha \simeq 1.2 \times 10^7$ V/m, which means that a displacement of 10^{-16}m gives for V the measurable value of 1 nV. For the

Frascati-Munich experiment[7] $\alpha \simeq 2 \times 10^8$ V/m, achieved by increasing the value of d and of ζ.

THE NOISE

The signal given by (18) has to be compared with the noise. There are the following kinds of noises:

a) The electronic noise of the FET amplifier. With the present technology the best amplifier is that used by Weber[6] which has a voltage noise $V_o \backsim 0.3$ $nV/\sqrt{HZ}$ and a current noise of about 25×10^{-15}A. With the large value of C, the ceramic capacity obtained by Weber with 25 piezoelectric ceramics of 36 cm^3 each one, the current noise is negligible. In other cases[7] where a different mounting of the ceramic is used in order to have a larger α the capacity C is smaller (of the order of 1.5 nF) and therefore the current noise is not negligible any more.

b) The Johnson noise in the piezoelectric ceramic at temperature T given by

$$W_J = \frac{4KT\ tg\delta}{\omega_o C} \tag{19}$$

where $tg\delta$ is the dissipation factor of the ceramic. At room temperature for $tg\delta \backsim 3 \times 10^{-3}$ and $C \backsim 7 \times 10^{-8}$ F[6] we have $W_J = 0.3$ $nV/\sqrt{HZ}$ which is comparable with V_o due to the electronics in the Weber case. We notice that this noise is much larger, about 2 $nV/\sqrt{HZ}$, for the Frascati-Munich experiment with small C. It is important to notice that at liquid helium temperatures W_J becomes negligible and therefore we shall not consider it for the cryogenic antennas and, in particular, we do not need to require, for this noise, that C be large.

c) The Brownian noise of the antenna which produces a voltage V(t) whose rms value is given by

$$V^2_{rms} = \frac{2KT\alpha^2}{m\omega_o^2} \tag{20}$$

If the quantity $V_B(t) = V(t)-V(t-\Delta t)$ is considered for a given value of Δt then its rms value is given by[8]

$$V^2_{B,rms} = \frac{2KT\alpha^2}{m\omega_o Q}\ \Delta t \tag{21}$$

Therefore, if Δt is small compared to Q/ω_o the Brownian noise associated to $V_B(t)$ is small compared to that associated to V(t).

The minimum noise is obtained, for our low temperature antenna, by choosing Δt such that $V^2_{B',rms} = V_o^*\Delta\nu$, where $\Delta\nu = 1/\Delta t$ and V_o^* includes both the voltage and current noise of the FET amplifier. We thus get the optimum bandwidth

$$\Delta\nu = \frac{\alpha}{V_o^*}\sqrt{\frac{2KT}{m\omega_o Q}} \tag{22}$$

and the minimum noise

$$V_N^2 = 2\alpha V_o^*\sqrt{\frac{2KT}{m\omega_o Q}} \tag{23}$$

THE PREDICTED SENSITIVITY FOR THE LOW TEMPERATURE ANTENNAS

At low temperature, a gravitational antenna in the Weber type configuration has a sensitivity to a gravitational pulse which can be computed by equating the signal $\alpha^2\xi_o^2$, with ξ_o given by (15), to the noise V_N^2 given by (23). Thus we get the minimum value of $h_o\tau$ which can be observed:

$$(h_o\tau)'_{min} = \left[2^{3/2}\pi^4\frac{V_o^*}{\alpha v^2}\sqrt{\frac{KT}{m\omega_o Q}}\right]^{1/2} = \left[2^{3/2}\pi^6\frac{V_o^*}{\alpha\omega_o^2L^2}\sqrt{\frac{KT}{m\omega_o Q}}\right]^{1/2} \tag{24}$$

where we notice the effect of the various parameters.

In the hypothesis that a noiseless transducer[9] is developed then we have to compare $\alpha^2\xi_o^2$ with $V^2_{B,rms}$ only and thus we get

$$(h_o\tau)''_{min} = \left[\frac{2\pi^4 KT\Delta t}{v^2\, m\omega_o Q}\right]^{1/2} = \left[\frac{2\pi^6 KT\Delta t}{mL^2\omega_o^3\, Q}\right]^{1/2} \tag{25}$$

which shows that one can use a small mass if Q is sufficently large. One of the advantages to go to low temperatures is that it has been found[2,9,10] that Q increases when T decreases below 50K.

For our low temperature antenna, we compute the sensitivity using the following values: $V_o = 0.3$ nV/√HZ[11], $\alpha = 1\times10^8$ V/m, $T=10^{-2}$K, m = 5000 Kg, $\omega_o = 5\times10^3$ rad/sec, $Q=10^6$, L = 3m and we get

$$(h_o\tau)'_{min} = 1.6\times10^{-21}\ \text{sec}$$

For a comparison, we compute the sensitivity for the Weber antenna, using (24) and therefore neglecting here the ceramic Johnson noise

$$(h_o\tau)^{Weber}_{min} = 1.1 \times 10^{-19} \text{ sec}$$

which, in energy, is about a factor 5000 larger than that of our cryogenic antenna.

It is interesting to notice that we can get, at least theoretically, a sensitivity better than that of a large room temperature antenna already with a 25 Kg antenna cooled to 1K. Much more can be gained with the low temperature antennas if a noiseless transducer is available because, as shown by (25), the minimum energy flux which can be detected depends on KT/Q instead than on $\sqrt{KT/Q}$.

Acknowledgments

The gravitational wave experiment in Rome has been suggested by Prof. E. Amaldi. The cryogenic part is being developed under the direction of Prof. I. Modena and the electronics under the direction of Prof. G.V. Pallottino. The experimental work is also being carried by Drs. M. Cerdonio, F. Bordoni, U. Giovanardi, C. Barbanera, C. Cosmelli and F. Ricci. Many interesting discussions were done with Profs. N.Cabibbo, A. De Gasperis and R. Ruffini. The Italian group is indebted to Profs. W.M. Fairbank and W.O. Hamilton who began to develop the low temperature gravitational wave antennas.

References

1. P. Carelli, A. Foco, U.Giovanardi, I. Modena. "The Magnetic Levitation of a Gravitational Antenna at Low Temperatures" Cryogenics, pag. 77, February 1976.

2. P. Carelli, A. Foco, U. Giovanardi, I. Modena, D. Bramanti, G. Pizzella. "Q Measurement Down to Liquid Helium Temperatures for a Gravitational Wave Aluminum Bar Antenna". Cryogenics, pag. 406, July 1975.

3. L. Adami, M. Cerdonio, F. Ricci, G.L. Romani. "A Superconducting Strain Transducer". Submitted to Applied Physics Letters, 1976.

4. J. Weber. "Proceedings of the International School of Physics E. Fermi". Varenna, Course 55, 1972.

5. G. Pizzella. "Gravitational Wave Experiments". Rivista del Nuovo Cimento, 5, 369, 1975.

6. D. Gretz, M. Lee, J. Weber. "Gravitational Radiation Experiments in 1973-74". Conference Proc. GR7, Tel Aviv, Israel, June 1974.

7. D. Bramanti, K. Maischberger, D. Parkinson. "Optimization and Data Analysis of the Frascati Gravitational Wave Detector" Lett. Nuovo Cimento 7, 665 (1973).

8. G.W. Gibbons, S.W. Hawking. "Theory of the Detection of Short Burst of Gravitational Radiation". Phys. Rev. D4, 2191, 1971.

9. V.B. Braginski . "Proc. of the International School of Physics E. Fermi". Varenna, Course 55, 1972.

10. G. Papini. "V Cambridge Conference on Experimental Relativity". Cambridge, Mass., 1974.

11. F. Bordoni, G.V. Pallottino. "Electronic Instrumentation for the Detection of Gravitational Waves". Colloque International sur l'Electronique et la Mesure. Pag. 183, Paris, May, 1975.

THE LSU LOW TEMPERATURE GRAVITY WAVE EXPERIMENT*

W. O. Hamilton, T. P. Bernat, D. G. Blair,† W. C. Oelfke+

Department of Physics and Astronomy
Louisiana State University
Baton Rouge, Louisiana 70893

INTRODUCTION

The field of experimental general relativity has come a long way in the past 15 years. In 1960 the tests of relativistic theories were generally felt to be the exclusive domain of the observational astronomer; only a relative few were even thinking about possible laboratory tests. One of these relative few is now the director of this course and the fact that there are so many attendees indicates that the field has indeed developed. In this paper we shall discuss but a small corner of the experimental gravitational effort: the design and construction of a cryogenic gravitational radiation detector. We have an unusual problem when we attempt to build a gravitational wave detector. Unlike detection schemes in spectroscopy, communications and radar where one searches for a known signal in unknown noise we must do the opposite and attempt to identify an unknown signal in the presence of noise, the properties of which are well defined.

PROPERTIES OF THE ANTENNA

The properties of thermal noise and its effect on the measurement have been covered by Kafka[2] and Bonazzola.[3] For detailed calculations of noise performance of a given system their papers should be consulted. For the purposes of this paper we use the results of Wang and Uhlenbeck[4] for the theory of a harmonic oscillator driven by a random fluctuating force $F_{f\ell}$ of spectral density $P_\omega = 4kT\beta$ where β is the damping constant of the harmonic oscillator. The equation of motion for the oscillator is

$$m\ddot{x} + m\beta\dot{x} + m\omega_o^2 x = F_{gw} + F_{f\ell}.$$

$F_{gw} + F_{f\ell}$ are the forces exerted on the oscillator. Clearly we mean F_{gw} to be the effective tidal force exerted by gravitational radiation and $F_{f\ell}$ is the Langevin fluctuating force (the force which causes Brownian motion).

In the absence of a gravitational wave we may use the Fokker-Planck equation to find an expression for the probability of measuring an amplitude y of our antenna at time t, given that at time t = 0 we measured an amplitude y_o:

$$P(y|y_o) = \sqrt{\frac{m\omega_o^2}{2\pi kT(1-e^{-\beta t})}} \exp -\left[\frac{m\omega_o^2(y-\bar{y})^2}{2kT(1-e^{-\beta t})}\right].$$

$\bar{y}$ is the amplitude that would be predicted at time t in the absence of random forces (i.e. $\bar{y} = y_o e^{-\beta t}$).

We should note that this description demonstrates the effective temperature concept very clearly. It has not always been obvious to everybody how the noise temperature can be less than the ambient temperature. As the time between the measurements increases, the above probability approaches the Boltzmann probability for ambient temperature T. If βt is small, however, the effective temperature (the denominator of the exponential) becomes

$$T_{eff} \simeq T\beta t \qquad \text{for} \qquad \beta t << 1.$$

Recalling that the definition of the Q of an oscillator is

$$\beta = \frac{\omega_o}{Q}$$

we see that

$$T_{eff} = \frac{T}{Q}\,\omega_o t \qquad \text{for} \qquad \frac{\omega_o t}{Q} << 1.$$

Thus the prime physical requirements for the design of our antenna in order to maximize our knowledge of the antenna deflection are demonstrated to be:

1) Low antenna temperature.
2) High mechanical Q (to lower T_{eff}).
3) Large mass (to raise the energy associated with a given y, note the position of m in the exponential).
4) Short time between measurements (to lower T_{eff}).

It should be obvious that we are discussing only the antenna in this section. When we couple the readout transducer and electronics to the system modifications will be in order. These considerations are further discussed for specific experiments by Kafka and by Bonazzola.[5] Physically we know that as we shorten the sampling time we must increase the bandwidth of the electronics. Hence we add noise and raise the effective temperature of the system. Coupling a transducer may lower the Q and further raise the effective temperature. It will do no good at all to improve the antenna effective temperature if we are unable to improve our transducer and amplifiers by a like amount.

The considerations discussed by Wang and Uhlenbeck are really more general than may be at first apparent. They demonstrate that a multi-variate Gaussian probability of the same form we have displayed is obtained for a general network of resistors, inductors and capacitors. Thus the probability for a given voltage output of the amplifier will have the form of $P(y|y_o)$ above when we consider the normal modes of the combined system. The effective temperature will be modified to include the damping of each normal mode observed.

THE EXPERIMENT DESIGN

The Mechanical Apparatus

Our experiment is still in the final stages of construction so we are afforded the luxury of discussing what it is supposed to do rather than the brutal facts of what it actually has done or is doing. There is however a substantial margin of improvement promised by low temperature detectors and the technology is sufficiently new that we will discuss it here.

A cross section of the experimental apparatus is shown below. A picture of the apparatus is also shown. The innermost shell, designated the 50 mK shell, is supported completely separately from the upper framework through isomode vibration filters similar to those used by Weber and through springs which are at the temperature of the shell. These are connected by stainless steel aircraft cable. The measured resonant frequency of the loaded shell-spring system is approximately 1.5 Hz. The antenna is supported in a cylindrical cradle which is bolted inside the 50 mK shell.

The remaining shells are supported from the outermost shell by stainless steel aircraft type cable. Their purpose is simply to hold the necessary refrigerants to keep the antenna cold. The 2 K shell is cooled by liquid helium, and the boiloff gas from this helium cools the 30 K shell. Liquid nitrogen cools the next outer shell and the outermost shell is the vacuum container. 125 layers

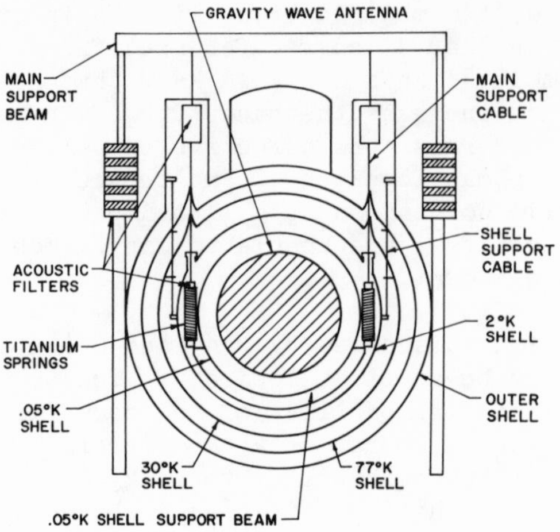

Figure 1. A cross section of the cryogenic gravity wave detector.

of NRC-2 superinsulation surround the 77 K shell, 20 layers the 30 K shell and 3 layers the 2K shell. A soldered lead sheet is used around the 2 K shell as a superconducting shield, preventing electromagnetic disturbances of any frequency from influencing the antenna.[6]

The 50 mK shell may be cooled by a He^3-He^4 dilution refrigerator which is designed to remove 500 erg/sec at 50 mK. Substantial engineering work has gone into this refrigerator design since with the large amount of gas being circulated (10^{-3} mole/sec max) heat exchange problems arise which do not affect a smaller design. Most of these problems seem to have been anticipated by the preliminary design since all of our experiments seem so far to give operating pressures and temperatures in close agreement with the design values.

One problem in the design of the large dewar's superinsulation was recognized but we underestimated its seriousness. The NRC-2 superinsulation is usually wound without a spacer between layers. Thermal contact between the layers of the superinsulation is generally unimportant because the crinkling of the superinsulation minimizes the contact area and the Mylar is not a very good thermal conductor. We followed the conventional technique but we wound each of the 125 layers independently to avoid the compressing of the inner layers of material by the outer. We should have used a nylon or fiberglass spacer between the layers because the weight of the NRC-2 has caused it to pull tight across the top of the shell. We do not know yet whether this will prove to be a serious problem but it will be a lengthy problem to fix if we have to. M. S. McAshan

Figure 2. The cryogenic container during final assembly. The antenna is seen in the foreground. The rails are for insertion of the antenna.

has suggested that Monsanto's Cerex nylon is an excellent spacer material and we will use it if the compression is too great.

The antenna is coated on the bottom with Nb-Ti sheet .020 inches thick. This sheet is bonded to the bar with Scotchcast Resin #8XR5236. The cradle in which the antenna rests is the support for superconducting magnets, wound as pancake coils and capable of being operated in the persistent mode. By passing a current through the pancake coils the bar can be supported without material contact with the cradle. The eddy currents induced in the Nb-Ti sheet on the bottom of the bar are repelled by the currents in the magnets and thus the bar is supported. We anticipate that this magnetic support will greatly aid our efforts to obtain a high mechanical Q in the bar since mechanical energy cannot be lost through the support. We have made extensive measurements of the support forces which can be expected from readily available materials and have every reason to believe that they will scale exactly to support our large bar. Our measurements agree well with those reported by Modena and the Rome group[7] and with the

Figure 3. Gluing the NbTi to the bar. The bands are removed when the epoxy is cured.

experiences of the Stanford group.[8] In the LSU experiment the magnets are charged by a rather simple flux pump.[9,10]

One additional benefit may accrue from the low temperature environment and the magnetic support.[11] If the entire antenna becomes superconducting the probability for electron scattering from phonons decreases as T^4 and that channel for losses becomes closed. Thus one may expect to see an increase in the Q of the antenna when it becomes superconducting. We have some evidence for this in our measurements of the Q of a small, magnetically supported, niobium bar. At 4.2 K the bar showed a Q in excess of 62×10^6. The bar was in helium gas during this measurement so we believe it to be a lower limit on what may be expected. A tracing of the experimental data is shown below.

The Transducer and Electronics

Treatment of electrical signals in the low temperature environment offers both advantages and disadvantages. The advantages reside in the use of superconducting elements, the possibility

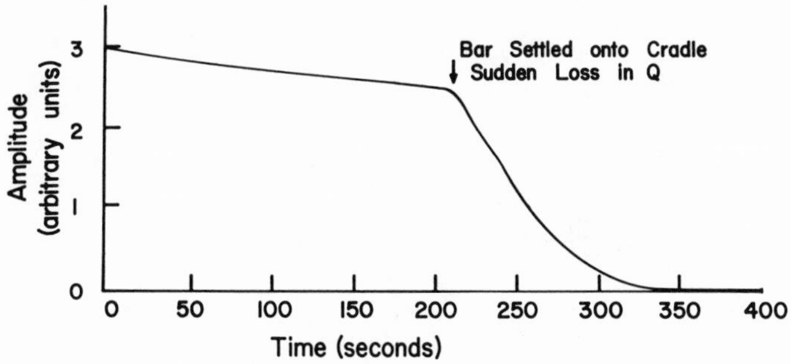

Figure 4. Decay of a magnetically supported Niobium bar at 4.2°K.

of low noise, and the existence of a very stable thermal environment. The disadvantages are all associated with the extreme change in the environment experienced by the apparatus between room temperature and the operating temperature. The apparatus is usually not available for adjustment after it is cooled down and yet it is mechanically very much changed because of thermal contraction. Much of the electronics, if it depends on superconductivity for operation cannot be tested until it is at helium temperatures. In spite of these drawbacks so much can be gained by low temperature signal processing that it should not be dismissed out of hand.

As an example of superconducting instrumentation in our application we may consider two versions of a superconducting accelerometer which may be used to determine the motion of the end of the gravitational antenna.

A diagram of the first accelerometer is shown below. The spool-shaped niobium test mass is supported on a superconducting wire through which a persistent current is flowing. The support field is the circular field around the wire due to the current. The test mass is therefore supported tightly against lateral motion on the wire but is free to move along the wire. The motion along the wire is virtually unimpeded and friction free. There are some small end corrections which are not terribly important.

A superconducting inductance is arranged so that its magnetic field is in contact with one end of the spool. In our experiment the inductance is a niobium spiral evaporated onto a sapphire substrate. This inductance forms a part of a tuned circuit at 40 MHz. Because of the superconductivity of all the circuit elements we should anticipate a high electrical Q for this parallel tuned circuit. Motion of the accelerometer case causes the inductance to move with respect to the spool-shaped test mass. This in turn

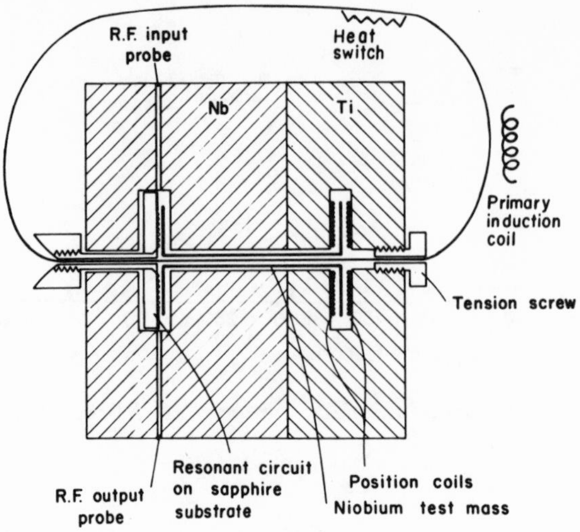

Figure 5. The free mass superconducting accelerometer.

causes the inductance to change and will vary the impedance of the tuned circuit. The low losses of the circuit cause the impedance to be an extremely rapidly varying function of frequency. Hence, if we feed the circuit by a constant current generator near its resonant frequency, the inductance variation due to motion of the accelerometer case results in a modulation of the voltage across the circuit or of the phase of the voltage with respect to the generator current. For a Q of 10^6 and an ambient temperature of 4 K we estimate that a relative displacement of 10^{-17} cm should be detectable.

The second accelerometer is a modification of the one just described, modified for higher frequency operation and for detection of accelerations of higher frequency. We utilize a re-entrant cavity configuration operated in the lowest TM mode. In this mode the electric field is largely confined to the region between the post and the end of the cavity. Van deGrift[12] has pointed out that this configuration looks a great deal like a resonant single turn inductance and a capacitance. The capacitance is the space between the center post and the end of the cavity. If the end of the cavity to which the re-entrant post is attached is made of a flexible superconducting diaphragm then motion of the cavity body will cause the spacing between the post and the cavity end to change, thus changing the cavity's resonant frequency. The modulation scheme is seen to be the same as that previously described except that here we are allowing the motion of the case to modulate the capacitance of the tuned circuit. To discriminate between output variations caused by frequency drift of the generator and the impedance change

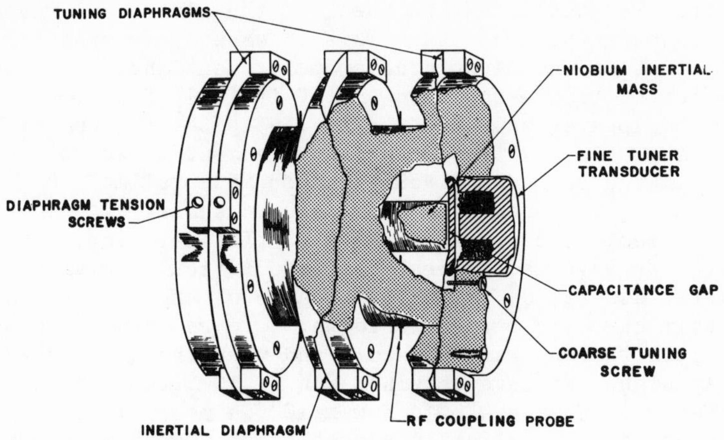

Figure 6. Superconducting re-entrant cavity accelerometer.

caused by acceleration we may use a dual cavity configuration as is shown in the drawing below. An acceleration along the axis of the cavities will cause the capacitance in one cavity to increase, the other to decrease. The output signal is the difference between the two cavities. A frequency drift of the generator will thus not be seen to first order while the effect of an acceleration is additive.

The difficulties associated with these superconducting readout circuits are very simply those involved with the low temperature

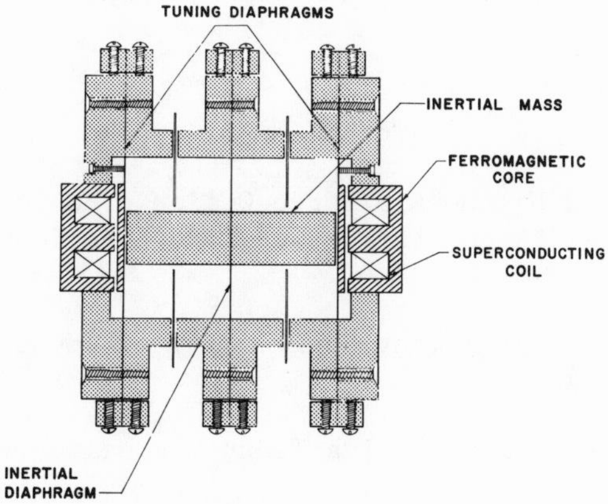

Figure 7. The r.f. cavity is resonant at 600 MHz.

environment. We cannot measure the Q of the electrical resonators before the apparatus is cooled. This is especially true for the thin film resonators of the first accelerometer mentioned. The levitation cannot be tested before cooldown and if it fails for some reason we must warm the apparatus up to fix it. In spite of these problems the benefits of the low temperature environment outweigh the disadvantages. If the resonant circuit can be made reliably and has a reliably high Q, the potential increases in sensitivity overbalance the difficulties. The operative word in this description is "reliably." We have had a great deal of difficulty with the first accelerometer described, primarily because of the high resistance of the thin film circuits before cooldown. The resonance can barely be observed until the apparatus is cold and then the Q is not necessarily high. We have obtained Q's of 10^5 from time to time but our experience is that a Q of a few x 10^4 is more likely. It is partly because of the reliability problem that we have adopted the second, re-entrant cavity type of accelerometer for our readout. We have, when coupling is proper, obtained Q's in excess of 10^5 repeatedly and others obtain 10^7 with proper cleaning. Much of the instrument alignment can be done at room temperatures and we obtain extra parametric gain by operating at 600 MHz.

The analysis of any resonant circuit will lead to the description of Wang and Uhlenbeck that we have already displayed. Here the electrical Q of the circuit assumes a dominant role. Again the crucial parameter will be the Q when the electrical circuit is suitably coupled for maximum signal to noise ratio and, as Kafka shows, this depends on the amplifier characteristics. We have demonstrated that a reduction in temperature must be only one facet of the effort to obtain a more sensitive detector. However, once the decision is made to operate a low temperature apparatus tremendous gains in sensitivity are possible. Hopefully our experience will be useful to any others attempting to make the new generation detectors.

REFERENCES

*Research supported by the Air Force Office of Scientific Research Grant No. AFOSR71-2054 and the National Science Foundation Grant No. PHY71-03166.

†Now at the Physics Department, University of Western Australia, Nedlands, Western Australia.

+On sabbatical leave from Florida Technological University, Orlando.

1. A good text for this study is Thomas, Statistical Communication Theory, Wiley, New York, 1969. A more advanced study is in

Whalen, Detection of Signals in Noise, Academic Press, New York, 1971.

2. P. Kafka, Erice Lectures, this volume.

3. S. Bonazzola, Erice Lectures, this volume.

4. M. C. Wang and G. E. Uhlenbeck, Rev. Mod. Phys. 17, 323 (1945).

5. See also the Ph.D. Theses of S. P. Boughn and H. J. Paik, Stanford University, 1975.

6. W. O. Hamilton, "Superconducting Shielding," in Methods of Experimental Physics, R. V. Coleman, ed., Academic Press, New York, 1973.

7. P. Carelli, A. Foco, U. Giovanardi, I. Modena, Cryogenics 16, 77 (1976).

8. W. M. Fairbank, Erice Lectures, this volume.

9. T. P. Bernat, D. G. Blair, W. O. Hamilton, Rev. Sci. Instr. 46, 582 (1975).

10. D. G. Blair, H. J. Paik, R. C. Taber, Rev. Sci. Instr. 46, 1130 (1975).

11. P. Carelli, A. Foco, U. Giovanardi, I. Modena, D. Bramanti, G. Pizzella, Cryogenics 15, 406 (1975).

12. The use of a re-entrant cavity as a displacement transducer is discussed by C. Van deGrift, Rev. Sci. Instr. 45, 1171 (1974).

OPTIMAL DETECTION OF SIGNALS THROUGH LINEAR DEVICES WITH THERMAL NOISE SOURCES, AND APPLICATION TO THE MUNICH-FRASCATI WEBER-TYPE GRAVITATIONAL WAVE DETECTORS

Peter Kafka
Max-Planck-Institut für Physik und AStrophysik
Föhringer Ring 6, 8000 München 40, W-Germany

CONTENTS

Probability distribution of λ in the presence of a signal of known strength.
The "signal content function" $\lambda(t)$, and its construction in the time domain.

4. Analysis of the Signal Content Function for Signal Pulses in a Resonant Detector.

Autocorrelation functions.
Almost monochromatic processes. The "co-rotating phase plane".
Statistics of the phase plane components. Rayleigh distribution.
The envelope $\Lambda(t)$ of the signal content $\lambda(t)$, and its normalized square $S(t)$, the "signal function".
Superposition of a known signal with the noise.
Threshold crossing.
Detectability at threshold S_o.
Optimal thresholds for limited and unlimited observation time.
Minimal detectable event rate as a function of signal strength. "Sensitivity and Detectability".

5. Optimal Procedure with Several Detectors.

Optimal multiple-detector sensitivity.
Loss of sensitivity when only the envelopes are combined.
"Coincidence experiments" for discrimination against local non-thermal disturbances.
Long-term search for rare events.

CHAPTER II: APPLICATION TO THE MUNICH-FRASCATI WEBER-TYPE GRAVITATIONAL-WAVE DETECTORS.

1. Summary of the Experiment.

2. Characterization of the Detectors.

Experimental details (supplied by W. Winkler).
The equivalent circuit and its noise sources.
Composition of the output U: Transfer functions; Spectrum; the kernel $\kappa(\omega)$.

3. Optimal Filtering for Short Gravitational Wave Pulses.

The reaction of the cylinder to a gravitational pulse.
Correspondence between gravitational signal strength and the equivalent electromotive force.
Calculation of the optimal sensitivity ϕ.
Optimalization of experimental parameters.
Calculation of the optimal filter for short pulses.
Optimal filtering in the components x and y in the "co-rotating phase plane".

INTRODUCTION

The first part (chapter I) of these lectures - a theory of signal detection in the thermal noise of a linear apparatus - is based on work done together with Friedrich Meyer between 1970 and 1973, when we first tried to understand the data processing in WEBER's (1969-1973) gravitational wave experiments, and then prepared the evaluation procedure for the Munich-Frascati Weber-type experiment. The essential ideas have been sketched earlier in various talks and in KAFKA (1973, 1974) and BILLING et al. (1975), where also results of the Munich-Frascati experiment have been reported.

Although we only attacked the special problem of the Weber-type experiment, it turned out that the theory applies to all devices with purely thermal noise sources, if both noise and signal appear linearly in the output. This certainly embraces a wide class of experiments, particularly if occasional non-thermal disturbances can be circumvented by the use of several detectors.

The aim of these lectures cannot be a general treatment of the problems connected with the detection of signals in noise. The author has not even studied the literature in this field. (Some general references, discovered afterwards, are given in the end.) Hence, experts may feel annoyed by a lack of agreement with possibly existing standard terminology. On the other hand, an independent approach to a special problem may justify an independent language, if this leads to a short but consistent presentation. This might be helpful for an experimentalist who does not want to spend too much time on beautiful mathematical theory but would rather know how to calculate the optimal evaluation algorithm and sensitivity for the detection of certain signals in his own apparatus.

In the second main part (Chapter II) we shall illustrate our general theory by detailed calculations for the Munich-Frascati experiment, which was built by H. Billing and W. Winkler in Munich and by K. Maischberger and D. Bramanti in Frascati. The optimal and the actual evaluation, as done by L. Schnupp and myself, are described, and the final null result of the search for short pulses within 350 days of observation is reported. The limits set by this result are the lowest achieved so far in gravitational wave experiments.

Finally, the present state of the art will be confronted with the far aims of gravitational pulse astronomy (Chapter III)

CHAPTER I

OPTIMAL DETECTION OF SIGNALS THROUGH LINEAR DEVICES WITH THERMAL NOISE SOURCES!

1. Formulation of the Problem

Consider a physical apparatus built for the detection of signals of a certain kind. In the absence of such signals its output is some function of time, e.g. a voltage U(t), due to the noise sources in the apparatus. In our treatment we make three assumptions about the apparatus and its noise: 1) The noise output U(t) is only due to thermal fluctuations of the constituents of the apparatus which are in thermodynamical equilibrium with "heat baths" of given temperatures. 2) The apparatus is linear in the sense, that its output noise is a sum of contributions which are linear functionals of the internal white noise sources. 3) The signal which one is looking for, causes in a known way an output $U_g(t)$ which is linearly superposed on the noise output.

An observation leads to a "piece of output" U(t) during some time interval, say $-\tau/2 \leq t \leq +\tau/2$. (We shall always assume that it is recorded with a time resolution sufficient for the mathematical procedures applied.) The signal which we are looking for, may for instance be a pulse which leads to a transient output $U_g(t)$, or a steady flux of radiation leading to a permanent addition $U_g(t)$ to the noise.

Our problem is to define the "signal content" of the observed output U(t) with respect to the assumed sort of signal $U_g(t)$. Of course, this can only be done in a statistical way, because the noise is a random process and will simulate any kind of signal with a certain probability. The smaller the mean square signal content simulated by pure noise, the more sensitive we shall call the apparatus. Expressing this concept of sensitivity in terms of the constituents of the apparatus, one will find the building instruction for a better apparatus.

With the construction of an optimal filter for the measurement of signal strength at each instant of time, the problem of optimal signal detection will not yet be solved. In actual survey experiments the task will not be to recover in an optimal way the strength of an individual signal, but rather to find the most likely rate-strength distribution of external signals which can explain the observed deviations from thermal statistics over a long observation time. For that purpose, or to judge the significance of null results, it is not enough to know the sensitivity, but one needs the full probability distributions in noise and in signal-plus-noise of the signal content as a function of time. We shall see that in principle the optimal procedure does not only depend on the type of the assumed signal, but even on assumptions about the rate-strength distribution. These questions will be discussed under the head-line "detectability" at the end of section 4.

Finally, we have to treat the case of more than one detector, and the exclusion of local disturbances in co-incidence experiments.

2. Statistics of Thermal Noise

A "piece of noise" as a vector in the "noise space" - An observation will usually mean the registration of some function $U(t)$ during a finite interval of time, say $-\tau/2 \leq t \leq +\tau/2$. $U(t)$ may e.g. be the output voltage of a detector, or the more indirectly defined electromotive force of an internal source of noise.

We want to consider such a piece of output as one entity and attribute to it a probability, assuming it is due to the noise sources of a known apparatus.

A real piece of observation can only contain a finite amount of information. Hence, we can avoid the difficulties with the definition of probability measures in infinitely dimensional spaces. We rather consider our functions $U(t)$ as vectors in a linear vector space of finite dimensions. A natural base will be given by the discrete instants of time at which the values of the output have been recorded. Let us assume they are $-\tau/2 = -N\,\Delta t, \ldots, -\Delta t, 0, \Delta t, \ldots N\Delta t = +\tau/2$. Then our space has the dimension $2N+1$. Let us further assume that the apparatus has smoothed the function $U(t)$ over the interval Δt. Then, nothing of importance happens below our time resolution and we can be sure that the representation of $U(t)$ by the $2N+1$ values contains "nearly all" information about it.

Next, let us consider the Fourier transform of U(t). Throughout these lectures we shall use the same functional symbol for a function of time U(t) and its Fourier transform U(ω):

$$U(t) = \frac{1}{\sqrt{2\pi}} \int_{-\infty}^{+\infty} U(\omega)\, e^{i\omega t} d\omega$$

$$U(\omega) = \frac{1}{\sqrt{2\pi}} \int_{-\infty}^{+\infty} U(t)\, e^{-i\omega t} dt$$

For these definitions to be applicable, we shall set U(t) = 0 outside the observation interval, and U(ω) = 0 for $|\omega| > \omega_c \approx 2\pi/\Delta t$. Because of the assumed smoothness on the scale Δt, the cut-off at ω_c is justified. In the ω-representation there will no essential information be contained on a smaller scale than $\Delta\omega = 2\pi/\tau$, corresponding to the finite observation time τ. Hence, U(t) will also be properly represented by the values of its Fourier transform at the frequencies $-\omega_c = -N\,\Delta\omega, \ldots -\Delta\omega, 0, \Delta\omega, 2\Delta\omega, \ldots N\Delta\omega = \omega_c$. Since $U(-\omega) = U^*(\omega)$, we can also use the real and imaginary parts of U(ω) at $\omega \geq 0$ only.

We see that with these conventions the dimension of the ω-base is the same as in the t-base. Therefore we shall from now on think of U as a vector in a "nearly infinite dimensional" linear vector space, in which we can go from one base to another. With our assumptions about smoothness etc., this will be a good approximation. In the same spirit we shall write integrals over t or ω (instead of sums over the points of support), and use e.g. Parseval's theorem. (We beg the pardon of mathematicians.)

The aim of this section will be to study the structure introduced in our "noise space" by the statistics of the output noise.

White noise of a resistor. Nyquist's formula - Thermal noise is the result of an extremely large number of "micro-events" which do not distinguish any frequency, except for a high frequency cut-off due to their quantum nature. Because of this lack of spectral features one speaks of white noise. The "height" of this constant spectrum is determined by the equipartition law of thermodynamics. Usually one defines the spectral density (or simply: "the spectrum") U^2 of a process U(r) by

$$U_\omega^2 = 2 \lim_{\tau\to\infty} \frac{1}{\tau} \overline{|U(\omega)|^2}$$

because then $\int_0^\infty U_\omega^2 d\omega = \overline{U^2(t)}$.

In order to practice our procedures, and because we shall represent the noise sources of a detector by resistors in an equivalent electric circuit, let us consider the fluctuating electromotive force (e.m.f.) u(t) of a resistor R in an L-C-circuit, and

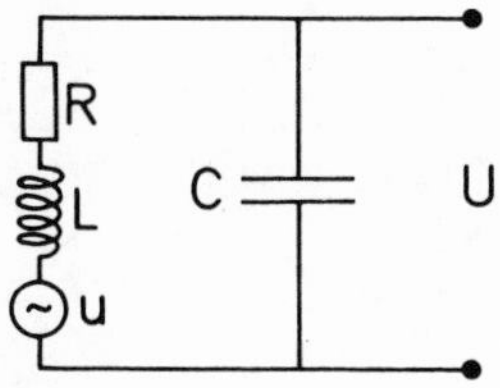

Figure 1: Output of a noisy L-C-circuit.

the corresponding output voltage as shown in figure 1. From thermodynamics we know that in this situation the mean energy of the L-C-oscillator has to be E=kT. Hence, the long-term mean square value of U(t) will be given by $E = LI^2 = CU^2 = kT$. The fluctuation current due to the e.m.f. u has to flow back through C. Therefore the Fourier transforms of pieces of noise U and u will fulfill

$$(U-u)/(R+i\omega L) = -U\cdot i\omega C$$

With $\omega_o^2 \equiv 1/LC$ this gives

$$u/U = 1-\omega^2/\omega_o^2 + i\omega RC$$

If we think of the limit of a long interval τ, or take the average $\langle\ldots\rangle$ over many pieces of noise, we write (using Parseval's theorem):

$$\overline{U^2} = \frac{1}{\tau}\int_{-\tau/2}^{+\tau/2} U^2(t)dt = \frac{1}{\tau}\int_{-\infty}^{+\infty} |U(\omega)|^2 d\omega =$$

$$= \frac{\Delta\omega}{2\pi}\int_{-\infty}^{+\infty}\frac{|u(\omega)|^2 d\omega}{|1-\omega^2/\omega_o^2+i\omega RC|^2} = \frac{\Delta\omega}{2\pi}|u(\omega_o)|^2\int_{-\infty}^{+\infty}\frac{\omega_o^4\, d\omega}{(\omega^2-\omega_o^2)^2+\omega^2(\omega_o^2 RC)^2}$$

In the last step we made use of the fact that the white noise e.m.f. u(t) will have no spectral features near ω_o, and $|u(\omega)|^2$ can be taken out of the integral. The remaining well known integral gives π/RC, and from $kT = CU^2 = \Delta\omega/2R\,|u(\omega)|^2$ we obtain

$$\langle |u(\omega)|^2\rangle = 2kTR/\Delta\omega \qquad (1)$$

This is our form of Nyquist's formula.
With the spectrum $u_\omega^2 = 2/\tau\,|u(\omega)|^2$ this reads

$$u_\omega^2 = 2kTR/\pi$$

or in the usual appearance, with $\omega = 2\pi\nu$:

$$u_\nu^2 \, d\nu = 4kTR \, d\nu.$$

<u>Probability density in the noise space</u> - According to the "Central Limit Theorem" of probability theory (see text books) a superposition of a very large number of independent events will lead to a normal distribution ("Gaussian"). Therefore, the real and imaginary parts of $u(\omega)$ at a fixed value of ω will have independent normal distributions around mean value zero with the same variance σ^2:

$$\sigma^2 = |u(\omega)|^2/2 = kTR/\Delta\omega$$

For instance, the probability of finding the real part of $u(\omega_1)$ in the range $R_1 \leq Re(u(\omega_1)) \leq R_1 + dR_1$ is

$$W(R_1)dR_1 = (2\pi\sigma^2)^{-1/2} \exp(-R_1^2/2\sigma^2)dR_1.$$

Because of the independence of real and imaginary part of $u(\omega)$, the joint probability to find $Re\ u(\omega_1) \approx R_1$ and $Im\ u(\omega_1) \approx I_1$, will be the product

$$W(R_1, I_1)dR_1 dI_1 = (2\pi\sigma^2)^{-1} \exp(-|u(\omega_1)|^2/2\sigma^2) \; dR_1 dI_1$$

because $R_1^2 + I_1^2 = |u(\omega_1)|^2$.

Now we can write down the joint probability distribution to find a whole vector u such that the components are in a small volume element around a fixed vector. Since the components at all frequencies ω_i are again independent, we find

$$W(u)dVol = \exp(-\Delta\omega/2kTR \cdot \sum_i |u(\omega_i)|^2) \cdot K \cdot dVol \; .$$

Here dVol is an abbreviation for the volume element $dR_O dR_1 dI_1 dR_2 \cdot \cdot dI_2 \ldots dR_N dI_N$ in our noise space in the base $\omega = 0, \Delta\omega, 2\Delta\omega, \ldots N\Delta\omega$. K is the normalization faction which makes $\int_{space} W(u)dVol = 1$.

In the limit of large τ and small Δt we can replace the sum in the exponent of the last equation by an integral and write

$$W(u)dVol = \exp(-(1/2kTR)\int |u(\omega)|^2 d\omega) \cdot K \cdot dVol \qquad (2)$$

W(u) is the probability density for finding a certain vector u as an observed piece of noise.

This basic concept will be needed to find the probability for the simulation of signals by detector noise. The Gaussian character of the probability distribution is essential for the following. For white noise this character will also be manifest in the time representation:

$$W(u)dVol \quad \exp(-\int_{-\infty}^{+\infty} u^2(t)dt/2\,\overline{u^2}\Delta t)\ dVol\ \ldots$$

(Remember $u(t) = 0$ outside an interval, and Δt = smoothing time.) However, this form will not be preserved after filtering.

Superposition of independent noise sources in the output of a linear device - We consider an apparatus with independent sources of white noise $\overset{1}{u}, \overset{2}{u}, \ldots$. By "linearity" we mean that the output U is a sum $U = \overset{1}{U}+\overset{2}{U}+\ldots$, where $\overset{i}{U}$ is a linear functional $\overset{i}{F}(\overset{i}{u})$ of the corresponding $\overset{i}{u}$. This condition will guarantee the Gaussian character of the total output noise. In general a linear functional in our noise space will be of the form $U_i = \sum_j F_{ij} u_j$ with a matrix F_{ij} given e.g. in the base $\omega_i = 0, \Omega, 2\Omega, \ldots N\Omega$ or any other. In the limit in which we write integrals this would e.g. appear as $\overset{1}{U}(\omega) = \int_{-\infty}^{\infty} F(\omega,\omega')\overset{1}{u}(\omega')d\omega'$. In our example of a gravitational wave detector we shall deal with an apparatus which can be represented by an equivalent circuit containing only resistors, capacitors and inductances. For such devices the relation between $\overset{i}{U}$ and $\overset{i}{u}$ will reduce to $\overset{i}{U}(\omega) = \overset{i}{F}(\omega)\cdot\overset{i}{u}(\omega)$ in the ω-representation. The kernel $F(\omega,\omega')$ becomes a δ-function with respect to ω'. The Fourier components are the eigenvectors of F_{ij}. Hence, for this most obvious application the ω-representation is distinguished by nature. The statistical independence given by white noise in all bases originally, will be kept in the ω-base even after the filtering. The function $\overset{i}{F}(\omega)$ may be called the transfer function of the i-th noise source.

In the more general case of a linear functional, no essential differences seem to arise, except notational complications. (One would have to go to a diagonal form by a transformation of axes.) Therefore we shall restrict our presentation to the case where the Fourier components are the eigenvectors. This is so if the noise sources are "white" (or filtered white) and if the linear apparatus does not mix frequencies. Then, the total output noise will be written:

$$U(\omega) = \sum_i \overset{i}{F}(\omega)\overset{i}{u}(\omega)$$

<u>Probability density of the total noise. The kernel $\kappa(\omega)$</u> - In the probability density of $\overset{i}{u}$ in the noise space (eq. 2) we can insert $\overset{i}{u} = \overset{i}{U}/\overset{i}{F}$ and find the probability density of $\overset{i}{U}$:

$$W(\overset{i}{U})\,dVol \sim \exp\left\{ -\int_0^\infty \frac{|\overset{i}{U}(\omega)|^2}{2k\overset{i}{T}\overset{i}{R}|\overset{i}{F}(\omega)|^2}\, d\omega \right\} dVol$$

Here and in the following we only write the interesting part of the right hand side and use the proportionality sign. To derive the density $W(U)$, for $U = \sum_i \overset{i}{U}$, we can simply argue that for each Fourier component we have a sum of independent normally distributed variables (with mean value zero) which will again be normally distributed around zero with a variance

$$\sigma^2 = \sum_i \overset{i}{\sigma}{}^2 .$$

Then we obtain for the whole vector (the collection of its Fourier components)

$$\left.\begin{aligned} W(U)\,dVol &\sim \exp\left\{ -\int_0^\infty \frac{|U(\omega)|^2}{\sum_i 2k\overset{i}{T}\overset{i}{R}|\overset{i}{F}(\omega)|^2}\, d\omega \right\} dVol \\ &= \exp\left\{ -\int_0^\infty (\omega)\,|U(\omega)|^2 d\omega \right\} \\ &\text{with } \frac{1}{\kappa(\omega)} \equiv \sum_i 2k\overset{i}{T}\overset{i}{R}|\overset{i}{F}(\omega)|^2 \end{aligned}\right\} \qquad (3)$$

This probability distribution for the output noise of the whole apparatus will be the starting point for all further considerations. The kernel $\kappa(\omega)$ contains all relevant information about the statistics of the pure noise. Apart from a numerical factor, due to our conventions about Fourier transforms and spectra, $1/\kappa$ must be the spectrum of the total output. We have $U_\omega^2 = 1/\pi\kappa(\omega)$.

(As an excercise one can derive formula 3 from the preceding one if one writes down the joint probability density for U_1, U_2, U_3 ..., and integrates over all $\overset{i}{U}$ ($i=2,3,\ldots$) under the side condition $\sum_i \overset{i}{U} = U$. To this end one has to complete the squares in the exponent consecutively and "kill" one variable after the other by integrations which give only contributions to the constant normalization factor.)

We remark that in the more general case of Gaussian statistics, where $\overset{i}{F}(\omega,\omega')$ replaces $\overset{i}{F}(\omega)$, the exponent in (3) will be replaced by an expression of the sort

$$\iint K(\omega,\omega')U(\omega)U(\omega')d\omega d\omega' \ .$$

For instance, a similar structure will appear in the time-representation, because no device will be able to make the values of U at all moments of time our statistically independent variables.

3. Signal Content in Noise

Definition of the "unit signal" output U_g - The external physical influence which is to be detected, will cause a certain output voltage in our apparatus. We demand that this signal output be a linear functional of the arriving signal. Clearly the optimal detection procedure will depend on the characteristics of the signal. For instance, a short pulse or a nearly monochromatic wave have to be treated differently. If one has tried to optimize a detector for one kind of signal, it may be far from optimal for another. What we want to describe here, is the optimal treatment of the output of a fixed detector for an assumed kind of signal. (The detector optimization in a concrete example will be discussed in chapter II, 3). Hence, we assume a unit signal, which causes an output voltage U_g, linearly superposed on the noise. The relevant characteristics of the signal input will be contained in U_g as a function of time or frequency. We define U_g to contain also the information about arrival time (or relative phase). E.g. it may mean: the output voltage of a short gravitational wave pulse of a given unit strength, arriving at time t_o.

The κ-metric in the noise space: Orthogonal splitting into signal and noise - If we have observed a piece of output, say U, we shall ask how much of the "signal" U_g is contained in it. Figure 2 gives an example with U and U_g shown as functions of time t.

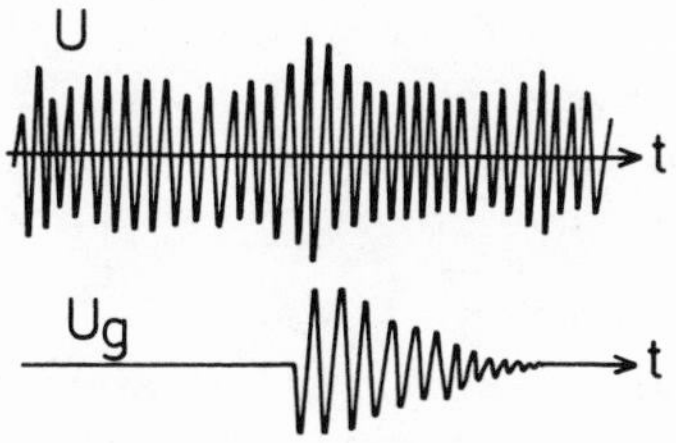

Figure 2: Output of noise and pure signal.

To answer our question we need a clear definition of the concept of "being contained". We can, of course, always write $U=V+\lambda U_g$ in an arbitrary way. In order to call λ the signal content of U with respect to U_g, we have to define the part V of U in such a way, that we can say: V does not contain any information about U_g.

Since the positive definite quadratic functional in the exponent of the noise distribution (3) defines a metric in our space, it seems plausible to use this metric to define orthogonality between the constituents V and λU_g of U. This metric introduces a scalar product $U \circ W$ and a norm $\|U\| = \sqrt{U \circ U}$ by

$$\left.\begin{aligned} U \circ W &\equiv \int_0^\infty \kappa(\omega)\, U^*(\omega)\, W(\omega)\, d\omega \\ \|U\|^2 &\equiv \int_0^\infty \kappa(\omega)\, |U(\omega)|^2 d\omega \end{aligned}\right\} \quad (4)$$

The meaning of this metric is: vectors with the same norm are equally likely found in pure noise; and mutually orthogonal vectors are statistically independent in pure noise. This statistical independence is in the sense that the probability of finding V+W as a piece of observed output is a product of the probabilities for V and W, because $\|V+W\|^2 = \|V\|^2 + \|W\|^2$ if $V \circ W = 0$.

We can use this metric to introduce an orthonormal frame in our space in such a way that one base vector E_o is the unit vector in the direction of the signal U_g, and the subspace orthogonal to U_g is spanned by an orthonormal frame $E_\nu(\nu=1,\ldots n)$. A vector U is then given by its components λ_ν in this frame:

$$U = \sum_{\nu=0}^{n} \lambda_\nu E_\nu$$

If we express the probability density of U (eq. 3) in this frame, it becomes a product of Gaussians for the components:

$$W(U)\, d\lambda_o d\lambda_1 \ldots d\lambda_n \sim e^{-\Sigma \lambda_\nu^2 \|E_\nu\|^2}\, d\lambda_o d\lambda_1 \ldots d\lambda_n$$

The linear transformation to this base introduces the Jacobian into W(U). However this is a constant which goes into the normalization.

For every single component λ_ν the probability distribution is a Gaussian with zero mean and variance $\sigma^2 = \frac{1}{2\|E_\nu\|^2} = \frac{1}{2}$:

$$W(\lambda_\nu)\, d\lambda_\nu = e^{-\lambda_\nu^2}\, d\lambda_\nu \cdot \frac{1}{\sqrt{\pi}} .$$

This is obvious because in the λ_ν-representation one can integrate over all but one components, producing only numerical factors.

We want to ascribe a real number λ to U, measuring the signal content with respect to U_g. If an external signal parallel to U_g

is added to a piece of noise, it will of course only influence λ_o. There is also noise in λ_o, which cannot be avoided. This noise is kept minimal if λ_o is defined by our orthogonal decomposition of U. With a non-orthogonal splitting, noise from other statistically independent components λ_ν would be admixed to λ_o and one would lose more information about the signal than necessary. Since the components λ_ν do not contain any information about each other, a functional $\lambda = \mathcal{L}(U) = L(\lambda_\nu)$ is a non-optimal measure of signal content, if the function $L(\lambda_\nu)$ contains any component but λ_o. However, if we have to choose a function of only one variable, it would be meaningless to distort the λ-scale by choosing a non-linear function $L(\lambda_o)$. Therefore, we can say:

An optimal measure of signal content is defined by orthogonal projection of U on U_g.

For non-Gaussian output statistics it may be helpful to make use of correlations between signal and noise, and thus use the noise as a sort of amplifier in some region of the noise space. For Gaussian noise this is impossible, because one can isolate the signal optimally from all noise except the one in the direction of the signal. Admixture of independent noise can never improve the signal recognition.

For the same reason, one should always try and build linear detectors, if the signal and the noise sources are linearly superimposed in the "input".

Calculation of the signal content λ: Definition of the "optimal filter" by the orthogonal splitting - To find out how the signal content λ of an observed piece of output U has to be calculated, let us multiply $U=V+\lambda U_g$ by U_g in the sense of the scalar product. Because of $V \circ U_g = 0$, the result is

$$\lambda = U \circ U_g / U_g \circ U_g \tag{5}$$

with the scalar product defined by equation (4). This algorithm for finding λ from U, we shall call the optimal filtering for the signal of type U_g. We have just seen, that no other algorithm can bring a better result in any average sense. If somebody finds that an individual, artificially induced signal of known strength comes out more clearly with a different algorithm, this is not a contradiction. On the average the algorithm described here will be optimal.

The sensitivity Φ: Definition through the probability density of signal content in pure noise - Integration of W(U)dVol over the subspace orthogonal to U_g gave

$$W(\lambda)d\lambda \sim \exp(-\lambda^2 U_g \circ U_g)d\lambda$$

We shall use the abbreviation

$$\Phi \equiv U_g \, U_g$$

for the squared norm of our unit signal. Since $W(\lambda)d\lambda$ has to be $= 1$, we can recover the correct normalization factor and wirte:

$$W(\lambda)d\lambda = e^{-\lambda^2\Phi} \cdot (\phi/\pi)^{1/2} \cdot d\lambda \tag{6}$$

$W(\lambda)$ is the probability density of the signal content λ in pure noise.

We see that λ has a normal distribution $e^{-\lambda^2/2\sigma^2}$ with mean value zero and variance $\sigma^2 = (2\Phi)^{-1}$. The bigger Φ , the less signal will be simulated by the noise on the average. Hence, we call $\Phi = U_g\, U_g$ the optimal sensitivity for the signal U_g. It has to be calculated from the parameters of the given detector and the properties of the signal:

$$\Phi = U_g \, U_g = \quad \kappa(\omega)\left|U_g(\omega)\right|^2 d\omega \tag{7}$$

<u>Probability distribution of λ in the presence of a signal of known strength</u> - For calibration, or for tests of a detector, one will apply artificial signals of known strength λ_g and study the distribution of the observed signal content as defined by the optimal filtering algorithm. Because of the independence of the noise and the artificial signal the observed signal content λ will be normally distributed around the strength λ_g of the applied signal, with the same variance as in pure noise:

$$W(\lambda|\lambda_g)d\lambda = (\Phi/\pi)^{1/2} \exp(-(\lambda-\lambda_g)^2\Phi)d\lambda \tag{8}$$

This distribution of λ, under the condition that a known signal of strength λ_g is present, will also be needed for the discussion of the rate-strength distribution of observed signals.

<u>The signal content function $\lambda(t)$ and its construction in the time domain</u> - So far we had thought of a U_g which includes the information about the arrival time or absolute phase of the signal. Unless the signal is a harmonic wave, we shall be interested in the signal content with respect to a signal arriving at time t. If we look back into figure 2, this means that we shift the lower curve along time t and calculate the signal content λ for each instant t. (By arrival time we mean of course the moment defined by some feature of the "shape of U_g. It does not have to be the beginning, if it starts smoothly.) Hence, we shall from now on use a unit signal $\overset{o}{U}_g(t)$ "arriving" at time $t=0$, and write $U_g(t) = \overset{o}{U}(t-t_o)$ for the signal of shape and strength U_g but "arriving" at time t_o. Then, from the definition of Fourier transforms, we know that

$$U_g(\omega) = \overset{\circ}{U}_g(\omega)\, e^{-i\omega t_o}$$

Instead of "the signal content of U with respect to $\overset{\circ}{U}_g$ arriving at time t" we shall say "the signal content at time t" and call it the signal content function $\lambda(t)$.
From formula (5) we find

$$\lambda(t) = \frac{1}{\Phi}\int_0^\infty \kappa(\omega) U_g^*(\omega) U(\omega)\, d\omega = \frac{1}{\Phi}\int_0^\infty \kappa(\omega) \overset{\circ}{U}{}_g^*(\omega)\, e^{i\omega t} U(\omega)\, d\omega \qquad (9)$$

Taking the integral from $-\infty$ to $+\infty$ and using Parseval's theorem we go to the time domain and find

$$\lambda(t) = \int_{-\infty}^{+\infty} G(\tau) U(t+\tau)\, d\tau \qquad (10)$$

where G(t) is the Fourier transform of

$$\frac{1}{2\phi}\, \kappa(\omega) \overset{\circ}{U}_g(\omega) .$$

$$G(t) = \frac{1}{\Phi} \cdot \frac{1}{\sqrt{2\pi}} \int_0^\infty \kappa(\omega) \overset{\circ}{U}_g(\omega)\, e^{i\omega t} d\omega \qquad (11)$$

Equation (10) shows that the signal content function $\lambda(t)$ is constructed from the observed output U(t) by convolution with the function $G(\tau)$.

Although the optimal filtering will usually not be possible by application of "hard-ware" filters in the frequency domain (it is not "physically realizable"), one can do it numerically by application of equation 10, if the output data have been recorded as a function of time. Of course, this may not be worth while, if a reasonable hard-ware approximation can be constructed, which recovers a sufficient fraction of the optimal sensitivity Φ. An analogue device for filtering would certainly be advisable in long-term survey experiments, if one knows what kind of signal one is looking for. Then one should develop an analogue installation (e.g. with delay lines which allow a simulation of equ. 10) for this kind of signal - or several ones for various classes of signals. Nevertheless, the knowledge of the optimal filters and their sensitivities Φ will be a valuable guide.

4. Analysis of the Signal Content Function for Signal Pulses in a Resonant Detector

In this section we shall restrict ourselves to the case of detectors whose output is dominated by a resonance, and to signals whose spectra do not show structure in the neighbourhood of this resonance.

Autocorrelation functions - For a statistical process f(t), like the output noise U(t) or its signal content λ(t), the autocorrelation function $C_f(t)$ is defined by

$$C_f(\tau) \equiv \overline{f(t)f(t+\tau)} \Big/ \overline{f^2(t)} \tag{12}$$

where the means are defined by $\lim_{T\to\infty} \frac{1}{T}\int_0^T \dots dt$. If one writes down its Fourier transform, one finds that the autocorrelation function $C_f(\tau)$ and the spectrum f_ω^2 are mutually connected by Fourier transformation:

$$C_f(\tau) = \int_0^\infty f_\omega^2 \cos\omega\tau \, d\omega \Big/ \int_0^\infty f_\omega^2 \, d\omega . \tag{13}$$

This relation and its inverse are sometimes referred to as Wiener-Khinchin relations.

Since the noise spectrum U_ω^2 defined our kernel κ(ω) and the metric in the noise space, we see that the knowledge of the autocorrelation function $C_u(\tau)$ would be sufficient as well.

Almost-harmonic processes. The co-rotating phase plane - The amplitude U of a damped harmonic oscillator, like the one considered in figure 1, follows an "equation of motion":

$$\ddot{U} + 2\beta\dot{U} + \omega_o^2 U = f(t)$$

where f(t) comprises the external and the "thermal" accelerations. The damping constant β, the resonance frequency ω_o and the "quality" Q are connected by $Q = \omega_o/2\beta$. The shape of the output spectrum is

$$U_\omega^2 \sim \frac{1}{(\omega^2-\omega_o^2) + (2\beta\omega)^2} \tag{14}$$

Applying (13) one finds the autocorrelation function

$$C_u(\tau) = e^{-\beta|\tau|} \cos \omega_1\tau \tag{15}$$

where $\omega_1 = \omega_o(1-\beta^2/\omega_o^2)$ equals the peak frequency of the spectrum up to 4th order in β/ω_o. The spectrum is symmetric around its peak in a similar approximation.

If the output of a resonant detector contains noise in addition to that introduced by the damping itself, the spectrum and autocorrelation function will be distorted. However, as long as the peak of the resonance remains dominant in the spectrum, the symmetry around the peak frequency and the "exponential cosine" character of the autocorrelation function will be approximately preserved. A random function of this kind may be called "almost harmonic".

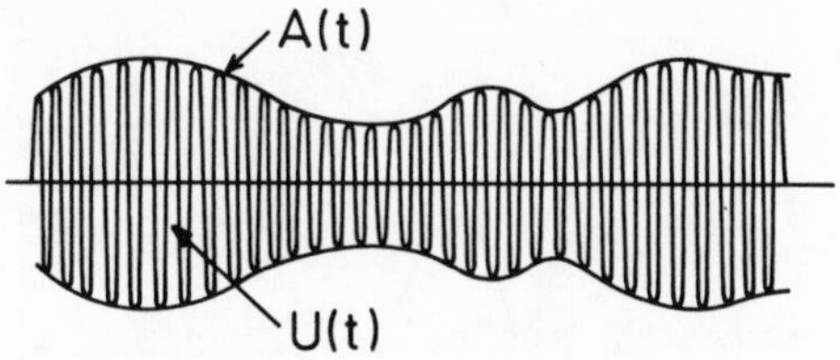

Figure 3: Envelope of an almost-harmonic process.

For any frequency Ω near the peak frequency of the spectrum we can write:

$$U(t) = A(t) \cos(\Omega t + \phi(t)) \tag{16}$$

where $A(t)$ and $\phi(t)$ are slowly varying functions of time. Slowness is with respect to the period $2\pi/\Omega$. If one wants $A(t)$ to be the envelope of $U(t)$ as in figure 3, one has to choose for Ω the frequency in the autocorrelation function. Instead of using the amplitude A and the phase ϕ as polar coordinates in the plane, one can also choose orthogonal coordinates x and y defined by

$$x(t) = A(t) \cos \phi(t)$$
$$y(t) = A(t) \sin \phi(t)$$

such that

$$\left.\begin{aligned} U(t) &= x(t) \cos \Omega t - y(t) \sin \Omega t \\ A(t) &= x(t) \cos \phi(t) + y(t) \sin \phi(t) \\ A^2 &= x^2 + y^2 \end{aligned}\right\} \tag{17}$$

The condition that A is the envelope of U, corresponds to x and y being uncorrelated. If one writes down the autocorrelation function of U, one sees that a beat between x and y is optimally avoided if Ω is chosen equal to the autocorrelation frequency of U.

In this "co-rotating phase plane" representation, the almost harmonic high frequency process U is approximated by a relatively slow random walk, driven back by the damping toward the center, in the plane spanned by A and ϕ or x and y.

Statistics of the phase plane components. Rayleigh distribution - We remember that at any fixed value of time the value of U has a probability distribution

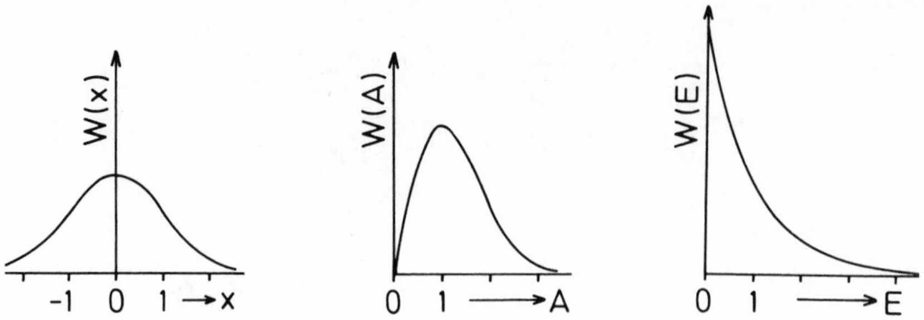

Figure 4: Gaussian, Rayleigh and Boltzmann distributions.

$$W(U)dU = (2\pi\overline{U^2})^{-1/2}\cdot\exp(-U^2/2\overline{U^2})dU.$$

If U is represented by mutually independent components x and y as in (17), we have $\overline{x^2} = \overline{y^2}$ and therefore $\overline{U^2} = \frac{1}{2}(\overline{x^2}+\overline{y^2}) = \overline{x^2} = \overline{y^2}$. Henze, x and y will have independent Gaussian distributions around zero with the same variance as that of W(U). The joint probability distribution, written with $x^2+y^2 = A^2$, will be

$$W(x,y)dxdy = e^{-A^2/2\overline{U^2}}\cdot dxdy/2\pi\overline{U^2} .$$

If we return to the polar coordinates A and ϕ, we have to substitute the phase plane element: $dxdy = AdAd\phi$. We integrate over the random phase, which gives a factor 2π, and find the "Rayleigh distribution" for the amplitude A:

$$W(A)dA = e^{-A^2/2\overline{U^2}}\cdot AdA/\overline{U^2} \quad (18)$$

For the "normalized energy" $E \equiv A^2/\overline{A^2}$ there follows the "Boltzmann distribution" $W(E)dE = e^{-E}dE$. Figure 4 shows the shapes of the distributions for x and y (Gaussian), for A(Rayleigh) and E (Boltzmann).

The envelope $\Lambda(t)$ of the signal content $\lambda(t)$ and its normalized square S(t), the "Signal-function" - From the construction of our signal content function $\lambda(t)$ in section 3. we can see that for an almost harmonic detector output U and a short duration of the signal U_g, the signal content $\lambda(t)$ will also be almost harmonic.

Shortness of the signal is demanded relative to the damping time $1/\beta$ in the autocorrelation function (14). Then the optimally constructed $\lambda(t)$ will have a similar autocorrelation function, except for a shorter correlation time. If one is not interested in the exact arrival time or absolute phase of a signal, but rather in its strength only, one will be satisfied with the observation of the slowly varying amplitude or envelope $\Lambda(t)$ in the co-rotating phase plane representation of $\lambda(t)$:

$$\begin{aligned} \lambda(t) &= \Lambda(t)\cos\left[\omega_c t + \phi(t)\right] \\ &= \lambda_x \cos\omega_c t - \lambda_y \sin\omega_c t \end{aligned} \tag{19}$$

where ω_c is the frequency in the approximate "exponential cosine" autocorrelation function $C_\lambda(\tau)$ of $\lambda(t)$, analogous to (15).

$$C_\lambda(\tau) \approx e^{-\mu|\tau|}\cos\omega_c\tau \tag{20}$$

The signal content envelope $\Lambda(t)$ will contain all information about the strength of arriving signals, while the information about the arrival time is only kept up to the width $1/\mu$ (with μ from $C_\lambda(\tau)$). The statistics of $\Lambda(t)$ are analogous to those of A in eq. (18). Remembering that λ had a Gaussian distribution with $2\sigma^2 = 1/\Phi$, we can write the Rayleigh-distribution of the envelope $\Lambda(t)$:

$$W(\Lambda)d\Lambda = e^{-\Lambda^2\Phi}\cdot 2\Phi\Lambda d\Lambda \tag{21}$$

This shows that it is reasonable to use Λ^2 instead of Λ: We define the "normalized squared signal content envelope"

$$S(t) \equiv \Phi\cdot\Lambda^2(t) \tag{22}$$

which then has the simple "Boltzmann distribution"

$$W(S)dS = e^{-S}\,dS \tag{23}$$

at each instant of time. For shortness we shall call S(t) the "signal-function". Its mean value in noise is unity.

Superposition of a known signal with the noise - We saw in section 3. that in the presence of a known signal of strength λ_g, the normal distribution of the signal content is simply shifted from mean value zero to λ_g. In order to find the corresponding distribution of the envelope Λ, we have to integrate the λ-distribution over the random phase in the co-rotating phase plane, after addition of a signal of strength λ_g. Because we integrate over the random phase anyway, we can put λ_g in a fixed direction, e.g. the λ_x-direction, if we have represented λ by independent components $\lambda_x = \Lambda\cos\phi$, $\Lambda_y = \Lambda\sin\phi$. If we now call the signal strength Λ_g (instead of λ_g) we obtain

$$W(\Lambda)d\Lambda = \int_{\phi=0}^{2\pi} e^{-[(\lambda_x-\Lambda_g)^2 + \Lambda_y^2]\Phi} \cdot \frac{\Phi}{\pi} \Lambda d\Lambda d\phi =$$

$$= e^{-[(\Lambda\cos\phi-\Lambda_g)^2 + \Lambda^2 \sin^2\phi]\Phi} \cdot \frac{\Phi}{\pi} \cdot \Lambda d\Lambda d\phi$$

$$= e^{-(\Lambda^2+\Lambda_g^2)\Phi} \cdot \Phi \cdot 2\Lambda d\Lambda \cdot \int_{\phi=0}^{2\pi} e^{2\Phi\Lambda\Lambda_g \cos\phi} \frac{d\phi}{2\pi} .$$

The integral $\int_0^{2\pi} e^{Z\cos\phi} \frac{d\phi}{2\pi} = I_o(Z)$ is known as the modified Bessel function of order zero. Hence, if we know about the presence of the signal of strength Λ_g, the signal content envelope will be distributed according to

$$W(\Lambda|\Lambda_g)d\Lambda = e^{-(\Lambda^2+\Lambda_g^2)\Phi} \cdot I_o(2\Phi\Lambda\Lambda_g) \cdot \Phi \cdot 2\Lambda d\Lambda \tag{24}$$

For $\Lambda_g = 0$ one has $I_o(0) = 1$ and one recovers the Rayleigh distribution. For large values of the argument one has $I_o(Z) \rightarrow e^Z/2\pi Z$. Therefore, for large Λ_g, $W(\Lambda)$ will approximate a Gaussian around Λ_g: $W(\Lambda)d\Lambda \rightarrow e^{-(\Lambda-\Lambda_g)^2\Phi} \cdot \frac{\Phi}{\pi} d\Lambda$. (The influence of zero which makes the Rayleigh distribution asymmetric, gets lost for large Λ_g.)

If we write (24) with our signal function $S = \Phi\Lambda^2$ we obtain

$$W(S|S_g)dS = e^{-(S+S_g)} \cdot I_o(2\sqrt{SS_g})dS \tag{25}$$

For a rough approximation one can instead use a Gaussian with expectation value S_g+1, and variance $\sqrt{1 + 2S_g}$.

The value S_g is the value which the signal function S would show in the absence of noise at the arrival of a signal of strength $\lambda_g = \lambda_g$, i.e. when

$$U^2 = S_g \cdot U_g^2/\Phi \tag{26}$$

Now we see, why we called Φ , and not 2Φ, the sensitivity with respect to the signal U_g: The average value unity of the Boltzmann-distributed signal function S corresponds to a mean squared simulated signal U_g^2/Φ.

We should mention here, that of course formulae analogous to (24) and (25) are valid for the amplitude and the normalized energy of the output U(t) itself.

Our signal function S(t) behaves like the energy of a damped harmonic oscillator in thermal equilibrium, normalized with its mean value kT. The unit signal produces a peak value $S \approx \Phi+1$.

Threshold crossing - To detect real signals in the output of a detector, one has to construct its signal function S (belonging to a suspected kind of signal) and look for significant deviations from the Boltzmann statistics. For that purpose one can e.g. measure how long S(t) stays above some threshold value, and compare the result with the prediction. Doing this for many thresholds, one will be able to derive all possible information about the rate-strength distribution of signals of type U_g - assuming there are no other disturbances present.

First, we calculate the probability W_o of finding the signal function above threshold S in pure noise. Integration of (23) from S to infinity gives

$$W_o(S) = e^{-S} \tag{27}$$

The duration τ of a peak above threshold S will be of the order $1/2\mu$, with the autocorrelation time $1/\mu$ from eq. 20. In general it will not be possible to derive a simple expression for the probability distribution of τ in pure noise. Even for the case of short pulses in a damped harmonic oscillator, where the energy itself would be our signal function, the derivation seems rather complicated (and perhaps remains to be done). However, it will always be easy to determine the distribution of the peak-width $\tau(S)$ experimentally.

Next let us calculate the probability W_1 of finding the signal function above threshold S at the "arrival time" of a signal of known strength S_g. Now we have to integrate (25) from S to infinity. The result, which cannot be expressed analytically is

$$W_1(S,S_g) = \int_S^{\infty} e^{-(X+S_g)} \cdot I_o(2\sqrt{XS_g})\,dX \tag{28}$$

Figure 5 shows this function of S for various values of the signal strength S_g. The curve for $S_g = 0$ is, of course, identical with $W_o(S)=e^{-S}$. If a peak in S(t) is caused or influenced by the presence of an external signal, the peak width τ will have a distribution which is different from that in pure noise. It can be found experimentally with series of artificial signals.

Detectability at threshold S - Suppose there are signals of strength S_g arriving at an average rate R. We shall demand that the signals are rare in the sense $R \ll \mu$, such that the signal function S is mostly due to noise, except for occasional peaks influenced by a signal. We want to discover this Rate R by observation of peaks above a threshold S.

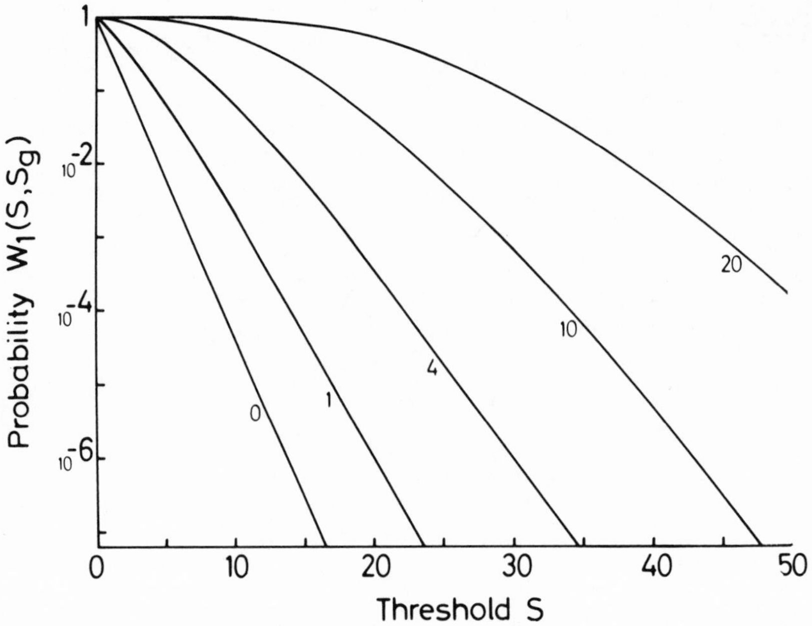

Figure 5: The probability $W_1(S,S_g)$ to find the signal function above a threshold S at the "arrival time" of a signal of known strength S_g.

Within a total observation time T, the signal function will be above S for a time

$$T_o = W_o(S) \cdot T$$

with an uncertainty ΔT_o. A fraction $W_1(S,S_g) \cdot RT$ of the arriving signals will be observed above S at the arrival time. If $\tau_1(S,S_g)$ is the average peak-width above S in the presence of S_g, the sum T_1 of the durations of all peaks containing external signals will be

$$T_1 = W_1 \cdot RT \cdot \tau_1 \ .$$

However, for a part $W_o T_1$ of this time, the signal function would have been above S even without external signals. Hence, the surplus of time above S will be

$$T_{plus} = (1-W_o)\cdot W_1 \cdot RT\cdot\tau_1 \ .$$

We neglect the uncertainty in T_{plus} caused by the distribution of peak-width around τ_1.

We want to compare this surplus with the expected fluctuation ΔT_o in pure noise. If $\tau_o(S)$ is the average peak-width above S in pure noise, the time T_o will consist of n_o peaks, with

$$n_o = T_o/\tau_o \ .$$

We can neglect the distribution of peak-width around τ_o in the fluctuation of n_o. The individual peaks can be considered as independent events, following Poisson statistics in time, as long as their duration is much shorter than their mutual distance. Therefore, the expected fluctuation in n_o will be $\sqrt{n_o}$, and the uncertainty in $T_o=n_o\tau_o$ becomes

$$\Delta T_o \approx \sqrt{T_o \cdot \tau_o} \ .$$

This corresponds to "one standard deviation".

In order to feel confident or at least suspicious about the existence of external signals, one will demand

$$T_{plus} > m\cdot\Delta T_o$$

with $m \not> 3$. Written out, this detectability condition for the rate R says:

$$R \geqslant \frac{m}{\sqrt{T\,\tau_o}} \cdot \frac{\tau_o}{\tau_1} \cdot \frac{\sqrt{W_o}}{W_1\cdot(1-W_o)} \tag{29}$$

where W_o and τ_o are functions of the threshold S, and W_1 and τ_1 functions of S and the signal strength S_g. Remember that $W_o(S)$ and $W_1(S,S_g)$ were given by equations 27 and 28, and that the average peak widths $\tau_o(S)$ and $\tau_o(S,S_g)$ can be determined experimentally, but will be of the order of autocorrelation time of the signal function S(t). Clearly this formula for the detectable rate becomes invalid at very low thresholds, where our concept of peaks as independent events breaks down. However, the divergence of R at $S \to 0$ is a correct feature..

Optimal thresholds for limited and unlimited observation time - At which threshold do signals of a given strength $S_g = \Phi\Lambda_g^2$ cause the most significant deviation from noise statistics? The answer is:

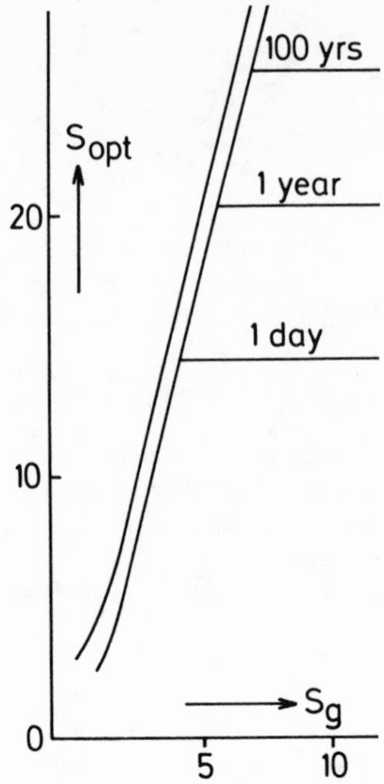

Figure 6: Optimal thresholds S_{opt} for limited and unlimited observation time as a function of signal strength S_g.

at the threshold S_{opt} where (29) has a minimum as a function of S for fixed S_g and T. The position S_{opt} of this minimum is determined mainly by the functions $W_o(S)$ and $W_1(S,S_g)$. Only for weak signals is there some influence of the average peak widths $\tau_o(S)$ and $\tau_1(S,S_g)$ on the value of S_{opt}. (But at low thresholds the notion of counting peaks becomes useless, anyhow. In figure 6, S_{opt} is plotted as a function of signal strength S_g. It has been found numerically. (The nearly linear character is a hint that it may be derivable analytically in a reasonable approximation.) The upper curve shows the position where the minimum of (29) occurs with τ_o and τ_1 set constant; the lower curve is with the approximation

$$\tau_o(S) \sim \frac{1}{S} \; ; \quad \tau_1(S,S_g) \sim \frac{S_g}{S} \; .$$

The steep rise of $S_{opt}(S_g)$ means that it is favourable to choose a high threshold, i.e. to suppress most of the external signals, because the number of accidental crossings drops even faster.

Of course, it does not make sense to search at thresholds, at which no accidental crossings are expected at all. For instance, the threshold S_T, at which the probability of observing accidental crossings within the total observation time T is about e^{-1}, will be found from the condition $e^{-S} . T/\tau_o(S) \approx e^{-1}$, or

$$S_T \approx 12 + 2.3 \cdot {}^{10}\log \cdot \frac{T[\text{days}]}{\tau_o[\text{sec}]} \tag{30}$$

For values of the signal strength S_g where S_{opt} is greater than S_T, it has to be replaced by some threshold near S_T. In figure 6 a few values of S_T are indicated, assuming $\tau_o(S) \approx 0.1$ sec. One can see that only for very weak signals (i.e. for $2 \leqslant S_g = \Phi\Lambda_g^2 \leqslant 7$) formula (29) can be relevant.

<u>Minimal detectable event rate as a function of signal strength. "Sensitivity and Detectability"</u> - The value of R in (29) at $S = S_{opt}$ is the minimal rate of signals of strength S_g which will be significantly detectable within the observation time T. However, if $S_{opt}(S_g) > S_T$ one cannot make full use of S_{opt}, because its choice would not only suppress all noise but also all real signals. In this case the minimal value of R in (29) has to be replaced by the larger rate, which still produces a few crossings at S_T. Hence, the minimal detectable rate of events of strength S_g within observation time T is

$$R_{min} \approx \begin{cases} R_{opt} \text{ from (29)} & \text{for } S_{opt}(S_g) < S_T \\ \dfrac{1}{TW_1(S_T,S_g)} & \text{for } S_{opt}(S_g) > S_T \end{cases} \tag{31}$$

This shows that the minimal detectable rate drops with $1/\sqrt{T}$ for weak signals, as one would expect. For strong signals it can drop as fast as $1/T$ if W_1 is already near 1.

The appearance of τ_o and τ_1 in the formulae for the minimal detectable rate, shows that a long correlation time of S(t) is wanted. Since this correlation time is determined by the time scale $1/\mu$ of the optimal filter, we see that the search for deviations from the noise statistics of the signal function introduces a new optimalization criterion which we did not use in our definition of optimal filtering. We see that the most general attack of optimalization is still more complicated than considered here. The "optimality" is different for the measurement of signal strength at a known instant of time and for the detection of signals in deviations from pure noise in the statistics of a "signal content function". In fact, there will not be a general concept of optimality for detection of a rate-strength distribution of signals. Like the type of the single signal appears in our optimal filter for the measurement of its momentary strength, the rate-strength distribution would enter, too. (However, if one already knows everything, one does no longer have to search.)

Fortunately,this problem will be unimportant in practice. The correlation time of S(t) enters the expression for the minimal detectable rate only in the square root. If one uses the optimal filter as defined by us, one will scarcely loose in the detectable rate. If the time scale $1/\mu$ in the optimal filter can be changed without much loss in the sensitivity, one may increase it somewhat and gain some "detectability" in spite of a little loss in "sensitivity". This can be checked in experiments with series of artificial signals, by doing the evaluation with various values of the filter time scale.

5. Optimal Procedure with Several Detectors

There are several reasons to use many detectors instead of one: a) Higher sensitivity, allowing for the detection of weaker signals; b) Telescope-character of any array, because of the directional information in the relations of arrival times; c) Discrimination against local non-thermal disturbances of single detectors.

A main advantage with two or more detectors is, that one can measure the rate of "accidental coincidences" directly, by repeating the common data evaluation, having shifted the data sequences relative to each other by arbitrary "time-delays". A deviation at zero time delay (or at a particular delay structure defined by the direction of a source) can then be found without exact knowledge of the statistics in thermal or additional non-thermal noise.

Optimal multiple-detector sensitivity - The optimal situation will be with purely thermal noise, and with known delays between the arrival times of an external signal.

If we have N detectors, and observe a piece of noise from each of them, we first have to shift them to a common time, using the known delays of arrival times for an external signal. (Obviously, this would have to be done separately for signals coming from various directions.) Let us call those adjusted pieces of output $\underline{U}_i$ (i=1,...N), and consider the collection

$$U \equiv (U_1, U_2, \ldots U_N)$$

as a vector in an enlarged space, which is the direct sum of the N individual spaces. Since the outputs U_i are all statistically independent in noise, the probability distribution of U will be a product of the individual contributions:

$$W(U)\,dVol \sim e^{-\sum_1^N U_i \circ U_i} \tag{32}$$

We have put the index i also under the symbols for the single-detector scalar products (defined by eq. 4 in I.3.2), because the kernels $\kappa_i(\omega)$ will be different for non-identical detectors. The sum in the exponent is positive definite and defines the "metric of statistical independence" in the combined space:

$$U \circ W \equiv \sum_1^N U_i \circ W_i \tag{33}$$

A pure signal would cause a total output

$$U_g \equiv (U_{1g}, U_{2g}, \ldots U_{Ng}) .$$

With the same arguments as in section 3 we shall measure the signal content λ of the total output U by orthogonal splitting with respect to U_g, using the metric (33):

$$U = V + \lambda U_g \quad \text{with} \quad V \circ U_g = 0.$$

As formerly, we integrate over the V-subspace and obtain the distribution of signal content in noise:

$$W(\lambda)\,d\lambda = (\Phi/\pi)^{1/2}\, e^{-\Phi\lambda^2}\, d\lambda \tag{34}$$

with the total sensitivity:

$$\Phi = \sum_i^N \Phi_i \,. \tag{35}$$

As before we find the prescription for the construction of λ:

$$\lambda = \frac{U \circ U_g}{U_g \circ U_g} = \frac{\sum \lambda_i \Phi_i}{\sum \Phi_i} \tag{36}$$

Thus, we have the plausible result: The combined optimal sensitivity of a collection of detectors is the sum of the individual optimal sensitivities, and the signal content is found by doing the optimal filtering separately in all detectors, and taking the sensitivity-weighted sum of the individually measured signal contents. In this way one would obtain a signal content function which could be further evaluated exactly like that of a single detector.

If we have resonant detectors, it will now not be sufficient to take the envelopes, because the knowledge of phase relations between the detectors allows one to predict how a given external signal will appear in λ_x and λ_y (cf. eq. 10). However, if one is not interested in the exact absolute arrival time, one can still save a lot of data recording: one only has to record λ_x and λ_y (or Λ and ϕ) for each detector with the time resolution better than the autocorrelation time of λ_x and λ_y. The high accuracy, to a small fraction of the resonance period, is only needed for the comparison of times (or phases of reference oscillators) between the various detector stations. The envelope Λ of the optimal combined signal content function $\lambda(t)$ can, however, be constructed from the slowly varying components $\underset{i}{\lambda}_x$, $\underset{i}{\lambda}_y$ in the properly defined common frame in the λ-phase-planes:

$$\Lambda^2 = \left(\sum_i \frac{\Phi_i}{\Phi} \underset{i}{\lambda}_x\right)^2 + \left(\sum_i \frac{\Phi_i}{\Phi} \underset{i}{\lambda}_y\right)^2 \qquad (37)$$

As in the single detector, the squared envelope Λ^2 will have a Boltzmann distribution with mean value $1/\Phi$, where Φ is the combined sensitivity (35). In the case of N identical detectors the autocorrelation of the total signal function will be the same as for each single one. Hence, the curve of minimal detectable rate versus signal strength will simply be contracted in the direction of the signal strength by a factor 1/N.

Loss of sensitivity when only the envelopes are combined - If one has several detectors and wants to save the expenditure for the measurement of their mutual phase relations, one cannot combine the individual signal content functions $\underset{i}{\lambda}(t)$ or their phase plane components, but only the envelopes $\underset{i}{\Lambda}$ or their normalized squares $\underset{i}{S} = \Phi_i \underset{i}{\Lambda}^2$. As we see from (37), one will then not be able to construct the optimally combined Λ or S. Since we don't have a common frame in the various phase planes, we will simply have to add the squares of all components and obtain

$$\Lambda'^2 = \sum_i \left(\frac{\Phi_i}{\Phi} \underset{i}{\Lambda}\right)^2 \qquad \text{or}$$

$$S' \equiv \Phi\Lambda'^2 = \frac{1}{\Phi} \sum_i \Phi_i \underset{i}{S} \qquad (38)$$

While the individual $\underset{i}{S}$ are Boltzmann distributed, S' is not. As an example we consider the distribution of $S' = 1/N \sum_i \underset{i}{S}$ in the case of N identical detectors. For N=2, we write $2S' = S_1 + S_2$ and $W(S')dS' = \left(\int_0^{S'} e^{-S1} e^{-(2S'-S1)} dS_1\right) 2dS' = 2S' e^{-2S'} dS'$. With complete induction we find in the case of N detectors:

$$W(S')dS' = \frac{(NS')^{N-1}}{(N-1)\,!} \; e^{-NS'} \; dS'$$

Because of our normalization the mean value of S' is still unity. Integrating this from a threshold S_o to infinity we obtain the probability of finding S' above some value S (corresponding to the probability $W_o(S) = e^{-S}$, defined by eq. 27:

$$W'_o(S) = e^{-NS} \cdot \left[1 + NS + \frac{1}{2}(NS)^2 + \ldots + \frac{1}{(N-1)\,!} \; (NS)^{N-1} \right]$$

For larger N, this becomes more and more concentrated around the mean value 1. In a similar way one can construct the corresponding probability $W'_1(S_o|S_g)$ in the presence of a signal of given strength, however, it would have to be done numerically. But because of $S' = 1/N \; \Sigma S_i$ we know that the expectation value of S' in the presence of the unit signal will be about Φ+1, as in the case of optimal combination of the detectors. The appearance of the bracket [...] shows the relative loss of "detectability".

In practical cases the combined signal functions S' will rarely be used, because for detectors in the same laboratory one can easily observe the phase relations and use the optimal procedure if there are no non-thermal disturbances. If such disturbances are present, one will in any case prefer the procedure of the next section, in order to discriminate against them.

"Coincidence experiments" for discrimination against local non-thermal disturbances - In this section, too, we restrict ourselves to resonant detectors and envelopes. The optimal N-detector signal function, or the sum S' of the single detector functions, are very sensitive to occasional local events which simulate a signal. Especially if the number N of detectors is not large, one will, therefore, if local disturbances are known to occur, demand that all detectors show an effect in coincidence. This means, that the signal functions of all detectors have to be above certain thresholds simultaneously, in order to define an event. (If travel times play a role, one has to use a corrected simultaneity.)

If we choose a set of thresholds,

$$\underline{S} \equiv (S_1,\; S_2, \ldots S_N) \;,$$

one for each detector, we can discuss the detectability at $\underline{S}$ in the same way as in section 4. For a total observation time T we calculate the expectation value of the time T_o during which all detectors are above threshold simultaneously. It will be

$$T_o = \underline{W}_o(\underline{S}) \cdot T \;,$$

where $\underline{W}_o(\underline{S})$ is simply the product of the individual $W_o(S_i)$. The expected deviation from T_o will again be

$$\Delta T_o \approx \sqrt{T_o \cdot \tau_o(\underline{S})} \quad ,$$

where τ_o is now a function of the N thresholds, and measures the average duration of the intersection of all peaks in an event in noise.

An actual rate R of external signals of strength Λ_g will cause about $\underline{W_1}(\underline{S},\Lambda_g)\cdot RT$ simultaneous peaks. $\underline{W_1}(S,\Lambda_g)$ is the product of the N individual values of $W_1(S_i,S_{gi})$ with $S_{gi} \equiv \phi_i \Lambda_g^2$. The average duration of the intersection of such simultaneous peaks will be a function $\tau_1(\underline{S},\Lambda_g)$ of the N thresholds and the signal strength. The sum of the durations of peak intersections containing external signals will be

$$T_1 = \underline{W_1} \cdot RT \cdot \tau_1 \quad .$$

A fraction W_o of this time would have been observed without signals, too. Hence the surplus due to the signals is

$$T_{plus} = (1-\underline{W_o}) \cdot \underline{W_1} \cdot \tau_1 \cdot RT \ .$$

The condition for detectability with m standard deviations is

$$T_{plus} > m \cdot \Delta T_o$$

and we arrive at a formula analogous to (29) for the detectable rate, with W_o, τ_o, W_1, τ_1, replaced by the corresponding underlined quantities.

The expression (29) is now a function of N thresholds, and the signal strength. Further, it contains the N sensitivities. It will again have a minimum as a function of the N thresholds for any fixed signal strength. It occurs at the "optimal set of thresholds". In general the thresholds in this set will not be identical with the optimal single detector thresholds for the same signal strength, because the correction factors $(1-W_o)$ and the functions $\underline{\tau_o}(\underline{S})$ and $\underline{\tau_1}(\underline{S},\Lambda_g)$ bring some "mixing".

For N identical detectors, one will have $\underline{W_o}(\underline{S}) = \left[W_o(S)\right]^N$ and $\underline{W_1}(\underline{S})\Lambda_g) = \left[W_1(S,\Lambda_g)\right]^N$. The average duration of peak coincidence in signal plus noise will not be strongly influenced by the demand of coincidence, and one will have $\tau_1(\underline{S},\Lambda_g) \approx \tau_1(S,S_g)$. The duration of chance coincidences, however, will drop with the number of detectors. For an estimate we can take $\underline{\tau_o}(\underline{S}) \approx \tau_o(S)/N$. Then we will obtain for the detectable rate at common threshold S:

$$R > \frac{m}{\sqrt{T\tau_o}} \cdot \frac{\tau_o}{\tau_1} \cdot \frac{W_o^{N/2}}{W_1^{N} \cdot (1-W_o^{N})} \cdot \frac{1}{\sqrt{N}} \tag{39}$$

where $W_o(S)$, $W_1(S,S_g)$, $\tau_o(S)$ and $\tau_1(S,S_g)$ are the one-detector quantities as in eq. 29 of I.4.7.

The common optimal threshold is where (39) becomes a minimum as a function of S for fixed signal strength S_g. We see that the optimal threshold is nearly the same as for the single detector, as long as the influence of the difference between the correction factors $(1-W_o^N)$ and $(1-W_o)^N$ is still negligible. This is the case at thresholds S with $Ne^{-S} \ll 1$.

Like in the one-detector case the optimal threshold from minimalization of (39) is only valid for unlimited observation time. As we see from the derivation of equation 30, the optimal threshold S_T for fixed observation time T will nearly come down to 1/N of the corresponding one-detector threshold. Looking again into figure 6, we see that with many detectors one would always work at rather low thresholds, no matter how long the observation time. The minimal detectable rate of external signals would then be of order 1/T for all strengths above a certain value, which can be brought down to about $S_g \approx 2$ by taking enough detectors.

Long term search for rare events - If one wants to find very rare events, one has to be sure, that within the observation time each real event would be detected, say with a probability $1-\varepsilon$, and that the probability of "false alarm" is very small, say ε.

For a fixed one-detector sensitivity ϕ_1 and a known signal strength Λ_g one has to combine N detectors in the optimal way, such that $S_g = \phi\Lambda_g^2 = N\phi_1\Lambda_g^2$ is sufficiently big to prevent false alarm in thermal noise. In order to exclude local non-thermal disturbances, one has to use at least two such detector groups in co-incidence.

If τ is the autocorrelation time of the optimally combinded signal function S (the same as for the single detector!), we can again use

$$\tau_o(S) \approx \tau/S.$$

The number θ of peaks above threshold S within the total observation time T will be

$$\theta = T/\tau_o(S) \ .$$

The condition that no accidentals occur at a threshold S reads

$$e^{-S} \cdot T/\tau_o(S) \approx Se^{-S} \cdot \theta \lesssim \varepsilon.$$

The threshold S_θ, defined by the equality, allows to calculate the minimal value of S_g, needed for certain detection of each event within T, from

$$1 - W_1(S_\theta, S_g) \lesssim \varepsilon.$$

For the whole interesting range of $1 < \theta < 10^{13}$ we obtain nearly linear relations between $\log\theta$ and S_θ or S_g respectively, shown in figure 7. For $\varepsilon = 10^{-3}$ we find:

$$S_\theta \approx 2.42 \log\theta + 9.2 \qquad\qquad S_g \approx 3.58 \log\theta + 26.$$

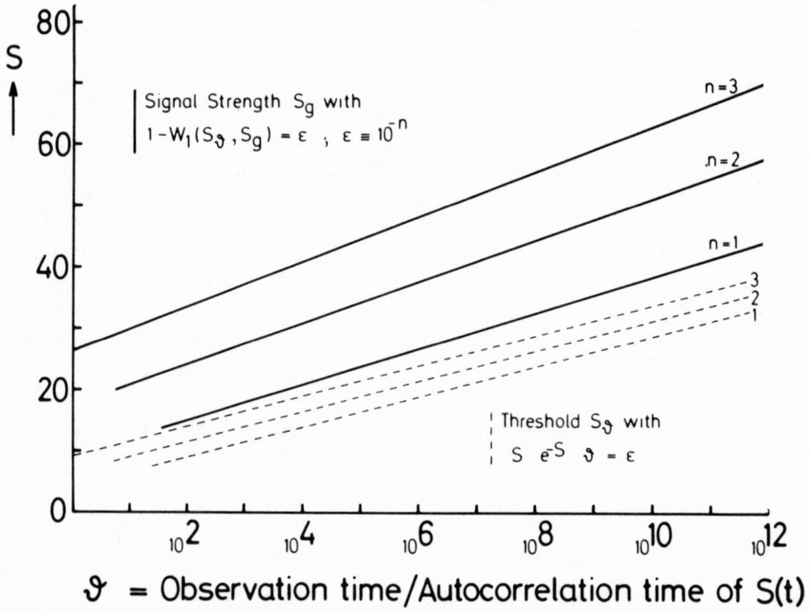

Figure 7: As functions of the number θ of accidental peaks in the signal function, there are plotted 1) the threshold S_θ at which accidental crossing has the probability $\varepsilon = 10^{-n}$ $(n = 1,2,3)$, 2) The strength S_g of signals which would cause a crossing at the threshold S_θ with probability $W_1(S_\theta, S_g) = 1-\varepsilon$.

The number N of detectors needed in each group (i.e. the necessary improvement in sensitivity) is then

$$N = S_g/\phi_1\Lambda_g^2. \tag{41}$$

If one wants to detect events which are already known to have occurred within a shorter interval - e.g. an optically observed supernova - the number θ has to be calculated with this interval in place of the total observation time. We see that this effect scarcely reduces the requirements: even with exact knowledge of the arrival time up to 0.3 seconds, we reduce the number in (41) only by about a factor 1/2, compared with the value for θ = 1 year/0.1 sec = $3\cdot10^8$.

CHAPTER II

APPLICATION TO THE MUNICH-FRASCATI WEBER-TYPE GRAVITATIONAL WAVE DETECTORS

1. Summary of the Experiment

In 1972, groups at the Max Planck Institute for Physics and Astrophysics in Munich and at the ESRIN Institute in Frascati started independently a repetition of J. Weber's gravitational wave experiment (WEBER 1969-73). Later the Frascati group was incorporated into the Munich institute.

The aim of the experiment was to verify or to exclude the existence of gravitational wave pulses of the kind proposed by Weber between 1969 and 1973 as an explanation for his results. The easiest and cheapest way toward that aim was to take over as much as possible from Weber, who agreed on this and cooperated very helpfully. Hence, the mechanical and electrical set-up of both detectors is very similar to Weber's. Only a few obvious improvements were built in, to increase the sensitivity. The data processing used achieved about 3/4 of the optimal sensitivities Φ for short signal pulses. No positive results were found. The results of 150 days have been reported in KAFKA (1974) and BILLING et al. (1975). The upper limits for the rate of short pulses were the lowest achieved so far in any gravitational wave experiment.

The Frascati detector has been moved to a site near Munich in February 1975. The total common observation time up to that time was nearly 1 year. The final result is described in section II 4.

2. Characterization of the Detectors

Experimental details (as given to me by W. Winkler) - The detectors follow Weber's principles. Using piezo-electric transducers, one observes the fundamental longitudinal mode of a cylinder and tries to discriminate excitations through gravitational waves from the thermal noise. The cylinders consist of the ALCOA alloy 6061-O. Their properties are given in table 1:

	length cm	diameter cm	mass g	frequency Hz	Original quality
Munich	153.7	62.2	$1.26\cdot10^6$	1654	$6.4\cdot10^5$
Frascati	153	70	$1.58\cdot10^6$	1659	$5.7\cdot10^5$

Table 1: Properties of the two cylinders

Details of the Frascati detector have been given in BRAMANTI, MAISCHBERGER and PARKINSON (1973), BRAMANTI and MAISCHBERGER (1972), and MAISCHBERGER (1973). Some minor changes were made, before we started the long-run coincidence experiment in July 1973.

The suspension and the shielding against mechanical and acoustical disturbances in Munich are as in Weber's experiment. The cylinder is suspended around its middle on steel wire of 4 mm diameter, on an iron yoke. This rests on acoustic filters of iron and (relatively vacuum-proof) rubber in an iron tank with walls which are 1 cm thick. The vacuum is always better than 10^{-3} Torr. The tank is isolated against electric or mechanic disturbances from the pump. The power supply is through an isolating transformer with low direct capacitance, in order to isolate the grounding of the amplifier from fluctuations. The input stage of the amplifier is powered by batteries.

The piezo-electric material is G 1408 (Glennite/Gulton). The producers quote the following properties: coupling coefficient K_{33}=0.63; coefficient of piezo-electric charge $d_{33}=2\cdot10^{-10}$ m/V; module of elasticity $y^E_{33}=8.2\cdot10^{10}$ N/m^2; mechanical quality Q_m = 1200; tgδ = 0.003. The measured value of tgδ near 1654 Hz is in fact 0.0015.

The transducers are fixed to the cylinder as shown in figure 8. The dimensions of an individual crystal are 1.5" x 1.5" x 0.75". Two columns, glued together from 6 transducers each, are mounted in the middle of the cylinder between 3 blocks of aluminium. The distance between the columns and the surface is 1.5 mm. The polarization points towards the middle of each column. The transducer endfaces near the aluminium blocks are grounded, the central sections

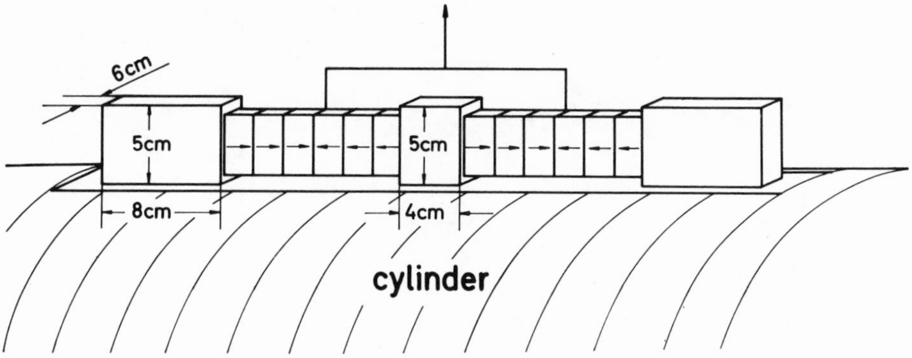

Figure 8: Mounting of the transducers in Munich.

of both columns are connected in parallel and supply the signal. The glue is Araldite AW 134 B, the hardener Hy 994; 100:40 (Ciba).

The gain in sensitivity as compared to Weber's detectors up to 1974, or to the Munich equipment before May 1973, is mainly due to three reasons: Higher quality and coupling of the new transducer material, the gain in coupling with parallel directions of strain and polarization, and the optimal adaptation (see section 3) to the high-impedance input stage of the amplifier, accomplished by the partial series connection. This important point, namely to minimize the amplifier noise by adaptation, was also stressed by MAEDER (1973) and by PAIK (1975).

The further apparative data processing in Munich (similar in Frascati) is shown in figures 9 and 10, supplied by W. Winkler. Figure 9 is a circuit diagram of the amplifier, figure 10 sketches the further data processing, up to the writing on magnetic tape.

The bandwidth of the Munich amplifier is 11.8 Hz. Two phase sensitive rectifiers achieve the decomposition of the output U(t) into the two slowly varying components x(t) and y(t) - the "phase-plane components in the co-rotating frame" as defined by equation (17) in section I 4. Both x and y are then smoothed with RC = 0.09 sec. (The total band width is then 2.7 Hz.) They are to be written on magnetic tape at intervals of 0.1 sec. Smoothing of x and y over more than the discretization interval would disturb the optimal evaluation, which has to be done on the computer. A still better way to define the values written on tape would be to take the mean values of x and y over the discretization interval (cf. LEVINE and GARWIN, 1973).

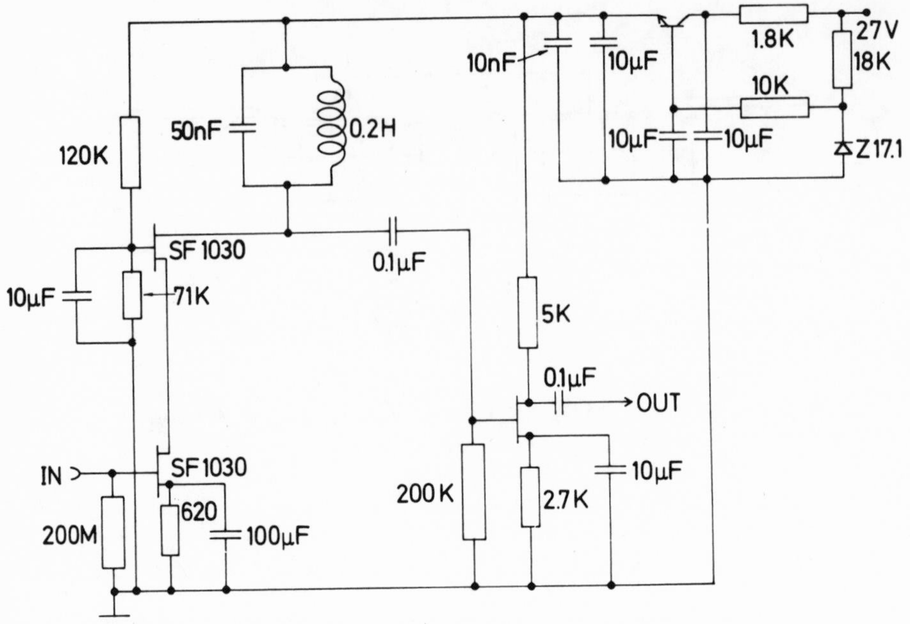

Figure 9. Circuit diagram of the Munich amplifier.

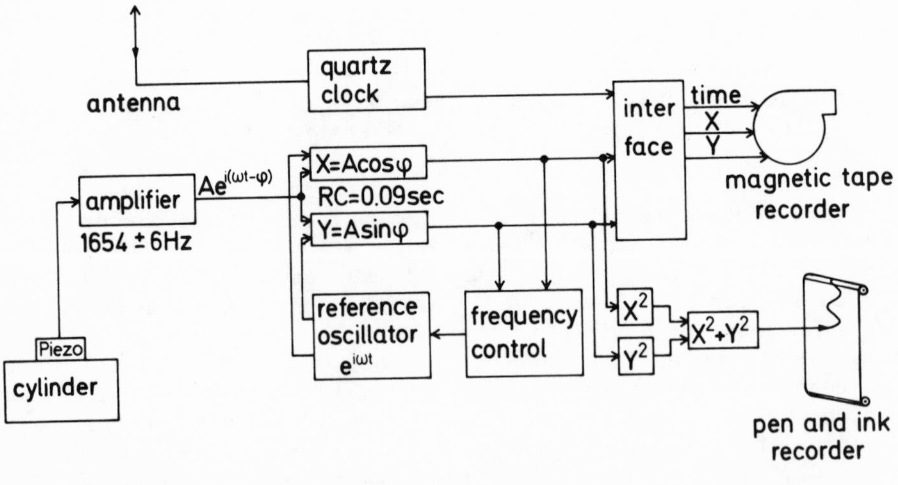

Figure 10. Sketch of the apparative data processing for the Munich detector.

A beat between the output frequency and the frequency Ω of the reference oscillator (-"the co-rotating frame"-) would be accounted for in the computer processing. However, it is time-consuming to measure a beat from the x,y-data on tape. Since the cylinder frequency varies by about 0.4 Hz per degree centigrade, the temperature would have to be kept constant to about 10^{-2} degrees if Ω were fixed. Instead, it is easier and cheaper to control the frequency of the reference oscillator by comparison with the output frequency. This was done in Munich from the beginning, and in Frascati when we started the long run experiment. For this purpose one measures in periods of high cylinder amplitude the mean tendency of phase drift and uses it to control the frequency of the reference oscillator. In this way the beat period could always be kept larger than 6 minutes (which is better than needed).

An interface brings the values of x and y into digital form, and they are recorded on magnetic tape with 8 bits per value, at intervals of 0.1 seconds. The data of one minute, i.e. 1200 values, are written in one block, at the beginning of which the time is written. Time information is supplied by a quartz clock which deviates by not more than a few milliseconds from Central European time (as broadcasted by the station DCF 77). Basis for this time, valid in the German Federal Republic, is the co-ordinated world time TUC.

In the same way the Frascati data were written on magnetic tapes which were mailed to Munich. Any physical interaction between the data from the two detectors was excluded.

Simultaneity with the time standard used in Frascati was tested by telephone communication and artificial excitation of both detectors. The simultaneous artificial pulses were detected by the evaluation program.

Artificial pulses could be applied with known reproducible strengths at moments accurate to milliseconds. The artificial excitation was first produced over a condensor plate opposite to the cylinder end-face. Instead of calculating the pulse strength from the applied voltage on a condensor plate, it is much safer to observe the result of excitations in the phase plane, and thus gauge the pulser.

The pen-and-ink recorder shown in figure 10 was only used to check the operation of the system. It plotted the "energy" $E = x^2+y^2$.

The equivalent circuit and its noise sources - The output before data processing can be described by an equivalent circuit. Figure 11 shows this circuit for Weber-type experiments. The values of the constituents are collected in table 5 for the Munich- and Frascati experiment and, for comparison, also for Weber's and Tyson's

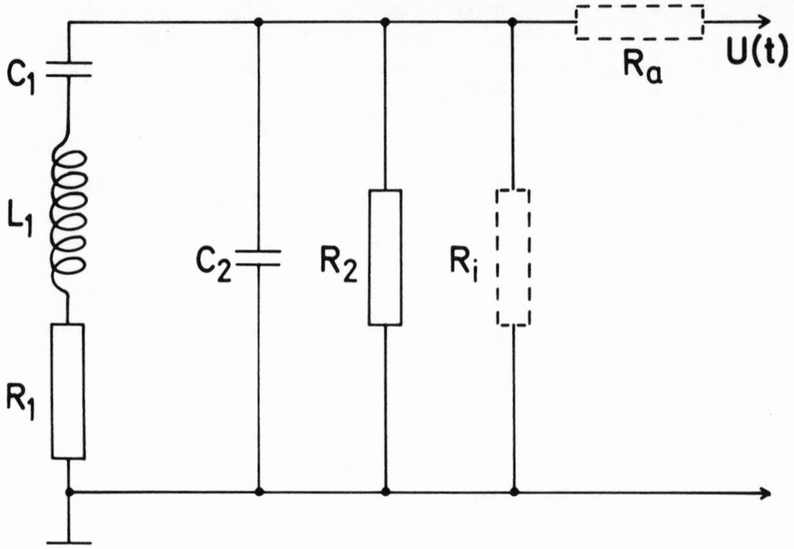

Figure 11: Equivalent circuit for Weber-type experiments.

(Bell Telephone Lab.) detectors. For the table see section II 6.

C_1, L_1, R_1 represents the elasticity, inertia and mechanical losses of the fundamental mode, C_2 and R_2 the capacity and electrical losses of the transducers. The thermal noise sources connected with the losses in R_1, R_2, R_a and R_i are the electromotive forces (e.m.f.) u_1 from the fundamental mode, u_2 from the transducers and u_a from the amplifier; and a current noise i_a from the amplifier. The source i_a can be replaced by an e.m.f. $u_i = i_a \cdot R_i$ considered as the noise of a resistor R_i. (R_i includes an actually connected resistor which stabilizes the working point of the pre-amplifier.)

The material properties fix the frequency and the mechanical and electrical loss factors. One still has the freedom to change the amount and the connection of the transducers and the properties of the amplifier. We do not derive the equivalent circuit here, but shall say a bit more about it in section 3.

In order to measure the quantities in the equivalent circuit, one can connect a relatively lossfree capacitor C_M in parallel to C_2. C_2 is measured directly. Measuring the damping time τ and resonance frequency ω_r with or without connection of C_M one can determine R_1, C_1, L_1 and R_2 from the equations

$$\omega_r^2 = \left(\frac{1}{C_1} + \frac{1}{C_2+C_M}\right)/L_1$$

$$\tau = \frac{2Q}{\omega_r} = 2L_1/\left(R_1 + \frac{1}{\omega_r^2 R_2 (C_2+C_M)^2}\right)$$

(For the quality Q cf. equation 50 in the next section.) R_2 can also be determined from measurements of noise power at frequencies sufficiently remote from the cylinder resonance. The value of R_a is determined from noise measurements with the amplifier entrance cut short. The equivalent current noise i_a, represented by R_i can be calculated from measurements of the noise with the transducers replaced by a known capacitance parallel to the input. The input capacity of the amplifier - to be included in C_2 - is measured by determination of the amplification with samll capacities connected in series with the input. In Munich it is 30 pF.

The noise sources R_2 and R_i can be combined into R_2' for the calculations, except in the treatment of optimal adaptation.

Of course, there will be a limitation of band width in the amplifier (and afterwards), which we have not included in the equivalent circuit. The band width has to be wide in comparison with the optimal filter which we are going to describe. On the other hand it is kept so low, that the mean square voltage $\overline{U_1^2}$ from the noise source R_1, i.e. the Brownian noise of the fundamental mode, dominates the output. One can calculate this mean square from the equivalent circuit and finds

$$\overline{U_1^2} = kT \cdot C_S / C_2^2 \qquad (41)$$

where $C_S \equiv \left(\frac{1}{C_1} + \frac{1}{C_2}\right)^{-1} = C_1 \left(1 + \frac{C_1}{C_2}\right)^{-1}$

Writing $\overline{U_1^2} \cdot C_2 = kT \cdot \frac{C_S}{C_2}$ we see why C_S/C_2 is called the coupling. Many authors used the abbreviation $C_S/C_2 \equiv \beta$, after the paper of GIBBONS and HAWKING (1973). Instead, we shall call it γ, since β was used for the damping constant (width of the resonance).

Measuring the output $\overline{U_1^2}$ over long periods one confirms (41) with an accuracy of a few percent (in Munich and Frascati).

The consistency of the various determinations of all parameters is important, because the theory of optimal evaluation starts from the equivalent circuit.

Composition of the output U: Transfer functions; Spectrum; the kernel $\kappa(\omega)$ - For a fixed equivalent circuit, we can treat R_i and R_2 together, writing $1/R_{2'} = 1/R_2 + 1/R_i$. Then we have only three noise sources R_1, R_2' and R_a, producing electromotive forces u_1, u_2 and u_a. Because of high impedance of the amplifier the total current in the circuit will be zero. For abbreviation we call the impedance of the branch L_1-C_1-R_1 in figure 11:

$$Z_1 \equiv R_1 + i\omega L_1 + \frac{1}{i\omega C_1} = R_1 \cdot \left[1 + iQ_1 \cdot \frac{\omega_o}{\omega} \left(\frac{\omega^2}{\omega_o^2} - 1 \right) \right] \qquad (42)$$

$$Q_1 \equiv \frac{1}{\omega_o C_1 R_1} = \text{"mechanical quality"}$$

Writing the total current (in frequency representation) we find

$$\frac{U - u_a - u_1}{Z_1} + (U - u_a) \cdot i\omega C_2 + \frac{U - u_a - u_a'}{R_2'} = 0.$$

With the total impedance Z, defined by

$$\frac{1}{Z} = \frac{1}{Z_1} + i\omega C_2 + \frac{1}{R_2'} \qquad (43)$$

this becomes

$$U = \frac{Z}{Z_1} u_1 + \frac{Z}{R_2'} u_2 + u_a \qquad (44)$$

Hence, our transfer functions are

$$F_1(\omega) \equiv \frac{Z(\omega)}{Z_1(\omega)} \; ; \; F_2(\omega) \equiv \frac{Z(\omega)}{R_2'} \; ; \; F_a(\omega) \equiv 1 \qquad (45)$$

Since an external signal acting on the cylinder corresponds to an e.m.f. in the same place as u_1, the transfer function for the signal is also F_1.

Now we can write down the kernel $\kappa(\omega)$ (cf. equ. 3 in II 2) or the total noise spectrum $U_\omega^2 = \frac{1}{\pi\kappa(\omega)}$. If all parts of the system are at the same temperature T, we obtain

$$\frac{1}{\kappa(\omega)} = 2kT \left(R_1 |F_1(\omega)|^2 + R_2' |F_2(\omega)|^2 + R_a \right). \qquad (46)$$

With the help of (42-45) and using the substitution

$$x \equiv \frac{\omega^2}{\omega_o^2} - 1$$

we find

$$|F_1(\omega)|^2 R_1 = \left(\frac{C_1}{C_2}\right)^2 \cdot R_1 \cdot \frac{x+1}{P(x)}$$

$$|F_2(\omega)|^2 R_2' = \left(\frac{C_1}{C_2}\right)^2 \cdot \frac{R_1^2}{R_2'} \cdot \frac{Q_1^2 x^2 + x + 1}{P(x)} \qquad (47)$$

with

$$P(x) \equiv x^3 + Bx^2 + Cx + D$$

$$B \equiv 1 - 2\gamma + (\alpha_1 + \alpha_2)^2$$

$$C \equiv -2\gamma + \gamma^2 + 2\alpha_1(\alpha_1 + \alpha_2)$$

$$D \equiv \gamma^2 + \alpha_1^2$$

with the definitions (and Munich values):

$$\gamma \equiv \frac{C_1}{C_2}\left(1 + \frac{R_1}{R_2'}\right) \approx \frac{C_1}{C_2} \approx 4.5\cdot 10^{-4}$$

$$\alpha_1 \equiv \frac{1}{Q_1} = \omega_o C_1 R_1 \approx 2.85\cdot 10^{-6} \qquad (48)$$

$$\alpha_2 \equiv \frac{1}{Q_2'} = \frac{1}{\omega_o C_2 R_2'} \approx 2.73\cdot 10^{-3}$$

The zeros of $P(x)$ are to good approximation $x_3 \approx -B \approx -1$ and $x_{1,2} \approx \gamma \pm i\alpha$ where the total loss factor α is the sum of α_1 and "coupling $\times\alpha_2$":

$$\alpha \equiv \alpha_1 + \gamma\alpha_2.$$

Hence, both $|F_1(\omega)|^2$ and $|F_2(\omega)|^2$ have poles in the complex ω-plane at

$$\omega = \pm\,\omega_r \pm i\beta$$

with $\omega_r^2 = \omega_o^2\,(1+\gamma)$ (49)

$$\beta = \omega_o \cdot \frac{\alpha}{2} = \frac{\omega_o}{2Q}$$

We see that the total quality factor Q is

$$Q = \left(\frac{1}{Q_1} + \frac{\gamma}{Q_2'}\right)^{-1}. \qquad (50)$$

ω_r is the peak frequency of the spectrum, and β measures the width of the resonance.

However, the spectrum has slightly more structure than a simple resonance curve, because $|F_2(\omega)|^2$ has complex zeros in the neighbourhood of the poles, as seen from (47). These zeros are at $x = -1/2Q_1 \ (1 \pm i\sqrt{3})$. This means, that the transducer noise makes its minimal contribution to the spectrum at some frequency different from ω_o and ω_r, but near ω_o.

The total spectrum is shown in figure 12, with some features, which will be discussed later.

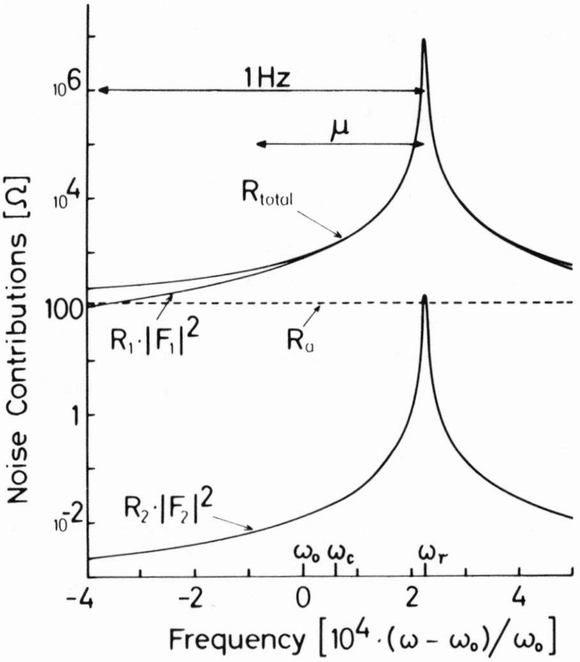

Figure 12: Spectrum of the noise sources R_1, R_2' and R_a (in Munich). R_{total} is the sum of those three contributions. ω_c is the frequency at which one would optimally detect sudden jumps in the state of the cylinder. $1/\mu$ is the time constant of the optimal filter for such jumps. (See further discussion in section 3.)

3. Optimal Filtering for Short Gravitational Pulses

The reaction of the cylinder to a gravitational pulse - Far from the source, every wave is nearly plane. Let the wave travel in z-direction, and x and y span the orthogonal plane. Then the structure of the plane wave (see text books) will be described by two functions of a retarded time t-z/c, namely the dimensionless amplitudes h_{xx} and h_{xy} of two independent states of polarization. Two free test particles separated by a spatial vector $\vec{r}$ suffer a relative acceleration $\vec{f}$ given by

$$f_x = \frac{1}{2} (\ddot{h}_{xx} \cdot x + \ddot{h}_{xy} \cdot y)$$
$$f_y = \frac{1}{2} (\ddot{h}_{xy} \cdot x - \ddot{h}_{xx} \cdot y) \qquad (52)$$

The wave is transversal, i.e. $f_z = 0$.
A solid body, which does not itself introduce relativistic effects, will behave under the action of such a wave, as though its parts suffered accelerations (52) with respect to the center of mass. (This has to be the result of a rigorous general relativistic theory of elasticity, applied to bodies in a laboratory.)

Let us consider a linearly polarized wave with $h \equiv h_{xx}$, and $h_{xy} = 0$, and an elastic bar, lying in the x-direction. The end-face displacement X of its fundamental longitudinal mode will follow the equation of motion

$$\ddot{X} + 2\beta\dot{X} + \omega_o^2 X = f(t) \qquad (53)$$

Internal thermal noise and external excitation are linearly superposed in f(t). To study the effect of a pure signal, we set the noise equal to zero. The fundamental mode of a cylinder cannot be represented analytically. An approximative treatment (PAIK 1975) shows that the effect of the thickness is negligible for the Weber-type cylinder. Therefore, like for a thin bar, the mode will be nearly sinusoidal in the x-direction, and the term on the right-hand side of (53) can be found by folding the mode and the acceleration (52). The result is a correction factor $8/\pi^2$ to the force term $f=1/2\ a\ddot{h}$, which would appear for a vibrator consisting of two point masses at relative distance 2a. Hence

$$f(t) = \frac{1}{2} a\ddot{h} \cdot 8/\pi^2 \qquad (54)$$

This means that the continuous bar reacts like a dumbbell with a length reduced by $8/\pi^2 = 0.81$. The small additional correction, due

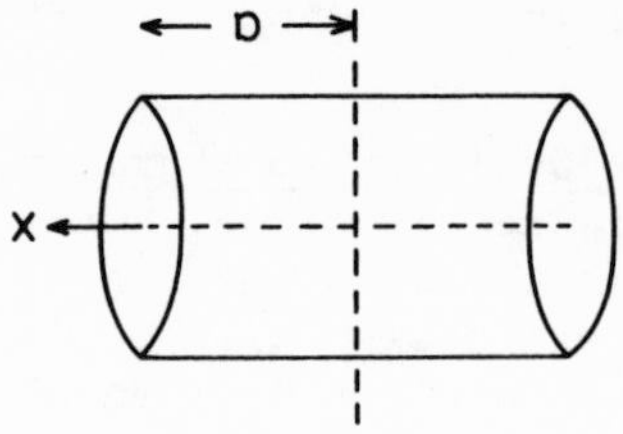

Figure 13: Schematic view of the cylinder.

to the thickness of the cylinder is only about 1.3 % for our cylinders, according to PAIK. (In BILLING et al., 1975, and some earlier publications, I had tacitly assumed that this correction bring the the factor $8/\pi^2$ nearly back to unity, and therefore used the dumbbell formula.)

Assuming that $f(t) \neq 0$ only during a finite pulse duration, we take the Fourier transform of (53):

$$X(\omega) = \frac{f(\omega)}{-\omega^2+\omega_o^2 + 2i\beta\omega}$$

and find

$$X(t) = \frac{1}{\sqrt{2\pi}} \int_{-\infty}^{+\infty} \frac{-f(\omega)e^{i\omega t}d\omega}{(\omega-\omega_1)(\omega-\omega_2)}$$

with the poles $\omega_{1,2} = \pm\,\Omega + i\beta$, $\Omega^2 \equiv \omega_o^2-\beta^2$.

If the pulse duration is much smaller than the damping time $1/\beta$, (about 1 minute in Munich) and $f(\omega)$ has no pole near ω_o, one obtains for the end-face displacement:

$$X(t) = \frac{\sqrt{2\pi}}{\Omega} \cdot \mathrm{Im}\left\{ f(\omega_1)e^{i\omega_1 t}\right\} = \frac{\sqrt{2\pi}}{\Omega} \cdot \left|f(\omega_1)\right| \cdot \cos(\Omega t+\phi)e^{-\beta t}$$

We see that the amplitude of X after a short pulse is to good approximation

$$A \equiv X_{max} = \frac{\sqrt{2\pi}}{\omega_o} \left|f(\omega_o)\right| \qquad (55)$$

Using (54) and $\ddot{h}(\omega) = i\omega\dot{h}(\omega)$, we find

$$A = \frac{4\sqrt{2\pi}}{\pi^2} \; a \; |\dot{h}(\omega_o)| \tag{56}$$

Since the energy in the cylinder (of mass M), corresponding to the end-face amplitude A is

$$E = \frac{1}{4} M \omega_o^2 A^2$$

and the velocity v of longitudinal sound is connected with the wave length $\lambda=4a$ by

$$v = \lambda \cdot \omega_o/2\pi = \frac{2}{\pi} a \omega_o \; ,$$

we obtain for the energy, which the pulse has given to the cylinder (with initial amplitude zero):

$$E = \frac{2}{\pi} Mv^2 \; |\dot{h}(\omega_o)|^2 \; . \tag{57}$$

For our gravitational wave (of most favourable direction and polarization) the energy flux vector (energy per area and time), analogous to the Poynting vector, is (see text books):

$$S = \frac{c^3}{16\pi G} \dot{h}^2 \tag{58}$$

We integrate this over the duration of the pulse, to obtain the energy per area in the pulse, and use Parseval's theorem to convert the time integral into a frequency integral:

$$\int_{\text{Pulse}} S(t)dt = \frac{c^3}{16\pi G} \int \dot{h}^2(t)dt = \frac{c^3}{16\pi G} \int_{-\infty}^{+\infty} |\dot{h}(\omega)|^2 \, d\omega$$

This shows that the spectral density of the energy in the pulse is

$$\frac{c^3}{8\pi G} \; |\dot{h}(\omega_o)|^2 \quad .$$

If we use the frequency scale $\nu = \frac{\omega}{2\pi}$, and define the spectral density $Z(\nu)$ by

$$\int_{\text{Pulse}} S(t)dt = \int_o^{+\infty} Z(\nu)d\nu$$

$$Z(\nu_o) = \frac{c^3}{4G} \; |\dot{h}(\omega_o)|^2 \tag{59}$$

we obtain (with 57):

$$Z(\nu_o) = 1.6 \cdot 10^{38} \, \frac{E}{Mv^2} \quad \left[\text{erg}/\text{cm}^2 \cdot \text{Hz} \right] \tag{60}$$

Clearly, a resonance detector can only measure the spectral density, and supplies no information about other properties of the short pulse.

Many authors define a "cross section" by

$$\int_0^\infty \sigma(\nu)\,d\nu \equiv \frac{E}{Z} = \frac{8}{\pi}\cdot\frac{GM}{c}\cdot\left(\frac{v}{c}\right)^2 = 6.3\cdot 10^{-39}\left(\frac{Mv^2}{erg}\right)\left[cm^2\cdot Hz\right] \qquad (61)$$

It is important that the length of the cylinder does not appear in this cross section or in the energy picked up from a given wave pulse. Since this energy has to be compared with the thermal noise energy kT, which is also independent of the length, a resonance detector does not make additional profit from the appearance of the detector length in (52) or (54). Whereas a free-mass antenna would simply indicate the dimensionless wave amplitude $\Delta l/l = 1/2\ h(t)$, a resonance antenna measures instead the Fourier component of its time derivative: When the pulse has passed through, the strain amplitude in the antenna is $\Delta l/l \sim \dot{h}(\omega_o)$. Only if the pulse spectrum is centered near ω_o, the dimensionless quantities h_{max} and $\dot{h}(\omega_o)$ will be comparable. E.g., with "a piece of a sine wave", i.e. $h(t)=h_{max}\cdot\sin(\omega_o t)$ from t=o to $t=n\cdot\pi/\omega_o$, one finds $\dot{h}(\omega_o) = h_{max}\cdot n\cdot\sqrt{\pi/8}$ and from (56): $(\Delta l/l)_{cylinder} = h_{max}\cdot n\cdot 2/\pi$.

Let us finally mention the size of the thermal fluctuations of the end-faces and the corresponding simulated gravitational "white-noise" strength: From 53,54 and $\overline{X^2} = 2kT/(M\omega_o^2)$ we find $f_\omega^2 = 8\beta kT/(\pi M)$ and $h_\omega^2 = f_\omega^2\cdot\pi^4/(16a^2\omega_o^4)$, or using the "quality" $Q = \omega_o/2\beta$, and the cylinder length L:

$$h_\omega^2 = (\pi^3/Q)\cdot(kT/M\omega_o^2L^2)\cdot(1/\omega_o).$$

For the Munich cylinder (with the values from table 1) we find

$$(h_\omega^2)^{1/2} \approx 10^{-20}\ Hz^{-1/2}$$

for the noise-simulated gravitational wave spectrum. The corresponding mean square value of $\Delta l/l$ is about $3\cdot 10^{-16}$, as follows from $\overline{X^2}/a^2 = 8kT/M\omega_o^2 L^2 \approx 10^{-31}$.

<u>Correspondence between the gravitational signal-strength and the equivalent electro-motive force</u> - In I 3 we introduced the unit signal output U_g. In our experiment, U_g is the output U of the equivalent circuit when the electro-motive forces of the noise sources are equal to zero, and the unit gravitational signal excites the cylinder. This excitation will correspond to an e.m.f. u_g in the equivalent circuit (cf. fig. 11 in section II 2). Let us, for convenience, choose our unit signal in most favourable direction and

polarization. A treatment of a special distribution of direction and polarization (e.g. with a source near the galactic center and detectors fixed to the rotating earth) would require only some integrations over transformed components of the tensor $h_{i\kappa}$.

We restrict ourselves to ignal pulses with a duration short compared to the damping time $1/\beta$ of the cylinder and with a flat spectrum in the neighbourhood of the resonance. The wave amplitude h(t) will cause an e.m.f. $u_g(t)$, which drives a current $I_g(t)$ through L_1 and $C_s = (\frac{1}{C_1} + \frac{1}{C_2})^{-1}$. After the wave pulse has passed, and u_g is zero again, the system has picked up an energy (if it was zero before):

$$E = \int_{\text{Pulse}} dt\; u_g(t) I_g(t) = \int_{-\infty}^{+\infty} u_g^*(\omega) I_g(\omega) d\omega$$

Writing

$$I_g(\omega) = u_g(\omega)/(i\omega L_1 + \frac{1}{i\omega C_s}) = i\omega C_s - u_g(\omega)/(1 - \frac{\omega^2}{\omega_r^2})$$

with $\omega_r^2 = 1/L_1 C_S$, we find

$$E = \frac{\pi}{L_1} \cdot |u_g(\omega_r)|^2 \tag{62}$$

Since the equivalent circuit was modeled after the real system, we have to compare this with eq. (57) and obtain the correspondence between u_g and h for a short pulse:

$$|u_g(\omega_r)|^2 = E \cdot \frac{L_1}{\pi} = \frac{2}{\pi^2} \cdot L_1 \cdot Mv^2 \cdot |\dot{h}(\omega_r)|^2. \tag{63}$$

L_1 was the inductance in the equivalent circuit (also replaceable by $L_1 = 1/C_s\omega_r^2$). M was the mass of the cylinder, and v the longitudinal velocity of sound.

It is convenient to define the "unit pulse" in a detector-dependent way as the pulse which would excite the cylinder from zero energy to 1kT, i.e. to its mean energy in pure noise. Then the Fourier component of the corresponding e.m.f. is simply given by

$$|u_g(\omega_r)|^2 = \frac{kT\, L_1}{\pi} . \tag{64}$$

The spectral density $Z_1(\omega_r)$ corresponding to this unit gravitational pulse will be (from 63):

$$\begin{aligned} Z_1(\nu_r) &= \frac{\pi}{8} \cdot \frac{c^3}{G} \cdot \frac{kT}{Mv^2} \\ &= 2.54 \cdot 10^7 \cdot \left(\frac{T}{300^\circ K}\right) \cdot \left(\frac{10^6 g}{M}\right) \cdot \left(\frac{v_{Al}^2}{v^2}\right) \left[\frac{erg}{cm^2 Hz}\right] \end{aligned} \tag{65}$$

where we used $v_{Al} = 5.1 \cdot 10^5$ cm/sec. Thus, to excite the Munich detector from zero to 1kT (at 300°K) we need a pulse with

$$Z_1(1654\ \text{Hz}) = 2.10^7\ \text{erg/cm}^2\ \text{Hz}.$$

Calculation of the optimal sensitivity - In section I 3 we found the optimal sensitivity with respect to a unit signal output U_g

$$\Phi = U_g \circ U_g = \int_0^\infty \kappa(\omega)\, |U_g(\omega)|^2 d\omega$$

The kernel $\kappa(\omega)$ for our equivalent circuit is defined by equation 46. The unit signal output U_g will be

$$U_g(\omega) = u_g(\omega) \cdot F_1(\omega),$$

with the transfer function $F_1(\omega)$ of the cylinder noise, defined in equations 45 and 57.

For a short pulse without spectral structure within the resonance, we can write

$$\Phi = |u_g(\omega_r)|^2 \cdot \int_0^\infty \kappa(\omega)\, |F_1(\omega)|^2 d\omega \tag{66}$$

For the integral we can write

$$\int_0^\infty \kappa(\omega)\, |F_1(\omega)|^2 d\omega = \frac{\omega_o}{4kT} \cdot (C_1/C_2)^2 \cdot \frac{1}{R_a} \cdot J \tag{67}$$

with

$$\left.\begin{aligned} J &\equiv \int_{-\infty}^{+\infty} \frac{z^2 dz}{Q(x)} \\ x &\equiv z^2-1, \quad z \equiv \omega/\omega_o \\ Q(x) &\equiv x^3+bx^2+cx+d \end{aligned}\right\} \tag{68}$$

The coefficients of the polynomial $Q(x)$ are defined by

$$\left.\begin{aligned} b &\equiv (R_2'/R_a + 1)/Q_2'^2 + 1 + c - d \\ c &\equiv -2\, C_1/C_2 + 1/Q_1^2 + d \\ d &\equiv (C_1/C_2)^2 \cdot (1 + R_1/R_a + R_1/R_2') \cdot (1 + R_1/R_2') + 1/Q_1^2 \end{aligned}\right\} \tag{69}$$

with the quality factors Q_1 and Q_2' defined in (48). We introduce two further Q-type quantities Q_a, expressing the noise of R_a, and Q_{el} combining all electrical noise sources, i.e. R_2, R_i and R_a.

$$\left.\begin{aligned} \frac{1}{Q_a} &\equiv \omega_o C_2 R_a \\ \frac{1}{Q_{el}} &\equiv \frac{1}{Q_2'} + \frac{1}{Q_a} \end{aligned}\right\} \qquad (70)$$

For our own equivalent circuits we can write (69) in a very good approximation:

$$\left.\begin{aligned} b &\approx Q_a/Q_{el} && \approx 3.7 \\ c &\approx -2\gamma && \approx -9\cdot 10^{-4} \\ d &\approx \gamma^2(1+Q_a/\gamma Q_1) && \approx 1.5\cdot 10^{-6} \end{aligned}\right\} \qquad (71)$$

The numerical values are for the Munich detector. They show that

$$d \ll c \ll b.$$

Therefore, the zeros of the denominator Q(x) are in good approximation:

$$\left.\begin{aligned} & x_3 \approx -b \qquad\qquad x_{1,2} \approx \xi \pm i\eta \\ & \xi \equiv -\frac{c}{2b} \quad , \quad \eta^2 = \frac{d}{b} - \xi^2 \end{aligned}\right\} \qquad (72)$$

To these 3 zeros of x there correspond 6 zeros of $z = \sqrt{1+x}$, or of $\omega = \omega_o z$. They are

$$\left.\begin{aligned} \omega_o z_{1,2,3,4} &\approx \pm\,\omega_c \pm i\mu \\ \omega_o z_{5,6} &\approx \pm\, i\omega_o \sqrt{Q_a/Q_2'} \\ \omega_c &\equiv \omega_o\left(1+\frac{\xi}{2}\right) \; ; \; \xi = \gamma\cdot Q_{el}/Q_a \\ \mu &\equiv \omega_o \frac{\eta}{2} \; ; \qquad \eta = \gamma\cdot Q_{el}\sqrt{\frac{1}{\gamma Q_1 Q_{el}} + \frac{1}{Q_2' Q_a}} \end{aligned}\right\} \qquad (73)$$

The integral J can be evaluated with the residue theorem. The poles at $z_{5,6}$ give negligible contributions, and the result is

$$J \approx 2\pi i \cdot 2i\,\mathrm{Im}\left\{ \frac{z_1^2}{(x_1-x_2)(x_1-x_3)\cdot 2z_1} \right\} \approx \frac{\pi}{b\eta}$$

If we use our unit signal pulse, corresponding to 1kT, with $|u_g(\omega_r)|^2 = kTL_1/\pi$, and insert this and J into (66-67) we obtain the sensitivity with respect to short pulses of 1kT:

$$\Phi = \frac{\gamma Q_{el}}{4} = \frac{1}{4}\left(\frac{1}{\gamma Q_1 Q_{el}} + \frac{1}{Q_2' Q_a}\right)^{-1/2} \qquad (74)$$

or expressed directly by the constituents of the circuit:

$$\Phi = \frac{1}{4}\left(\frac{R_1+R_a}{R_2'} + \omega_o^2 C_2^2 R_1 R_a\right)^{-1/2} .$$

This is a pure number, telling us - according to sections I 3 and I 4 - the mean square value of the optimal signal content function λ or its envelope Λ,

$$\overline{\lambda^2} = \frac{1}{2\Phi} , \quad \overline{\Lambda^2} = \frac{1}{\Phi} ,$$

with respect to a unit signal pulse of 1kT. Using (65) we find that the spectral density of signal pulses, simulated by pure noise in the optimally constructed Λ^2, is

$$Z_\Phi(\nu_r) = \frac{1}{\Phi} \cdot \frac{\pi}{8} \cdot \frac{c^3}{G} \cdot \frac{kT}{Mv^2} \qquad (75)$$

For the Munich detector we find $\Phi = 49$, and the simulated pulse strength (for pulses of most favourable direction and polarization):

$$Z_\Phi(1654 \text{ Hz}) \approx 4\cdot 10^5 \text{ erg/cm}^2\cdot\text{Hz} = 4 \text{ GPU} \qquad (76)$$

if we use the abbreviation 10^5 erg/cm^{2}Hz $\equiv$ 1 GPU, introduced in MISNER (1973) as "gravitational pulse unit".

Optimalization of experimental parameters - If one wants to optimize the detector for the detection of short pulses, one has to maximize Φ as a function of the experimental parameters (for given Mv^2/T). With given material constants one has fixed values for $L_1C_1 = 1/\omega_o^2$ and $Q_2 = \omega_o C_2 R_2$. (The quality of the transducer material is often characterized by the "loss angle" δ with $tg\delta = 1/Q_2$.) One is still free to choose the amount of transducer material and its fixation to the cylinder as well as the electric connection of transducer pieces. This will fix the coupling and the mechanical quality $Q_1 = 1/\omega_o C_1 R_1$. In addition, one can choose R_a and R_i by choice of an amplifier. Hence we must use R_2 and R_i separately now, instead of combining them in parallel into R_2'.

If we again use the abbreviation $C_s/C_2 \approx C_1/C_2 \approx \gamma$ for the coupling (many authors called this quantity β), we can rewrite (74) as

$$\left.\begin{aligned}(\frac{1}{4\Phi})^2 &= \frac{1}{\gamma Q_1 Q_2} + \\ &+ \frac{1}{\gamma Q_1}\cdot\left[\frac{1}{\omega_o C_2 R_i} + \omega_o C_2 R_a\ (1 + \frac{\gamma Q_1}{Q_2})\right] + \\ &+ R_a/R_i\end{aligned}\right\} \tag{77}$$

This has to be minimized. For our and similar experiments we can neglect the last term and minimize the sum of the first two terms for fixed Q_2.

The definition of sensitivity given in GIBBONS and HAWKING (1971), apart from neglecting the amplifier noise, overestimated the achievable sensitivity by a factor $(8\pi)^{1/2}$.

If one starts with a cylinder of quality Q_o, corresponding to a damping constant $\beta_o = \omega_o/2Q_o$, the application of transducers will reduce the quality to Q_1, or increase the damping to $\beta_1 = \omega_o/2Q_1$. (The small change of the resonance frequency can be neglected here.) If many transducers are fixed in such a way that each of them introduces the same amount of additional damping and the same contribution to the coupling, the additional damping will be proportional to the coupling:

$$\gamma \sim \beta_1 - \beta_o$$

or

$$\gamma Q_1 \sim (\beta-\beta_o)/\beta_1 = (Q_o-Q_1)/Q_o \tag{78}$$

From (77) it is obvious that we want a big value of γQ_1. From (78) we see that γQ_1 approaches a maximum rather quickly. Although the maximum would be reached if the whole resonator would consist of the transducer material (and have the corresponding high damping), we achieve already 90 % of the maximal sensitivity with $Q_1 = Q_o/5$, because $\Phi \sim ((Q_o-Q_1)/Q_o)^{1/2}$. Even for $Q_1 = Q_o/2$, we have 70 % of the maximal Φ. The maximum approached by γQ_1 with more transducers fixed in the same way, would be

$$(\gamma Q_1)_{max} = (\gamma Q_1)\cdot Q_o/(Q_o-Q_1) \tag{79}$$

We see that larger coupling and damping do not bring much advantage. They would only force us to study the output with higher time resolution, which would mean a waste of data tape and a loss of "detectability" (cf. section I 4).

Once we have fixed γQ_1 at a reasonable fraction of its maximal value, the remaining optimalization problem is mainly to minimize the bracket in the second term of (77) by proper choice of C_2 and the amplifier. The minimum will be achieved if the two terms in the bracket are equal, i.e. when

$$\frac{1}{\omega_o^2 C_2^2} = R_i R_a \left(1 + \frac{\gamma Q_1}{Q_2}\right) \quad . \tag{80}$$

To fulfill this condition of "optimal adaptation" between the amplifier and the other parts of the circuit, one can either change R_i and R_a by using several amplifiers in series or in parallel, or by connecting the individual piezo crystals partly in series and partly in parallel, such that (80) is fulfilled. The latter way was chosen in Munich and Frascati.

In table 2 the various contributions to the sensitivity have been collected for the detectors in Munich and Frascati (Lines 11, 14, 15). The optimal adaptation is seen from lines 12 and 13. The resulting sensitivities are given in line 16, the corresponding spectral density of a pulse which could deposit an energy 1kT/Φ in the cylinder, is given in line 17. (Remember: 1 GPU $\equiv 10^5$erg/cm^{2}Hz.) Line 18 is the sensitivity with a noiseless amplifier, i.e. taking only the first term in (77). Line 19 is the sensitivity with a noiseless amplifier and the formal maximum of γQ_1 from (79). If we would build the whole cylinder from the transducer material with its mechanical quality $Q_1 \approx 1200$, electrical density $Q_2 = 667$ and electromechanical coupling 0.3, we would have $\gamma Q_1 \approx 360$, while the value of $(\gamma Q_1)_{max}$ is 348 in Munich and 155-170 in Frascati. This shows that the crystals have been fixed in the optimal way (without additional losses in the glue etc.) in Munich, while in Frascati they introduce about twice the unavoidable losses, which means a loss $1/\sqrt{2}$ in sensitivity without amplifier noise. We see that both detectors have been built in a nearly optimal way. (In Frascati the mechanical quality deteriorated a little between July 1973 and August 1974). Afterwards the original values were nearly re-established by changing the suspension.)

Calculation of the optimal filter for short pulses - As we saw in section I 3, the signal content at time t is optimally measured by convolution of the output U(t) with the function G(t):

$$\lambda(t) = \int_{-\infty}^{+\infty} U(t+\tau)G(\tau)d\tau$$

$$G(\tau) = \frac{1}{2\phi\sqrt{2}} \int_{-\infty}^{+\infty} \kappa(\omega)U_g(\omega)e^{i\omega\tau}\, d\omega \; .$$

We are not interested in constant factors with $\lambda(t)$, because we shall normalize with the mean square value. Hence, for a short pulse, the only relevant contribution of $U_g(\omega) = u_g(\omega)\cdot F_1(\omega)$ will be the transfer function $F_1(\omega)$, and we can write:

$$G(\omega) \sim \kappa(\omega)\cdot F_1(\omega) \sim \frac{F_1}{|F_1|^2 R_1 + |F_2|^2 R_2 + R_a} \tag{81}$$

		Munich	Frascati
1	ω_o [sec^{-1}]	$1.04 \cdot 10^4$	$1.04 \cdot 10^4$
2	C_2 [F]	$8\ 10^{-10}$	$1.2 \cdot 10^{-9}$
3	$\gamma = C_1/C_2$	$4.5 \cdot 10^{-4}$	$7.6 \cdot 10^{-4}$
4	Q_o	$6.4 \cdot 10^5$	$5.7 \cdot 10^5$
5	$Q_1 = 1/\omega_o C_1 R_1$	$3.5 \cdot 10^5$	$(1.2-1.6) \cdot 10^5$
6	$Q_2 = \omega_o C_2 R_2$	670	670
7	$Q_2' = \omega_o C_2 R_2'$	366	386
8	$Q = (1/Q_1 + \gamma/Q_2')^{-1}$	$2.45 \cdot 10^5$	$(0.97-1.2) \cdot 10^5$
9	$Q_a = 1/\omega_o C_2 R_a$	1000	1000
10	$Q_{el} = (1/Q_2' + 1/Q_a)^{-1}$	268	278
11	$1/\gamma Q_1 Q_2$	$0.95 \cdot 10^{-5}$	$(1.65-1.23) \cdot 10^{-5}$
12	$1/\omega_o C_2 R_i' \equiv \alpha_i$	$1.2 \cdot 10^{-3}$	$1.1 \cdot 10^{-3}$
13	$\omega_o C_2 R_a (1+\gamma Q_1/Q_2) \equiv \alpha_a$	$1.2 \cdot 10^{-3}$	$(1.1-1.2) \cdot 10^{-3}$
14	$(\alpha_i + \alpha_a)/\gamma Q_1$	$1.52 \cdot 10^{-5}$	$(2.5-1.9) \cdot 10^{-5}$
15	R_a/R_i'	$1.2 \cdot 10^{-6}$	$1.1 \cdot 10^{-6}$
16	Φ	49	39 - 44
17	Z_Φ [G.P.U.]	4	4 - 3.5
18	$\sqrt{\gamma Q_1 Q_2}/4$	81	61 - 71
19	$\sqrt{(\gamma Q_1)_{max}\ Q_2}/4$	120	69 - 85
20	$\omega_r/\omega_o = 1 + \gamma/2$	$1 + 2.3 \cdot 10^{-4}$	$1 + 3.8 \cdot 10^{-4}$
21	$\omega_c/\omega_o = 1 + \xi/2$	$1 + 6.0 \cdot 10^{-5}$	$1 + 10.5 \cdot 10^{-5}$
22	$\Delta\omega/\omega_o = (\omega_r - \omega_c)/\omega_o$	$1.65 \cdot 10^{-4}$	$2.74 \cdot 10^{-4}$
23	μ [sec^{-1}]	3.2	7.0 - 6.2
24	Ψ	62°	68° - 65°
25	$1/\beta_o = 2Q_o/\omega_o$	123 sec	110 sec
26	$1/\beta_1 = 2Q_1/\omega_o$	67 sec	24 - 31 sec
27	$1/\beta = 2Q/\omega_o$	47 sec	19 - 23 sec
28	$1/\mu$	0.31 sec	0.14 - 0.16 sec

Table 2: Various numerical values for our detectors.

with the definitions from (47). Again using the abbreviations $z=\omega/\omega_o$ and $x=z^2-1$ we find

$$G(t) \sim G_o(t) \equiv \int_{-\infty}^{+\infty} \frac{z \cdot g(z) e^{i\omega_o z t} dz}{x^3 + bx^2 + cx + d} \tag{82}$$

with b,c,d from (59) and

$$g(z) \equiv (x - \frac{C_1}{C_2}) z + i \left[x(\frac{1}{Q_2'} + \frac{1}{Q_1}) + \frac{1}{Q_1} \right] \tag{83}$$

The poles in the integrand of (82) are the same as in (72). Using the theorem of residues we find for $t > 0$

$$G(t) \sim \mathrm{Re} \left\{ (b + \xi - i\eta \cdot g(Z_1) \cdot e^{i\omega_c t} \right\} \cdot e^{-\mu t} \tag{84}$$

with b, ξ, η, ω_c and μ defined in (69)-(71). For $t < 0$ the value of $g(Z_1)$ has to be replaced by its complex conjugate, as one can see from the definition of $G(t)$.

Evaluating the real part in (84) we find for the future and past:

$$G^{\pm}(t) \sim H^{\pm} \cdot \cos(\omega_c t + \psi^{\pm}) \cdot e^{\mp\mu t} \tag{85}$$

where to very good approximation $H^+ \approx H^-$ and $\psi^+ \approx -\psi^-$, because $-g(Z_1) \approx \gamma - \xi - i\eta$. Hence we find

$$G^{\pm}(t) \sim (\gamma-\xi) \cos \omega_c t \pm \eta \sin \omega_c t$$

The final result is, with $\psi \equiv \arctan \frac{\eta}{\gamma-\xi}$,

$$\left. \begin{aligned} &G(t) \sim e^{-\mu t} \cos(\omega_c t-\psi) \qquad \text{for } t > 0 \\ &G(-t) = G(t) \end{aligned} \right\} \tag{86}$$

Figure 14 shows this shape of the optimal filter in the time domain schematically. The value of the frequency ω_c is given by (73): The shift with respect to the peak frequency ω_r is

$$\Delta\omega \equiv \omega_r - \omega_c = \omega_o \cdot \frac{\gamma}{2} \cdot \frac{Q_{el}}{Q_2'} \tag{87}$$

(Remember, that Q_{el} combined R_2, R_i and R_a; Q_2' combined R_2 and R_i, i.e. R_2'.)
The frequency ω_c drops further away from the peak frequency ω_r as the amplifier noise becomes lower ($\omega_c \to \omega_o$ as $Q_{el} \to Q_2'$). This shift is due to the fact that the ratio between the noise contributions from R_2' and R_1 has a minimum at ω_o. Therefore, one gains a little sensitivity by observing at ω_c, which lies somewhere between ω_o and the peak frequency ω_r.

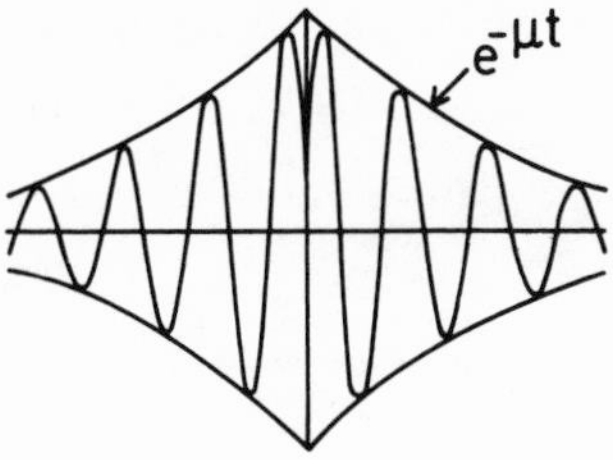

Figure 14: Schematical representation of the optimal filter in the time domain.

The value of the width μ in the optimal filter was given in (73). We see that

$$\mu = \frac{\omega_o}{2} \cdot \frac{\gamma Q_{el}}{4\Phi} \tag{88}$$

with the sensitivity Φ from (74). The width μ, the frequency shift $\Delta\omega = \omega_r - \omega_c$ and the phase shift ψ in the optimal filter (86) are related by

$$\left.\begin{aligned} \Delta\omega &= \mu \cdot \frac{4\Phi}{Q_2'} \\ tg\psi &= \frac{\mu}{\Delta\omega} \end{aligned}\right\} \tag{89}$$

The values of $\omega_c, \mu, \tau_\mu \equiv \mu^{-1}$ and are given in table 2 (lines 22 ff). The values were also shown in figure 12.

<u>Optimal filtering in the components x and y in the "co-rotating phase plane"</u> - So far we assumed that the output U(t) of our detector is recorded as a function of time, in a wide band around the resonance frequency. Actually, we only record the values of x and y (cf. equ. 17 in I 4) in a "co-rotating frame", ten times a second, and after smoothing over this discretization interval. Our control of the reference oscillator (see II 2) keeps the frequency of co-rotation very near the peak frequency ω_r (equ. 49). The beat period being kept $\geq$ 10 min, the error in frequency is always small compared to the width of the spectrum and to the shift $\Delta\omega$ between ω_r and ω_c, the frequency of optimal evaluation. As we saw in I 4, the components x and y, recorded ten times a second, will be sufficient to reconstruct the envelope $\Lambda(t)$ of our optimal signal content function $\lambda(t)$ with good accuracy, because the discretization interval of 0.1 sec is shorter than the time scale $1/\mu$ of the optimal filter, which de-

termines the autocorrelation time of our signal function $S \equiv \Lambda^2/\Lambda^2$, as defined in section I 4.

In order to construct S, we have to represent $\lambda(t)$ in a frame co-rotating at ω_c, i.e.

$$\lambda(t) = \lambda_1 \cos \omega_c t - \lambda_2 \sin \omega_c t$$

whereas x and y were gained at the co-rotation frequency $\omega_r = \omega_c + \Delta\omega$, i.e.

$$U(t) = x \cos \omega_r t - y \sin \omega_r t.$$

If we insert this into

$$\lambda(t) \sim \int_{-\infty}^{+\infty} G(\tau) \cdot U(t+\tau) d\tau$$

with $G(\tau)$ from (86), we find the signal content envelope Λ from

$$\Lambda^2(t) = \lambda_1{}^2 + \lambda_2{}^2$$

$$\lambda_1 = \int_{-\infty}^{+\infty} \{X(t+\tau)G_c(\tau) - y(t+\tau)G_S(\tau)\} \, d\tau \tag{90}$$

$$\lambda_2 = \int_{-\infty}^{+\infty} \{X(t+\tau)G_S(\tau) + y(t+\tau)G_c(\tau)\} \, d\tau$$

where we have defined the filter functions

$$G_c(\tau) \equiv e^{-\mu|\tau|} \cdot \cos(\Delta\omega|\tau| + \psi)$$

$$G_S(\tau) \equiv \pm e^{-\mu|\tau|} \cdot \sin(\Delta\omega|\tau| + \psi) \tag{91}$$

with "±" for future and past respectively. Figure 15 shows these functions for the Munich detector. If we define X_c, X_S, Y_c, X_S as the convolutions of x(t) and y(t) with the functions $G_c(\tau)$ and $G_S(\tau)$, we can write

$$\Lambda^2 = X_S{}^2 + Y_S{}^2 + X_c{}^2 + Y_c{}^2 + 2(X_S C_c - X_c Y_S) \tag{92}$$

If we would produce the components x and y with $\Delta\omega = 0$, i.e. by co-rotation with ω_c (or by numerical rotation with $-\Delta\omega$, if the reference frequency is fixed at ω_r), the two functions $G_c(\tau)$ and $G_S(\tau)$ would simply be $e^{-\mu|\tau|}$, weighted with $\cos\psi$ and $\pm\sin\psi$. This procedure would be far from measuring the jump in the co-rotating phase plane, because at ω_c one would have to correct for the fast beat $\Delta\omega$ (about 3.7 sec beat period for the Munich values). However, if we look at figure 15, where the filtering is shown for x and y produced with the reference frequency fixed at the peak frequency

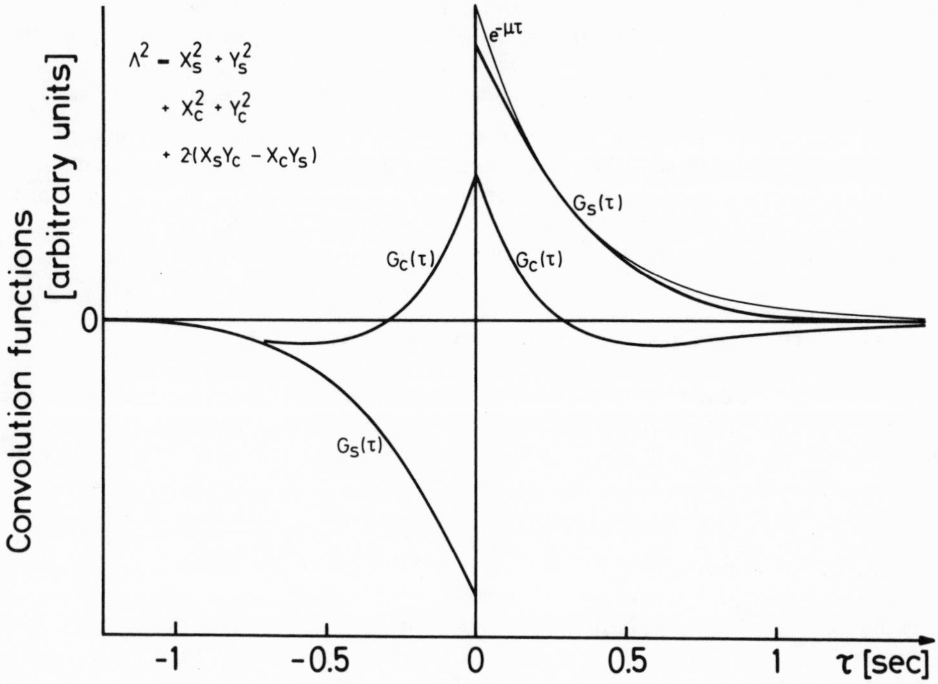

Figure 15: The convolution function for optimal determination of a cylinder jump in the co-rotating phase plane. (Munich detector.)

ω_r, we see that the contribution of G_C in (92) will be much smaller than that of G_S (for our detectors). And since $G_S(\tau)$ does not differ much from $e^{-\mu|\tau|}$, as seen in the figure, it will be a reasonable approximation to neglect the influence of ψ, and simply take

$$G_C(\tau) \approx 0, \qquad G_S(\tau) \approx \pm e^{-\mu|\tau|}. \tag{93}$$

The plausible meaning of this approximation would be: "Measure the jump between future and past in the phase plane, co-rotating at the peak frequency of the output spectrum. Define future and past by exponential smoothing with the time scale $1/\mu$, given by (88), which increases with the spectral contributions of disturbing electrical noise."

We see that our general approach of optimal evaluation tells us a bit more than this "plausibility": One can gain sensitivity by taking into account the spectral structure of the various noise contributions. The appearance of $G_C(\tau)$ in (90) and fig. 15 means a certain discrimination against jumps in the phase plane which do not have the cylinder characteristics. For different parameters of the detector these differences between the "plausible" and the optimal filtering may become important. In our case they are not.

We should stress again that so far we have only considered the optimal evaluation for short pulses, exciting the cylinder within a duration $1/\mu$ or less. If one wants to look for influences which last longer than $1/\mu$, the duration of the pulse will enter the optimal algorithm. An example will be considered in section II 5.

4. Actual Evaluation in the Munich-Frascati Experiment

The data processing: Construction of S(t) - In the last section we saw how the strength of short pulses in the cylinder would be measured optimally. There, it was assumed that x and y are determined without apparative filtering. Due to the discretization at 0.1 seconds, connected with a physical smoothing over this time scale, we had to expect some loss of sensitivity. On the other hand this smoothing will have the effect that (93) gives a better approximation of the signal function, because the "fast" noise in the co-rotating phase plane is suppressed. Hence, we could expect to approach a reasonable fraction of the optimal sensitivity, by simply taking the jumps between a mean future and past. Because of the smoothing it might be better to separate the past and the future by more than 0.1 seconds. Therefore, we tried the following approximations of Λ^2:

$$\left.\begin{aligned} \Lambda_N^2(t) &\equiv (x^+ - x^-)^2 + (y^+ - y^-)^2 \\ x^+ &\equiv \mu\Delta t \sum_{n=N}^{\infty} x(t+n\Delta t)\, e^{-n\mu\Delta t} \\ x^- &\equiv \mu\Delta t \sum_{n=1}^{\infty} x(t-n\Delta t)\, e^{-n\mu\Delta t} \\ &\text{and analogously for } y^+ \text{ and } y^-. \end{aligned}\right\} \quad (94)$$

$\Delta t = 0.1$ sec is the discretization interval. $N = 0$ means that we start x^+ and x^- at adjacent data points, $N = 1$ means that we leave out one point, etc. Because we smoothed only over $\Delta t=0.1$ sec, we tried N=0 and N=1. The corresponding signal functions $S_o \equiv \Lambda_o^2/\overline{\Lambda_o^2}$ or $S_1 \equiv \Lambda_1^2/\overline{\Lambda_1^2}$ were calculated from the data tape. Because of the exponential decay one finds $x^{\mp}(t\pm\Delta t)$ from $x^{\mp}(t)\cdot e^{-\mu\Delta t}$ + $x(t\pm\Delta t)$, and correspondingly for y. (Future and past are treated in the same way within a long block of data kept in the storage.) The appearance of only one multiplication and one addition makes the numerical evaluation rather fast. (A better approximation of equation 90 would increase the numerical work considerably without much gain in sensitivity.)

Calibration of actual sensitivity achieved with our approximation of the optimal filter - In order to measure the actual sensitivity we applied series of artificial pulses of known strength and studied the threshold crossing of our signal functions. The probabilities to find S_o or S_1 (from the last section) above a given threshold S are again found to be $W_o(S) = e^{-S}$ in pure noise, and $W_1(S,S_g)$ at the arrival of a signal pulse of strength S_g. The function $W_1(S,S_g)$ is the same as derived in section I 4, namely the expression (28). The only difference is that the sensitivity Φ has some value smaller than the optimum. It has to be determined from (28) experimentally with known artificial pulses.

Artificial mechanical exciation of the detector cylinder was first achieved with electric pulses in a condensor plate, fixed opposite to the cylinder end-face. However, this equipment tended to be mechanically unstable over long periods. Therefore, the excitation in the final set-up was done over piezo-electric transducers. In Frascati a little crystal was glued on especially for this purpose, in Munich the pick-up system itself was used and a compensation circuit took care that the exciting electric pulse was suppressed in the output signal. To produce short pulses, a rectangular voltage, lasting half a cylinder period, was applied. Since one is always within the linear regime, it is sufficient to gauge the pulser with strong pulses, and one can extrapolate very reliably to weak ones. The action of strong pulses was observed in the co-rotating phase plane, either directly on an oscilloscope or from the x,y-data on tape. The larger the jump in the phase-plane, the better is the relative accuracy of its observation, because the uncertainty of the mean past and future is given by the same noise. On the other hand, one must not make the pulses too strong because one has to stay within the range of linearity of the amplifier, and within the data range defined by the encoding procedure. Applying many equally strong pulses, we could gauge the pulser to an accuracy of a few percent.

Experimentally it is most natural to choose as the unit pulse one which would excite the cylinder from energy zero to 1kT. Since the contribution of the electrical noise to the long term value $\overline{E} \equiv \overline{x^2+y^2}$ is negligible in our detectors and data treatment, we have $\overline{E}$ = 1kT (plus less than 1 %). Thus, the radius $r \equiv \overline{(x^2+y^2)}^{1/2}$ is an experimentally defined unit in the phase-plane and a good measure of our unit pulse strength: A unit pulse causes a jump of this length (plus the random influence of the noise) in the co-rotating phase-plane, no matter where and in which direction in the plane it occurs. In our evaluation, a short pulse does not have any properties, except its strength.

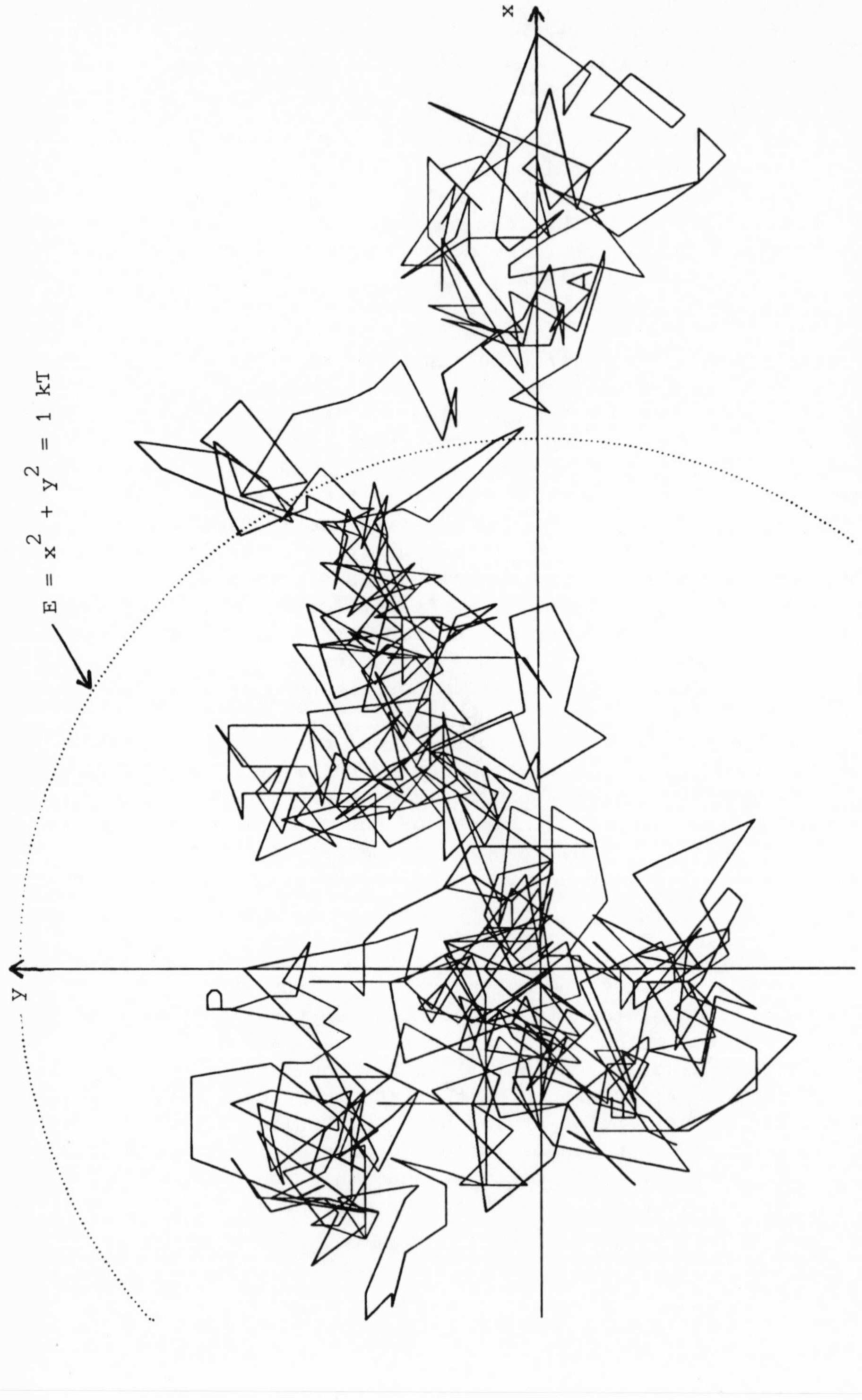

Figure 16: Munich raw data of one minute from A to E in the (x,y)-plane. Artificial pulse at P.

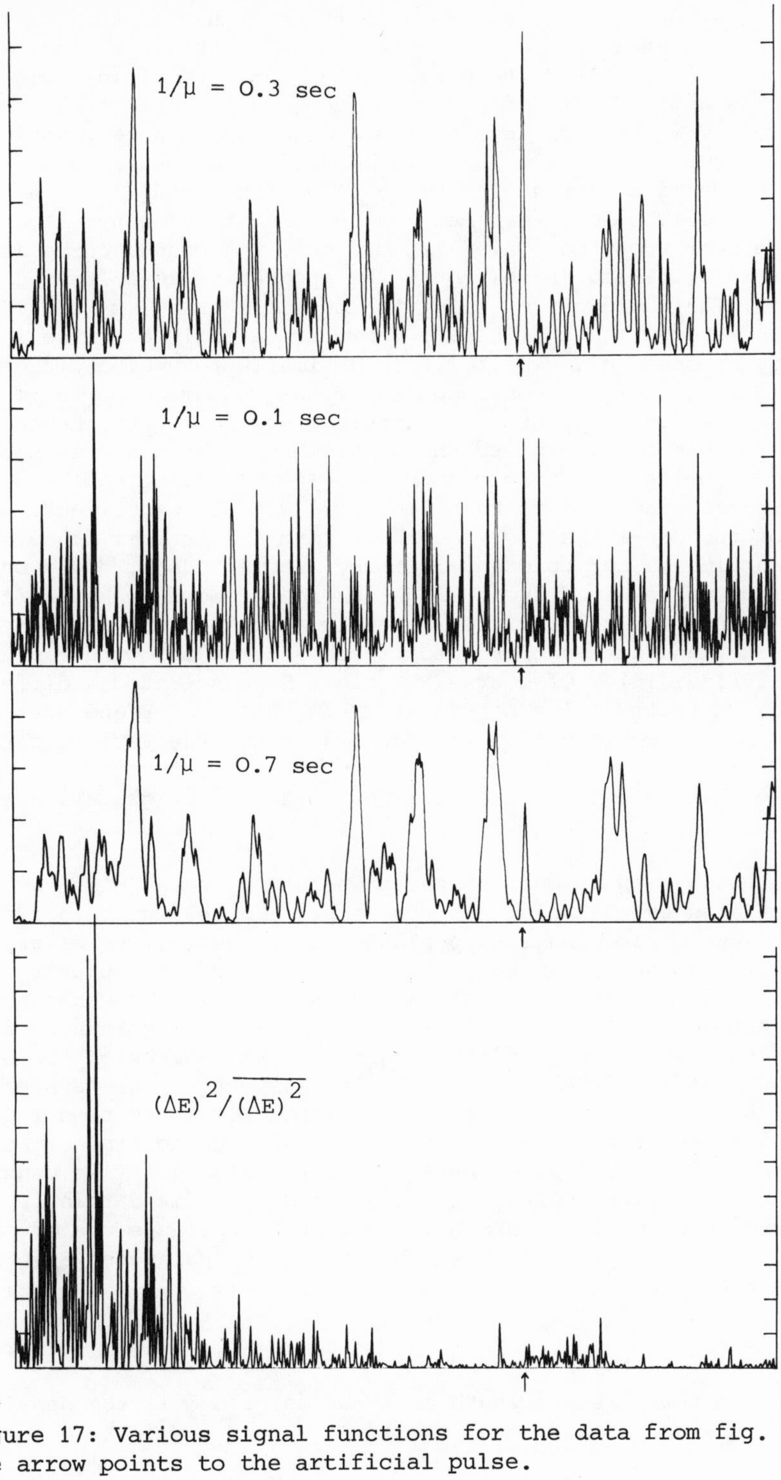

Figure 17: Various signal functions for the data from fig. 16. The arrow points to the artificial pulse.

Figure 16 shows the x,y-data of the Munich detector for 1 minute. The 600 data points are connected by a line. The start was at "A", the end at "E". At the moment "P" a weak artificial pulse of 0.1 kT was applied. It did not have much obvious effect on the data. In figure 17 we show the evaluation of the same minute with the algorithm from eq. (94) with N=0, for different values of the filter time $1/\mu$. We have plotted the signal function $S = \Lambda_o^2/\overline{\Lambda_o^2}$ as a function of time. The long term mean value is unity. The upper curve is with the nearly optimal value $1/\mu=0.3$ sec. The second curve is with too little filtering (namely only the apparative RC-filtering of 0.1 sec). The third curve is for too much filtering, namely $1/\mu=0.7$ sec. The arrow marks the instant where the artificial pulse was applied. As one would expect, it is optimally recovered with the nearly optimal filter. For comparison, the bottom curve in figure 17 shows the function $(\Delta E)^2$, that is the square of the change in x^2+y^2 from one data point to the next. This curve has also been normalized with its long term mean value. It is similar to the signal function recorded most of the time in J. Weber's experiment. Because of the non-linearity, the short-term statistical behaviour of this function is not constant in time, but depends strongly on the instantaneous value of the cylinder energy E. Most peaks of $(\Delta E)^2$ appear at high values of E. Our artificial pulse, which occured at low energy, and went rather into the phase than into the radius, is nearly invisible. Of course, we might have selected a different picture, where the pulse occurs at a high value of E and acts radially. But as we know from section I 3 in the theoretical part, the $(\Delta E)^2$ algorithm must be worse than the optimal one on the average. (We shall see that this difference is in fact quite considerable.)

Although even in our optimal signal function a pulse of 0.1 kT was simulated by the noise about once per minute, we could use a large number of such pulses to measure our sensitivity. We produced the artificial pulses at known instants (accurate to about a millisecond), and calculated the signal function in some neighbourhood. Due to the nearly Gaussian statistics in x and y (avoidance of a beat!) and the linearity of the algorithm, the values of the signal function S at any fixed time shift with respect to the "arrival time" follow the statistics of equations (25) or (28) in section I 4. Figure 5 in section I 4 showed the probabilities to find S above a given threshold for a given strength of the signal. If we count how often in a series of pulses the value of S at a fixed relative time was found above various thresholds, we must reproduce one of the curves in this type of figure. This fixes the parameter S_g. From (26) we know that

$$S_g = \Phi E_g \tag{95}$$

where E_g is the pulse strength in kT-units, and Φ is the sensitivity of the algorithm with respect to pulses of 1kT. Thus, we have determined the sensitivity of the algorithm at a given time-shift with respect to the arrival time of the signal. It must have a maximum at a certain value of the shift. Because of the apparative filtering in Munich and Frascati with a time constant of about 0.1 sec, we expect and find a positive shift of this order. The decay of sensitivity as a function of the shift is governed by the correlation time of S(t) which is of the order $1/\mu$. The maximum of the sensitivity (found by variation of the shift) will, of course, be called the sensitivity of the algorithm. Figure 18 shows the determination of this sensitivity for the algorithm Λ_o^2 in Munich and Frascati.

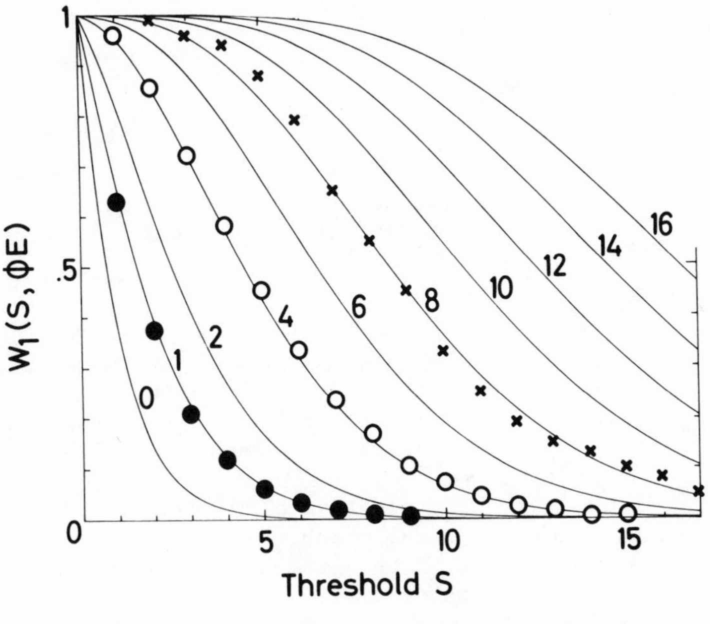

Figure 18: Gauging of actual sensitivity in Munich and Frascati.

● 3500 pulses with E_g=0.025 in Munich.
○ 1000 pulses with E_g= 0.1 in Munich.
× 100 pulses with E_g= 0.28 in Frascati.

In Munich for 3500 pulses with $E_g = 0.025$ and for 1000 pulses with $E_g = 0.1$; in Frascati for 100 pulses with $E_g = 0.28$. The results of the counts (at the most favourable time shifts) are plotted with the theoretical curves from fig. 5. Using (95) we see that the actual sensitivities of the two experiments were for the algorithm Λ_o^2:

$$\begin{aligned} \Phi_{\text{Munich}} &\approx 40 \approx 0.82\ \Phi_{\text{opt}} \\ \Phi_{\text{Frascati}} &\approx 28 \approx 0.72\ \Phi_{\text{opt}} \end{aligned} \tag{96}$$

as compared to the theoretical optimal values of 49 and 39-44 (in line 16 of table 2). With the algorithm Λ_1^2 the sensitivity was nearly unchanged in Munich and about 10 % higher in Frascati.

We have to remember, that these values of sensitivity are with respect to detector dependent unit pulses producing a jump of "$\sqrt{1\text{kT}}$" in the phase plane. Since the mass of the Frascati detector is larger by a factor 1.25, we have to multiply Φ_{Frascati} by 1.25 (obtaining 35), if we refer both detectors to the same unit pulse of 200 GPU, corresponding to 1kT in the Munich cylinder. Hence, we find for the spectral density Z_ϕ, defined in equation (75) of section II 3, but now using the <u>actual</u> sensitivities,

$$Z_\phi \approx \left\{ \begin{array}{l} 5 \text{ GPU in Munich} \\ 6 \text{ GPU in Frascati} \end{array} \right\} \tag{97}$$

From (59) we find the spectral density of the gravitational wave amplitude (or rather its time derivative), corresponding to 5 GPU:

$$|\dot{h}(\omega_o)| = 7\cdot 10^{-17} \ .$$

Because of the slight changes in the Frascati detector, we assumed a value of 30 for the Frascati sensitivity (with respect to the Munich 1kT-pulse) during the whole period of evaluation. But this is a lower limit which was in fact never reached.

The fact that with our evaluation algorithm we loose only about 20 and 30 percent of the optimally achievable sensitivities, shows that it would not have been worth while to try harder. In order to approach the optimum even better, we would have to sample the data at smaller intervals than 0.1 sec, and use correspondingly less apparative filtering.

<u>Common evaluation</u> - In both detectors there still occured nonthermal disturbances due to local effects. Whereas the Munich detector (housed in a separate, well isolated hut) showed such events very rarely, the Frascati detector (situated in a laboratory with much activity) was sometimes excited several times per day by elec-

trical or mechanical disturbances. Since the relative phases of the reference oscillators were not registered, the two-detector evaluation could not be done in the optimal way (section I.5.1) anyhow. It had to be done as described in section I 5 under "coincidence experiments", i.e. both signal functions were demanded to be above threshold simultaneously in order to count an "event".

Before April 1974 we used the signal functions with N=O in (94) for both detectors. Later we used N=1 for the Frascati detector, which slightly increased the actual sensitivity. Otherwise, the evaluation was always done in the same way. The signal functions were computed and written on tape again, every 0.1 seconds, normalized with their long term mean values. The lowest thresholds for these normalized signal functions S_1 and S_2 were set at 5 and 4. The lower value for Frascati was chosen because of the smaller sensitivity: For identical real pulses one would observe $S_1 > S_2$. Both functions are expected above threshold S with the probability e^{-S}, which is confirmed with high accuracy, if the rare local disturbances are neglected. The probability of finding both signal functions above thresholds S_1 and S_2 respectively, is of course $e^{-(S_1+S_2)}$, because of the independence. Since our aim was simply to confirm or disprove the existence of the pulses reported by Weber, we counted the number of pairs of data points from both S-tapes, where both values were above certain threshold values S_1 and S_2. These values were chosen in 18 steps of 0.5 above the lowest values 6 and 4. Hence, we counted for all combinations of the following thresholds:

$$S_1 = 5,\ 5.5,\ 6,\ \ldots\ldots\ldots,\ 13.5,\ 14$$

$$S_2 = 4,\ 4.5,\ 5,\ \ldots\ldots\ldots,\ 12.5,\ 13.$$

The resulting numbers $N(S_1,S_2)$ of coincident pairs were stored as a 19x19-matrix. This matrix was produced not only for coincidences at the same instant, but also for 40 values of time delay between the two data tapes. The neighbourhood of zero delay was studied at all delays between -1 sec and +1 sec, at intervals of 0.1 sec. In addition, delays between -5 min and +5 min were studied at intervals of 30 sec. Thus, the results of the evaluation for each day were stored on magnetic tape in the form of 41 matrices.

Since both cylinders were situated in east-west direction, nearly on the same geographic longitude (Munich-Rome), the plane of maximum sensitivity was given by the meridian. If one were aiming at a fixed direction in the sky - say the galactic center - one should make these counts separately for fixed fractions of a sidereal day.

The counts for time delays larger than a few tenths of a second should be uninfluenced by the kind of signal we were looking for. Therefore, we used the average over the 22 largest time delays

($\delta t = \pm 1$ sec, ± 30 sec, ± 60 sec, ... ± 300 sec) as a definition of the rate of accidental coincidences. The fluctuations within this sample could then be used to judge the significance of possibly observed deviations of the counts at $|\delta t| < 1$ sec from the mean value over all $|\delta t| \geq 1$ sec.

For an optimal detection of a deviation due to common external sources, one should construct the time delay histogram in a somewhat more sophisticated way. In section I 4 we discussed the question of "sensitivity and detectability" and saw that we could gain a little with a longer autocorrelation time of the signal function. If the average duration of a coincidence at a given pair of thresholds were much longer than the discretization interval, one should allow for a "coincidence window" i.e. combine the counts of coincidences within some range of time delay into one bin of the histogram. (Of course, this would have to be done consistently at all delays.)

In our experiment, where the autocorrelation times are near the discretization interval of 0.1 sec, it was not worth while to introduce such complications. We could even neglect the differences in the moments of maximum sensitivity for the algorithms Λ_o^2 and Λ_1^2 (one expects a shift of 0.05 seconds between them). Experiments with series of simultaneously applied artificial pulses (both triggered by the clocks, or with a count down via telephone) showed that the instants of maximum sensitivity are indeed separated by less than 0.1 seconds. Hence, we would have detected gravitational pulses nearly optimally by counting the coincidences at zero delay.

Overall results of 350 days of common observation - Between July 1973 and January 1975 the detectors in Munich and Frascati were active simultaneously for nearly one year. The set-up was not changed during this period. The Munich apparatus was very stable, the Frascati one was deteriorating a bit during the first year, but returned to the initial sensitivity after renewal of the suspension. (For the changes in the equivalence circuit cf. table 2 in section II 3.) Losses of observation time were mainly due to break-downs of the tape recorders, occasionally also in the interfaces, which converted the data into digital form, and rarely in other parts of the electronics. When longer disturbances in one of the detectors were clearly recognizable as non-gravitational, the data were not used for this period. Data containing short pulses, which did not have the cylinder characteristics, were usually kept, except for a few days when they became very frequent in Frascati. (This kind of disturbances was not observed in Munich.)

The Munich-Frascati experiment ended on January 27, 1975, when the Frascati detector was dismantled. (It was later rebuilt in Garching near Munich.) The total evaluated time was nearly one year, namely 504410 minutes ≈ 350.3 days. The counts over this whole period do not indicate the existence of common external excitation at any of the observed time delays. To set upper limits to the rate of infalling gravitational wave pulses of certain strengths, we calculate the expression analogous to (29) and (39) (in sections I.4.6 and I.5.2) as a function of the thresholds S_1 and S_2 for our sensitivities $\phi_1 = 40$, $\phi_2 \geq 30$. The average peak width as a function of thresholds was determined from the data. It is not important because even at our lowest observed pair of thresholds (5;4), i.e. $S_1 = 5$, $S_2 = 4$, the average duration τ of a chance coincidence is only 1.2 times the discretization interval $\Delta t = 0.1$ sec. Calculating the detectable rates we find the optimal pairs of thresholds for various signal strengths. For weak signals they would be below (5;4) where we started counting. For strong signals, we have to choose a pair of thresholds where no coincidences are observed, near the point in the (S_1,S_2)-plane where $W_1(S_1,S_2,\phi_1,\phi_2,E_g)$ - as defined in section I.5.3 - has a minimum along S_1+S_2 = const. Again, we measure the signal strength by E_g, which is the fraction of 1kT in the Munich Cylinder. $E_g = 1$ corresponded to a pulse of 200 GPU = $2\cdot 10^7$ erg/cm^2 Hz.

Table 3 shows the rate, detectable with 3 standard deviations within one year, as a function of signal strength E_g. For $E_g \geq 0.2$ the table shows instead of the rate the probability $1-W_1$. It tells how likely we have missed a pulse of given strength at the given pair of thresholds.

Signal Strength E_g	Thresholds $S_1;S_2$	Rate per year detectable with 3σ	Probability of missing, $1-W_1$
0.001	5;4	$> 10^7$	
0.005	5;4	$\approx 2\cdot 10^6$	
0.01	5;4	$6\cdot 10^5$	
0.025	5;4	10^5	
0.05	5;4	$1.7\cdot 10^4$	
0.1	12;8.5	1400	
0.2	12;9	55	0.94
0.3	12;9	11	0.70
0.5	12;9	4	0.15
0.7	12;9	2	$1.3\cdot 10^{-2}$
1	12;9	1	$1.8\cdot 10^{-4}$
2	12;9	1	$<10^{-11}$

Table 3: Detectable rates in the Munich-Frascati experiment (= Upper limits from the negative result over 350 days).

$S_1;S_2$	$N \pm \Delta N$	n_τ	$\langle N \rangle \pm \langle \Delta N \rangle$	$N-\bar{N}$ for $\delta t=0$	N
5;4	37559 ± 194	1.20	37349 ± 212	− 85	
5;5	13887 ± 113	1.15	13740 ± 126	−195	
6;4	13845 ± 110			+ 8	
6;5	5122 ± 69	1.11	5055 ± 75	−143	
7;4	5148 ± 85			− 22	
6;6	1895 ± 39	1.09	1860 ± 45	− 22	
7;5	1901 ± 52			− 87	
7;6	708 ± 22	1.08	684 ± 27	− 47	
8;6	713 ± 34			− 1	
7;7	267 ± 13			− 8	
8;6	267 ± 16	1.07	252 ± 16	− 25	
9;5	271 ± 19			+ 1	
8;7	101 ± 8.4	1.06	93 ± 10	− 7	
9;6	104 ± 12			− 10	
8;8	37 ± 5.7			+ 2	
9;7	39 ± 5.8	1.05	34 ± 5.9	+ 1	
10;6	42 ± 8.0			− 6	
9;8	15 ± 3.5	1.0	12.5 ± 3.6	± 0	15
10;7	16 ± 4.2			− 5	11
9;9	5.5 ± 2.3	1.0	4.6	+ 0.5	6
10;8	5.2 ± 2.4			− 0.2	5
10;9	1.7 ± 1.6	1.0	1.7	+ 0.3	2
11;8	1.8 ± 1.2			− 0.8	1
10;10	0.7 ± 1.0			− 0.7	0
11;9	0.7 ± 0.9	1.0	0.62	− 0.7	0
12;8	0.9 ± 0.9			+ 0.1	1
11;10	0.3 ± 0.6			− 0.3	0
12;9	0.4 ± 0.6	1.0	0.23	− 0.4	0
13;8	0.3 ± 0.5			+ 0.7	1
11;11	0.1 ± 0.5			− 0.1	0
12;10	0.1 ± 0.4	1.0	0.08	− 0.1	0
13;9	0.04 ± 0.2			− 0.04	0
12;11	0.1 ± 0.4	1.0	0.03	− 0.1	0
13;10	0			0	0
12;12	0			0	0
13;11	0		0.01	0	0
14;10	0			0	0

Table 4: Some of the counts for 350 days, and their comparison with thermal statistics. The columns show:

1) $S_1;S_2$ is a "reasonable" pair of thresholds.
2) $\bar{N}$ is the average count over 22 bins with $|\delta t| \geq 1$ sec; ΔN its dispersion.
3) n_τ is the average number of coincident data points (0.1 sec) in a peak above $(S_1;S_2)$.
4) $\langle N \rangle = 30264600 \cdot e^{-(S_1+S_2)}$ is the expected value of the count; $\langle \Delta N \rangle \approx \sqrt{\langle N \rangle \cdot n_\tau}$ is the expected dispersion.
5) and 6) The observed counts N at time delay $\delta t = 0$, or their deviations $N-\bar{N}$ from the average $\bar{N}$.

At time delays $|\delta t| \leq 0.1$ sec no coincidences were found at thresholds 12;9, and no significant deviation from the mean at lower thresholds. Hence, the rates given in the table are trustworthy upper limits for the arrival of short gravitational pulses. Any such pulse stronger than $E_g \gtrsim 0.5$ should have been detected, but there were none.

Table 4 shows how good the results fit the prediction for thermal noise. Obviously, no surplus at zero delay has been observed. The result looks similar for the other time delays and at other pairs of thresholds. Of course, there appear larger positive deviations at some delays, but they are not significant, even though deviations >3 appear (after normalization with the observed scatter ΔN).

Analysis of fluctuations in time. Search for intermittent sources or in special directions of the sky - It remains the question whether there exist significant deviations of the counts near zero delay within certain periods, or for certain directions in the sky. Of course, for signals stronger than $E_g \gtrsim 0.5$, it remains true that nothing was found. To set limits to rate-strength distributions of fluctuating sources, we analysed the counts for periods from a few days up to months and looked for significant features near zero time delay. The result was negative: No feature was discovered which would discriminate zero from arbitrary time delay. "Suspicious" peaks occured at all time delays. Clearly, the larger the number of thresholds, delays and time-intervals, the larger will be the biggest deviation found. But this does not mean the detection of a significant effect. The surplus near zero time delay, which appeared for a period of two weeks in March 1974, seems to have been accidental (KAFKA, 1974 and BILLING et al., 1975). The "accident" was, that we had inspected a small selection of pairs of thresholds for a short period of observation (defined by the length of a tape), and that up to this time, the largest deviation printed out was near zero delay. Now, after inspection of the full results (as stored on magnetic tape), this feature does no longer seem unlikely. We always stressed very carefully that we did not believe to have discovered real events! On the contrary: we wanted to show how easily one could be deceived by accidents. Nevertheless, some authors and translators seem to have misunderstood us. (Cf. WEBER, 1974 and the English translation of BRAGINSKY et al. 1974.)

An evaluation of our data with respect to certain directions in the sky could be done, if new evidence were found for the existence of any positive effect. It does not seem worth-while now, because it would reduce the upper limits for a source of weak pulses in a particular direction only by a factor of about 2 below the rates given in table 3. But we do already know that there were no pulses at all with $E_g \gtrsim 0.5$. Our original data will, however, be kept on tape for some time, for the case that other experimentalists might report evidence for a certain moment or direction.

5. Discussion of other Kinds of Signals.

The evaluation in the Munich-Frascati experiment was done for short pulses of gravitational radiation. The major part of a change in the state of the cylinder has to take place within the time $1/\mu$, in order to make our processing a good approximation of the optimal. Hence, our upper limits on the rates of pulses of given spectral density are valid for pulses which make their main effect within $\tau_p < 0.2$ seconds (and whose spectra do not show structure within the width β of our resonator).

A short pulse with only a few waves would be expected from some collapse or collision events, and the strongest possible events would certainly be the shortest. On the other hand, vibrations of transitorily formed neutron stars or fast rotation of strongly asymmetric configurations in collapse, may create longer pulses or even long trains of nearly monochromatic radiation. Of course, in the latter case a resonant antenna is not useful, unless the event happens near the frequency ω_r. But if the frequency of the source is changing in time, it may for a while also be near ω_r. A wide class of realistic physical processes will produce a gravitational wave amplitude as a function of time of the kind shown in figure 19. The time derivative is of the same sort, and it is the absolute square of its Fourier component at ω_r, which our detectors measure. In order to obtain a pulse acting longer than 0.2 seconds at 1660 Hz, the frequency of the process in fig. 19 would have to stay within our resonance width $\beta \approx 0.02\ \text{sec}^{-1}$ for a duration $\tau > 0.2$ sec. This means

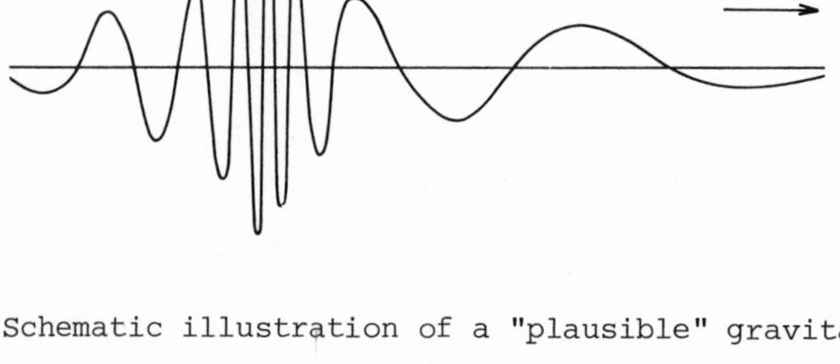

Figure 19: Schematic illustration of a "plausible" gravitational pulse.

that f(t) has to look like a sine at $\omega_r \pm \beta$ for several hundred waves. If we still want to have a chance to observe a reasonable fraction of this type of event, (without making our observation frequency a universal constant), we have to assume that the source sweeps through a bandwidth of several hundred Hz in a similar way. This might still be physically reasonable, but the total energy in this sort of event is then again obtained by multiplication of the observed spectral density with a bandwidth of nearly 1kHz.

The optimal evaluation for this type of long lasting pulses must be different from that for short pulses. One expects that one has to look for the change in the state of the cylinder over a time of the order of the duration. To confirm this, let us consider a special type of pulse, given by

$$u_g(t) \sim e^{-\alpha|t|} \cdot \cos \omega_S t \tag{98}$$

This looks nearly like the shape in figure 19, except for the constant space between the zeros in (98). We set the main signal frequency ω_S equal to ω_C. A frequency shift will reduce the sensitivity. (A fixed frequency $\omega_S \neq \omega_c$ would enter in the optimal filter, but we cannot expect such signals with a unique frequency near our resonance.)

$$\left.\begin{aligned} &\text{The Fourier transform of } u_g(t) \text{ is} \\ &u_g(\omega) \sim \frac{i}{\omega-\omega_{S1}} + \frac{i}{\omega-\omega_{S2}} - \frac{i}{\omega-\omega_{S3}} - \frac{i}{\omega-\omega_{S4}} \\ &\text{with } \omega_{S1,2,3,4} = \pm\omega_S \pm i\alpha \end{aligned}\right\} \tag{99}$$

The appearance of poles is due to our unphysical signal. A smearing of frequency would remove them, but the feature we are interested in would not be changed. For the calculation of the optimum filter, $u_g(\omega)$ has to be included as a factor in (82), introducing 4 additional residues. Then the result for G(t) becomes essentially:

$$\left.\begin{aligned} G^{\pm}(t) \sim\ & \mu e^{-\mu|t|} \cdot \left[\Delta\omega \cdot \cos \omega_c t \pm \mu \sin \omega_c t\right] \\ & + \alpha e^{-\alpha|t|} \cdot \left[\Delta\omega \cdot \cos \omega_c t \quad \alpha \sin \omega_c t\right] \end{aligned}\right\} \tag{100}$$

with $\mu, \Delta\omega$ and ω_c as in section II.3.5.

This is the expected result: For a pulse duration shorter than $\approx 1/\mu$ the optimal filter does not depend on the pulse, whereas for much longer pulses their duration simply replaces the time $1/\mu$ in the optimal filter. For our special pulse shape (98), the "duration" is in fact $2/\alpha$, which means that $2/\mu$ is the critical duration, beyond which the filter time would have to be adapted. (Remember that this was 0.6 sec in Munich and 0.3 sec in Frascati.)

Since we have not done an evaluation directed especially at longer pulses (except for a few days, and with negative result), we should estimate how the sensitivity for such events drops with the filter time. A nearly monochromatic wave train would superpose an amplitude on the cylinder, proportional to its duration. With longer filter times the main noise is that of the cylinder. The corresponding distance travelled in the phase-plane goes with the square-root of time. Hence, the sensitivity (with respect to the spectral energy-flux density) will roughly be proportional to the fraction of the pulse duration covered by the filter time. This means, that even for quasi-monochromatic waves lasting for one second, our algorithm must have achieved about 1/5 of the optimal sensitivity!

For signals without good phase relations over the duration of the pulse - e.g. a stochastic sequence of very short pulses over a duration of several seconds - the result would simply be an enhanced random walk in the phase plane. This would be discovered nearly equally well with our short-pulse filter as with a longer one. (The peaks in the signal function would of course be wide or multiple in this case, which would favour detection according to equations (29) or (39)!)

If the pulse duration becomes comparable to the damping time $1/\beta$ of the cylinder (say about half a minute), one has to distinguish again the quasi-monochromatic and the stochastic source. In the first case the sensitivity becomes proportional to $1/\beta$ (but the frequency has to lie within the small range $\omega_r \pm \beta$). In the second case, the random sequence of short pulses would be detectable with our algorithm if it is detectable at all. E.g. one might try to find this sort of event by the increase of energy in the cylinder. Let us assume that each "event" consists of 300 short random pulses of 1/100 kT each, spread over 20 seconds. On the average the energy at the end of this time would be about 2kT. The accidental number of cases where both detectors stay above 2kT for about 10 seconds will be about 430 000 per year. To discover a surplus with 3σ one would need about 2000 real events of that kind. However, the same events would cause changes of about 0.03 kT within 0.2 seconds. From table 3 we see that we would need about 10^5 pulses of this strength per year for a 3σ deviation. Since each event would cause about 100 such pulses, the 2000 real events would bring about 2.10^5, which is more than needed to reach the significance achievable with an algorithm using the energy. For instance, the algorithm applied by Weber most of the time, was sensitive to the energy, as one can see at the bottom of figure 17 in section II 4. It showed the square of the jump in the energy over a short interval (0.2 to 0.5 seconds). Clearly, the biggest jumps of this kind appear at high energy. But it does not matter very much, whether the energy is high due to the thermal noise of the cylinders or due to a common external source of the kind discussed. And although the number of common peaks in $(\Delta E)^2$ during one common period of high energy may be high, the siggnificance of detection will scarcely be increased over that achievable with observation of the energy itself (instead of its jumps). We see that even for the example of random series of very weak short pulses over a long time, our evaluation was more efficient than Weber's $(\Delta E)^2$-algorithm.

We may say, that our experiment has set the strictest limits, so far achieved in gravitational wave experiments, to all kinds of physically reasonable sources (i.e. which have not been cleverly designed by nature with the aim of better detection in a different algorithm).

6. Comparison with other Experiments

Weber's experiment has been repeated or modified by several groups. Coincidence experiments have been run at the University of Moscow (BRAGINSKY et al 1974), the University of Glasgow (DREVER et al. 1973), between the Bell Telephone Laboratories in Holmdel, New Jersey, and the University of Rochester (TYSON, 1973 and DOUGLASS et al. 1975), and between Munich and Frascati. Moreover, single detectors have been built at the observatory of Paris at Meudon (BONAZZOLA, 1973), and in the IBM laboratories (LEVINE and GARWIN, 1974).

As soon as the various detectors were working properly, and no group could find gravitational pulses, it became more and more obvious that Weber's findings must have had some other origin. This was no longer very surprising, because a closer look at Weber's evaluation procedure had made it clear that his detection efficiency was quite low and his results would have meant e.g. the removal of several million solar masses per year from the centre of our galaxy (KAFKA, 1972). At the end of 1972 we had reached a sensitivity which should have allowed the detection of at least several pulses per day, of the kind reported by Weber. Our first negative result of a few days of coincident data was reported at the Texas Symposium in New York, December 1972. At the same time, the negative results of Braginsky, Drever and Tyson also seemed to become significant evidence against the further occurence of events, as reported by Weber in 1969 and 1970. However, Weber had improved his sensitivity in the mean time, and still reported positive results. The fact that even the observation with single detectors with sufficiently pure thermal noise became significant, did not convince him. Therefore, we decided to start a long run of our experiment, without any further changes. This was done from July 1973 to January 1975. (See the results for the total time of 350 days of data, given in section II 4.) Again, the result was negative. No other group had done the experiment over a comparably long period, and none achieved the same sensitivity as ours. Hence the limits set by our experiment must be the lowest.

In order to compare different experiments it would be useful to have the equivalent circuits (and confidence in their determination). Then one could calculate their optimally achievable sensitivity for an assumed kind of signal. Of course, tests with artificial signals are even more important, but they clearly have to be done with large series and must give repeatable results (statistically). Full confidence in an experiment is only justified when the theory based on the equivalent circuit and the tests with artificial signals give the same results.

In table 5 we compare some detectors for which we had sufficient information to calculate the optimal sensitivity which might have been achieved with optimal data processing. (Cf. table 2 for more details of our detectors.)

	Munich	Frascati	Bell Lab.	Weber '73	Weber '70
$C_1\,(10^{-13}\ F)$	3.6	9.1	100	122	5
$L_1\,(10^3\ H)$	25	11	5	0.75	20
$R_1\,(\Omega)$	760	750	82	50	2000
$C_2\,(10^{-9}\ F)$	0.8	1.2	50	70	100
$R_2'\,(10^6\ \Omega)$	44	31	1	0.23	0.1
$R_a\ (\Omega)$	120	80	20	15	?
μ^{-1} (sec)	0.31	0.15	1.4	0.69	> 2.8
Φ	49	41	18	10	< 2
Z_Φ (GPU)	4	4	3.7	20	> 100

Table 5: Comparison of some detectors. Quantities in the equivalent circuits and corresponding sensitivities and filter constants.

The same table was given in Billing et al., 1975, but with a wrong value for R_2' in Frascati, giving a slightly worse number for Φ. As explained in section II 3, in connection with eq. 54, we had moreover used 120 GPU instead of the correct 200 GPU for the equivalent of 1kT in a cylinder of the Munich mass. This explains the different values in the last line of the former table. This last line gives the spectral flux density (1 GPU = 10^5 erg/cm^2 Hz) simulated by the detector noise on the average. We see that Tyson's detector was theoretically the best, with 3.7 "pulse units". This is due to the higher mass which cancels the effect of the lower Φ. (With better adaptation of the amplifier, the sensitivity in Tyson's experiment could still have been improved.) Since the sensitivity of Douglass' detector in Rochester is considerably lower, their coincidence experiment did not quite reach ours (cf. DOUGLASS et al. 1975). (It also worked at a lower frequency, namely 710 instead of 1660 Hz.)

For a comparison with the experiment just mentioned, we should also remark that these authors express the sensitivity Φ by an "effective noise temperature" T^*. It corresponds to the temperature of the cylinder, divided by Φ:

$$T^* = T/\Phi. \tag{101}$$

The authors reported a value $T^* = 16^oK$ for a 3 week period or 18.3^oK for about 3 months. This value was for the coincidence experiment, with detectors of different sensitivities. (One should make clear, how T^* is defined in this case!) In Munich, we reached $T^* = 7.5^oK$ for the whole period of 350 days!

For the coincidence experiments in Moscow and Glasgow, we did not have all details of the equivalent circuits. Both experiments modified Weber's and gave significant null results, but they cannot have reached our sensitivity.

A comparison with Weber's experiment was included in KAFKA (1974) and BILLING et al. (1975). We repeat here, why its actual sensitivity must have been even much smaller than the optimal one shown in table 5.

We tested the signal function $(\Delta E)^2/\overline{(\Delta E)^2}$ with the same series of test pulses which were used to gauge our actual sensitivity with our usual algorithm. For this purpose one simply had to count threshold crossings of the jumps in the quantity $E = x^2+y^2$, instead of the jumps in x and y. In figure 18 we showed how good the latter fitted the curves of thermal statistics. In figure 20 the same curves for $W_1(S,S_g)$ are shown, with a logarithmic scale for the probability of finding the signal function above threshold S for the series of artificial pulses with $E_g = 1/40$ and 1/10, i.e. $S_g = 1$ and 4 respectively. The curve labeled 0 is e^{-S}, the probability to find our signal function above S in pure noise. With Weber's preferred algorithm, taking the jump of $(\Delta E)^2$ over 0.1 seconds, the probability of finding $(\Delta E)^2/\overline{(\Delta E)^2}$ is experimentally found to be near the broken line labeled 0'. The corresponding probabilities in the presence of the same signals as before, are found near the curves labeled 1' and 4'. (We also included an estimate for still stronger pulses of $E_g = 1/4$ kT, i.e. $S_g = 10$.) The curves 1' and 4' show that the probability of finding $(\Delta E)^2/\overline{(\Delta E)^2}$ above a

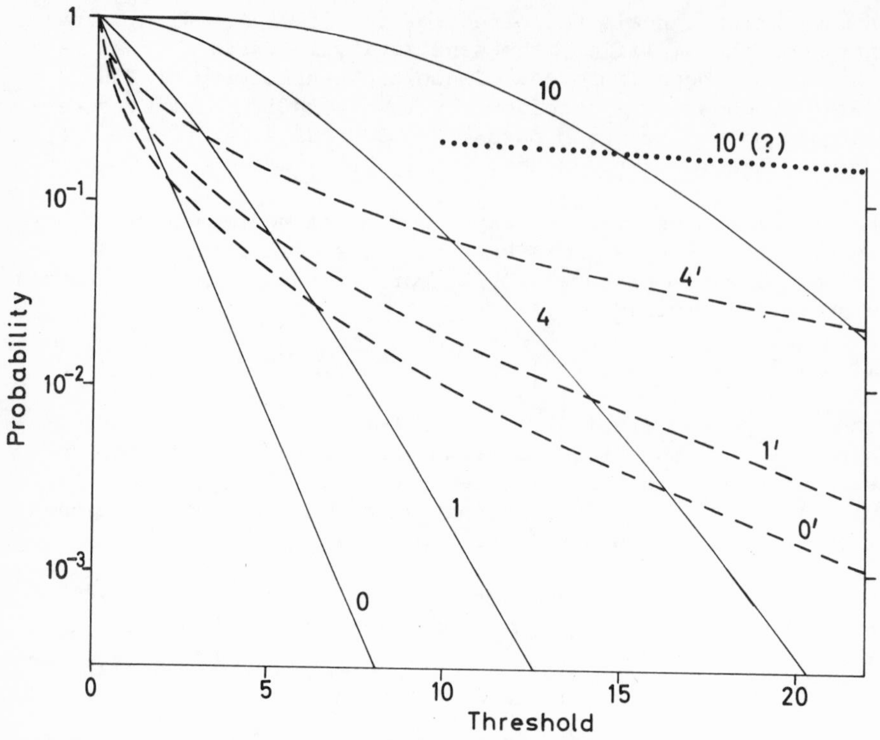

Figure 20: Threshold crossing with Weber's preferred $(\Delta E)^2$-algorithm, compared to the optimal one for short pulses. See text for detailed explanation.

threshold, drops only slowly with the threshold. However, this "high detection efficiency" is more than cancelled, because the curve 0' behaves in the same way. And we know that detectability is governed by the ratio $\sqrt{W_o/W_1}$ (cf. equation 29 in section I 4). Thus we could estimate the minimal detectable rates for Weber's algorithm with our detector. Extrapolating to Weber's lower sensitivity, we found the rates for Weber's coincidence experiment. The result was that we should have detected short pulses about 100 to 1000 times more significantly than Weber, in the range of pulse strength $0.05 < E_g < 1$.

We also applied the $(\Delta E)^2$-algorithm to about 1 month of our coincident data, selecting periods in which there was a positive deviation at zero time delay with our algorithm, but the result was random (as we could expect, because the counts for $(\Delta E)^2$ and $(\Delta x)^2 + (\Delta y)^2$ should scarcely be correlated in pure noise.

The positive deviations of the counts at zero delay in Weber's experiment were formerly confined to delays smaller than 0.1-0.3 seconds. Recently (cf. the panel discussion of DREVER, KAFKA, SCIAMA, TYSON and WEBER at the "GR7" conference in Tel Aviv, June 1974, Weber reported time delay histograms with a wide peak of the counts around zero delay. The width was of the order of the damping time of the cylinders. This would suggest that the effect was due to a slow excitation of both cylinders to high energy roughly at the same time. The excitation must have been "slow" (i.e. over about one or several seconds) because Weber usually did not find a positive effect if he used the "pulse algorithm" $(\Delta x)^2 + (\Delta y)^2$, which he also tried sometimes. Unfortunately, Weber did not record the data x and y, but always an already filtered signal function $(\Delta E)^2$ or $(\Delta x)^2+(\Delta y)^2$. Hence, he could not check whether the broad peak in the histogram for $(\Delta E)^2$ had been accompanied by a correlation of the energies E themselves. Therefore, the origin of Weber's results is still unknown, except for the fact that it is not gravitational radiation. At the conference in Paris (KAFKA, 1973) we had proposed to check the following mechanism: The analogue coincidence detector in Maryland, which registered coincidences with the Argonne detector over a telefone line, might excite the Maryland detector (through some electrical influence) at the detection of an accidental coincidence. This would then lead to an enhanced rate of coincidences over a period of the order of the damping time, without any influence spreading to Argonne. This or a similar hypothesis about a physical origin of Weber's results could only be rejected if the results would remain the same with the telefone line or any other connection of the two detectors permanently removed.

Note added for publication: We also tried Weber's preferred $(\Delta E)^2$-algorithm for two periods of about 1 month each, for which Weber told us he found especially significant results. We found no distinguished effects at all near zero time delay. This seems to exclude the hypothesis that Weber's positive results were due to very special gravitational wave pulses, which happened to be aimed particularly at the $(\Delta E)^2$-algorithm - which would be worse for "almost-all" signals.

CHAPTER III

AN OUTLOOK

Gravitational waves could not be detected at the present level of sensitivity. This was not surprising, but sad. What remains, are the questions which sources we would expect, and which experimental improvements can be made in order to detect them.

As a conservative astrophysicist one may expect that in supernovae the rest-mass energy of one to a few percent of a solar mass might go into gravitational radiation lasting for a millisecond or up to a second, in a bandwidth between a few kHz and, say, 10 Hz. The strongest possible source would be the collision of black holes, which may happen at a certain stage of a quasar remnant in a galactic nucleus, but is unlikely to happen now in one of the nearer galaxies. (KAFKA, 1970). For a survey of many kinds of sources cf. PRESS and THORNE (1972). One is not at all on safe ground, if one predicts rates of events of given kinds and strengths from sources in our own galaxy or the next clusters of galaxies. From the observation of supernovae in distant galaxies and from the statistics of supernovae remnants and pulsars, one estimates a rate of supernovae in our galaxy of about 2 to 4 per century (although no supernova became visible in our own galaxy for centuries). The center of our own galaxy is at a distance of nearly 10 kpc. The next big cluster of galaxies, the Virgo cluster, is about 1000 times more distant and contains several thousand galaxies (including the giant elliptical M87 with an active nucleus). Figure 21 shows these two main aims of gravitational pulse astronomy in the rate-strength plane. The stars represent the prediction of a moderate optimist (conversion of 0.01 $M_{\odot}$ into a pulse of gravitational radiation, going into a bandwidth of 1 kHz in the kHz-range). The strong part of the broken line near the star is an estimate of the same moderate optimist. The extensions around the question-marks may be too optimistic, but necessary for an experimentalist applying for money. The lower scale shows the fraction of 1kT at room temperature to which a cylinder of Weber's type and size would be excited, the upper scale is the corresponding spectral density in gravitational pulse units (1 GPU = 10^5 erg/cm^2 Hz), assuming a pulse of most favourable direction and polarization. (To make a survey, one would use enough detectors to be always sensitive for all suspicious directions and all polarizations.)

At the right hand side of the figure, the present limits of the Munich-Frascati experiment, and our estimate of the corresponding curve for Weber's experiment are shown. All other experiments done so far lie in between these two curves. We see that not much progress has been made in view of the aims.

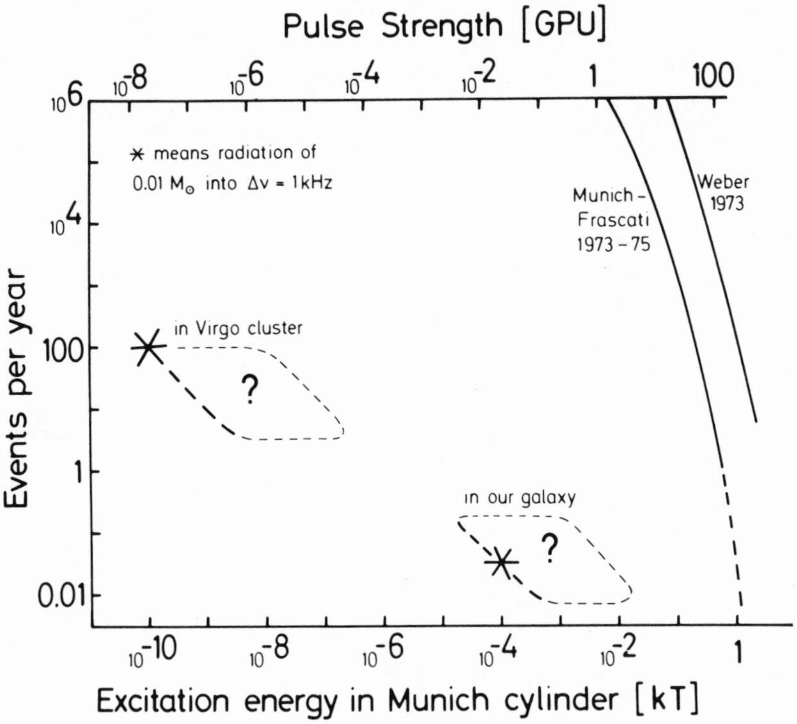

Figure 21: "The state of the art and the aims". See text.

Some possibilities to increase the sensitivity have been discussed here in Erice by BRAGINSKY, FAIRBANKS, HAMILTON and others. It seems obvious, that only a combination of extremely high quality and extremely low temperature will bring resonance detectors near the range where astronomical work is possible. (Another way which seems worth exploring, is laser interferometry with long "free-mass antennas".) The detection of the events in our own galaxy would be possible with equipment which could be developed within a few years. But it would be frustrating to wait for the next supernova. Nevertheless, taking into account man's curiosity, we dare predict that gravitational wave astronomy will be successful while Joe Weber, its founder, can still enjoy it.

General References:

1. On the detection of signals in noise:

J.S. BENDAT, Principles and Applications of Random Noise Theory, (John Wiley, New York), 1958.

A.D. WHALEN, Detection of Signals in Noise, (Academic Press), 1971

D.C. CHAMPENEY, Fourier Transforms and their Physical Applications, (Academic Press), 1973

(From the last book I just learnt, that the criterion for optimal detection of signals in noise is called "Wiener-Hopf-condition". This seems to be similar to what we have applied.)

2. On the theory of gravitational waves, possible sources, and detectors:

C.W. MISNER, K.S. THORNE and J.A. WHEELER, Gravitation, (W.H. Freeman & Co), 1973

W.H. PRESS and K.S. THORNE, Annual Review of Astronomy and Astrophysics 10, 355-374, 1972

List of References:

H. BILLING, P. KAFKA, K. MAISCHBERGER, F. MEYER and W. WINKLER, 1975, Lettere al Nuovo Cimento 12, 111

S. BONAZZOLA, 1973, in: Ondes et Radiations Gravitationnelles Proceedings of the Colloque C.N.R.S. Nr. 220, Paris

D. BRAMANTI and K. MAISCHBERGER, 1972, Lettere al Nuovo Cimento 4, 1007

D. BRAMANTI, K. MAISCHBERGER and D. PARKINSON, 1973, Lettere al Nuovo Cimento 7, 665

V.B. BRAGINSKY, A.B. MANUKIN, E.I. POPOV, V.N. RUDENKO and A.A. KHOREV, 1974, Zh. Eksp. Teor. Fix. 66, 801 (English Transl.: Sov. Phys. JETP 39, 387)

D.H. DOUGLASS, R.Q. GRAM, J.A. TYSON and R.W. LEE, 1975, preprint May 1975, to appear in Phys. Rev. Letters.

R.W.P. DREVER, J. HOUGH, R. BLAND and G.W. LESSNOFF, 1973, in Ondes et Radiation Gravitationnelles, Proc. of the Colloque C.N.R.S. Nr. 220, and Nature 246, 340

G.W. GIBBONS and S.W. HAWKING, 1971, Physical Review D4, 2191

P. KAFKA, 1970, Nature 226, 436
1970, in: External Galaxies and Quasistellar Objects (D.S. Evans ed., Proc. of IAU Symposium No 44)
1972, "Are Weber's Pulses Illegal?", unpublished essay for the "Gravity Research Foundation Award"
1973, in: Ondes et Radiations Gravitationelles, Proc. of the Colloque C.N.R.S. No 220, Paris
1974, Panel Discussion in the Proceedings of "GR7" the seventh International Conference on General Relativity and Gravitation, Tel Aviv. (G. Shaviv and J. Rosen, editors)

J. LEVINE and R. GARWIN, 1973, Phys. Rev. Letters 31, 176
1974, Phys. Rev. Letters 33, 794

D. MAEDER, 1973, in: Ondes et Radiation Gravitationnelles, Proc. of the Colloque C.N.R.S. No 220, Paris

K. MAISCHBERGER, 1973, in: Ondes et Radiations Gravitationnelles, Proc. of the Colloque C.N.R.S. No 220, Paris

C.W. MISNER, 1974, in: Gravitational Radiation and Gravitational Collapse, (C. deWitt ed., Proc. of IAU Symp. 64)

H.P. PAIK, 1975, Ph.D. Thesis, Stanford University, HEPL report 743

W.H. PRESS and K.S. THORNE, 1972, Annual Review of Astronomy and Astrophysics 10, 335

J.A. TYSON, 1974, in: Gravitational Radiation and Gravitational Collapse (C. deWitt ed., Proc of IAU Symp. 64)

J. WEBER, 1969, Phys. Rev. Letters 22, 1320
1970, Phys. Rev. Letters 24, 276
1970, Phys. Rev. Letters 25, 180
1970, Lettere al Nuovo Cimento 4, 653
1971, Nuovo Cimento 4B, 197
1972, Proc. of Internat. School of Phys. Enrico Fermi, Varenna, Course 55
1973, Phys. Rev. Letters 31, 779
1973, in: Ondes et Radiations Gravitationnelles, Proc. of the Colloque C.N.R.S. No 220, Paris
1974, Panel Discussion in the Proc. of GR7, Tel Aviv.
1974, reply to a letter of R. Garwin in Physics Today, December 1974.

SYNCHROTRON RADIATION AND ASTROPHYSICS

A. A. Sokolov

1. INTRODUCTION

Synchrotron radiation is produced by the ultrarelativistic electrons moving along some curvilinear trajectories possessing a microscopic curvature radius.

The radiation was first observed in cyclic electron accelerators. Three stages should be distinguished in the development of the physics of synchrotron radiation. At the first stage the radiation was considered as a very harmful phenomenon since the relevant energy loss determined the upper bound of the electron energy attained in the betatron regime: only the relativistic mass increase was compensated for in the latter but not the radiation energy loss. Therefore after the radiation had been discovered both the construction and designing of all new several hundreds Mev betatrons was stopped. The Veksler-Mc Millan autophasing principle made it possible to compensate both the mass increase and the radiation losses thus allowing the cyclic electron accelerators to be built operating not only at several hundreds of Mev but at some Gev as well.

At the second stage the physical properties of the synchrotron radiation itself were studied using the ultrarelativistic generalisation of classical and quantum electrodynamics. Experiments were done to check the theory.

Finally we came to the third stage of the synchrotron radiation physics. The point is that it is possible to vary the radiation spectrum in rather a broad range: by varying the electron energy the intensity maximum may be shifted to any part of the electromagnetic spectrum - from the radiofrequencies up to the x-ray regions. This is practically the sole calibrated source available which produces the radiation having both a prescribed polarisation and a prescribed shape of the spectral intensity curve. Thus a special problem arises of providing the synchrotron radiation sources. The best of these are the electron (or positron) storage rings.

At present the synchrotron radiation is being used to study the band structure of solids (vacuum ultraviolet range is needed 50-200 Å), molecules and molecular biology problems (2 - 10 Å). This is quite a new application of the storage rings which have already recommended themselves in their usual field. We recall the new particles discovered using the colliding electron positron beams. Thus the 760 Mev ρ-mesons were produced in the Novosibirsk storage ring at electron and positron energies equal to 2 x 380 = 760 Mev. The mesons were found to decay rapidly into two π-mesons. The subsequent resonances corresponded to the ω - and ϕ -meson production at 780 Mev and 1020 Mev respectively. All these particles fitted into the then existing table of the elementary particles. New resonances were discovered in November, 1974 at the total electron and positron energy of 3,1 and 3,7 Gev. This was done in Frascati using the colliding beams "Adone" and in Stanford (USA) at the storage ring "Spear," in Hamburg using the new storage ring "Doris" etc. The new particles called ψ_1 and ψ_2 decayed into hadrons, $e^- e^+$ and $\mu^- \mu^+$.

The new particles did not fit into the existing conventional schemes of elementary particles including the well-known Gell-Mann-Nishi-jima scheme. It seems that along with the strangeness and baryon number some new quantum number should be introduced to describe the family of the so-called "coloured" or "charmed" quarks.

2. SYNCHROTRON RADIATION

It is believed at present that the non-thermal radiation of Galaxies is of a synchrotron origin (called sometimes a magneto-Bremmstrahlung). For example the magneto-Bremmstrahlung is emitted by the Crab Nebula. As contrasted to the synchrotron where the electrons move along almost circular trajectories the magneto-Bremmstrahlung of the Galaxies is due to the spiralling electrons.

For symplicity we consider the electron moving in a constant and homogeneous magnetic field of strength H directed along the z-axis. Then the trajectory is a spiral described by

$$z = v_{11}(t,x) = R\cos\omega t, \quad y = R\sin\omega t, \tag{1}$$

Here $v_{11} = c^2 p_z/E$ - is the (constant z - component of the velocity and p_z is the respective momentum. The angular velocity ω, and the radius of curvature R are given by

$$\omega = \frac{e_o cH}{E} \; ; \quad R = \frac{\sqrt{E^2 - m_o^2 c^2 - c^2 p_z^2}}{e_o H} \tag{2}$$

We treat the problem in some detail since many well-known theoreticians were repeating the same error when calculating the intensity of the radiation emitted by a spiralling electron. They were trying to obtain the energy flux across a spherical surface following the Shott (1912) calculation of the radiation intensity in case of the circular trajectory. The correct solution was found by myself and my coworkers [1]. We obtained a completely rigorous expression for the radiation intensity in the frame of a classical relativistic theory. To this end it was necessary to calculate the electromagnetic energy flux not across a spherical surface, relative to which the electron position changed in time but across a cylindrical surface of the radius r > R with an axis directed along the external magnetic field H. One obtains then the following expression for the radiation intensity

$$W = \frac{ce_o^2}{R^2} \sum_{\nu=1}^{\infty} \nu^2 \int_0^{\pi} \frac{\sin\theta d\theta}{(1-\beta_{11}\cos\theta)^3} \left[\left(\frac{\cos\theta - \beta_{11}}{\sin\theta} \right)^2 J_\nu (x) + \beta_1^2 J'^2_\nu (x) \right] \tag{3}$$

Here ν is the number of the harmonic, θ is the angle between the z-axis and the direction of radiation, $c\beta_{11} = v_{11}$ and $c\beta_\perp$ are the longitudinal and transverse velocity components respectively. The argument of the Bessel function is

$$x = \frac{\nu\beta_\perp \sin\theta}{1-\beta_{11}\cos\theta} \; . \tag{4}$$

Putting $\beta_{11} = 0$, $\beta_\perp = \beta$ in eq. (3) the Shott eq. is re-derived Integrating (3) with respect to θ and doing the sum over ν we obtain

$$W = \frac{2}{3} \frac{ce_o^2}{R^2} \frac{\beta_o^2}{(1-\beta_o^2)^2} \, , \tag{5}$$

where

$$\beta_o = \sqrt{\frac{\beta^2 - \beta_{11}^2}{1 - \beta_{11}^2}} \tag{6}$$

is the invariant velocity. Thus the radiation power (5) is also an invariant.

Note that in some papers on the magneto-Bremmstrahlung an eq. of the type (3) was obtained with the factor $(1 - \beta_{11}\cos\theta)^3$ in the denominator being replaced by $(1 - \beta_{11}\cos\theta)^4$ [2]. Hence an incorrect expression was found for the radiation intensity not satisfying the relativistic covariance requirements.

Doing just the θ - integral in eq. (3) we obtain the spectral distribution of the radiation intensity:

$$W(\nu)=\nu W \frac{\sqrt{3}}{2\pi}\gamma^6 \int_{\frac{2\nu}{3}\gamma^3}^{\infty} K_{5/3}(x)\,dx, \qquad (7)$$

Here $\gamma = \sqrt{1 - \beta_o^2}$. The right and left circularly polarised waves are emitted respectively at the acute $(0 < \theta < \pi/2)$ and obtuse $(\pi/2 < \theta < \pi)$ angles with respect to the magnetic fields. The

$$\sin\theta_o = \frac{\sqrt{1-\beta_{11}^2}\sin\theta}{1-\beta_{11}\cos\theta}, \quad \cos\theta_o = \frac{\cos\theta - \beta_{11}}{1-\beta_{11}\cos\theta}$$

$$W=W_\pi + W_\sigma = \frac{ce_o^2}{R^2}\sum_{\nu=1}^{\infty}\nu^2 \int_0^\pi \sin\theta_o\, d\theta_o (1+\beta_{11}\cos\theta_o)$$

$$\times \left[f_\pi(\nu,\theta_o)+f_\sigma(\nu,\theta_o)\right]$$

$$f_\pi(\nu,\theta_o)=\mathrm{ctg}^2\theta_o J_\nu^2(\nu\beta_o\sin\theta_o),$$

$$f_\sigma(\nu,\theta_o)=\beta_o^2 J_\nu'^2(\nu\beta_o\sin\theta_o).$$

ratio of the so called σ - component (electrical vector of the wave perpendicular to the z -axis) to the π-component (magnetic vector of the wave perpendicular to the z - axis) is equal to 7: 1.

Using eq. (7) one finds the harmonic corresponding to the max maximum of the radiation intensity:

$$\nu_{max} \backsim \frac{1}{2\gamma^3} \qquad (8)$$

The energy emitted is concentrated within a narrow cone with an angle γ directed along the electron velocity.

Observations of the spectrum and of the polarization effects showed the magneto-Bremmstrahlung of the galaxies to be of the synchrotron nature. Note that the longitudinal velocity component being absent ($\beta_{11} = 0$), eqs: (2), (5), (8) reduce to the formulae which describe the circular electron motion and form the basis of the theory of radiation emitted by the modern electron synchrotrons [3,4], then

$$\gamma = \sqrt{1 - \beta^2} = \frac{m_o c^2}{E}$$

3. QUANTUM EFFECTS

Quantum effects lead to the fluctuational character of the radiation. In modern synchrotrons the external magnetic field being of the order of 10^4 Gauss, this becomes noticeable at the energies about 500 Mev.

Quantum effects lead to the exitation of the betatron and synchrotron oscillations in modern synchrotrons as a result of which the storage rings may function [5]. Indeed, the growth is compensated for by a strong radiation damping. Hence the favourable regime of the stationary orbit broadening may be obtained which is necessary for the storage ring to work.

On the other hand at the fields and energies considered the quantum effects lead to a radiative self-polarization of the electrons.

This results in about 96% of electrons (positrons) having their spins directed against (along) the external magnetic field after about 2,5 hours work. This prediction was fully confirmed by observations at the storage rings ACO in Orsay (France) and in Novosibirsk.

The influence of quantum effects upon the radiation intensity is described by the parameter

$$\xi = \frac{3}{2} \frac{E}{m_o c^2} \frac{H}{H_o}, \qquad (9)$$

Here $H_o = m_o^2 c^3 / e_o \hbar = 4{,}41.10^{13}$ Gauss is the so called Schwinger magnetic fieldstrength.

If ξ is much less than one, the quantum effects do not change appreciably the classical eq. (5) for the radiation intensity. Such is the case in conventional synchrotrons and storage rings. However when dealing with the cosmic magneto-Bremmstrahlung, the values of ξ of the order of and even much higher than one might well be expected. This might be due to the high energy values at comparatively small magnetic fields ($H \ll H_o \gamma$).

Then the radiation intensity calculated by means of the quantum theory, W^{qu}, becomes much less than the classical value W:

$$W^{qu} = \frac{2^{8/3}}{\perp} \Gamma\,(2/3)\,\frac{W}{\xi^{4/3}} \qquad (10)$$

At $\xi \gtrsim 1$ the one photon processes of the pair creation and annihilation might become of importance. When the photon of the energy $\hbar\omega \gg m_o c^2$ enters the magnetic field region a shower of particles may be produced - just as the cosmic rays produce showers in the atmosphere. The probability of the pair creation by a photon moving perpendicularly to the magnetic field is given by [7].

$$w = w_o \begin{cases} \frac{3}{16}\sqrt{\frac{3}{2}}\, e^{-\frac{4}{\xi}} & \xi \ll 1 \\ \frac{5}{7}\, 2^{-4/3}\, \frac{\Gamma(\frac{5}{6})}{\Gamma(\frac{7}{6})} \left(\frac{2}{\xi}\right)^{1/3} & \xi \gg 1. \end{cases} \qquad (11)$$

Here

$$w_o = \alpha \frac{m_o c^2}{\hbar} \frac{H}{H_o}$$

The probability of the one-photon annihilation of an electron and positron moving in a plane perpendicular to the field and having equal energies E is

$$w_1 = \frac{4}{3}\alpha \frac{c}{L^3}\left(\frac{\hbar}{m_o c}\right)^2 \frac{H_o}{H}\left(\frac{m_o c^2}{E}\right)^4 \left[2K_{1/3}^2\left(\frac{2}{\xi}\right) + K_{2/3}^2\left(\frac{2}{\xi}\right)\right] \quad (12)$$

Here L^3 is the volume of the interaction region, K_μ are the modified Hankel functions.

Note that an essentially quantum region $\xi \gtrsim 1$, might be attained as well when the magnetic field H, is high enough. Such a situation might be realized, say, within the pulsars.

According to [7], at the magnetic fields near to the critical value H_o an important contribution to the probability of the processes considered comes from the transitions to the ground state or to some level near to it. For example the probability of the transition to the ground state at $H \sim H_o$ is given by [7]:

$$w_1 = \frac{1}{2}\alpha \frac{m_o^2 c^4}{\hbar E} \frac{H}{H_o} e^{-H_o/H} \quad (13)$$

Thus such transitions contribute noticeably to the radiation intensity. On the other hand at $H \ll H_o$ their probability is exponentially small. Analogous result holds as well for the probability of the pair creation by the photon having the energy $\hbar\omega \gg m_o c^2$ if one of the particles is created in the ground state. To get this probability one has but to replace E by $\hbar\omega$ in eq. (13).

Gravitational radiation of the charge moving in a homogeneous magnetic field was studied as well. The radiation consists of the two parts: one of them comes from the transformation of the electromagnetic radiation into the gravitational one in an external homogeneous field and is given by the following formula:

$$I_{trans.} = \frac{2}{3} \frac{G m_o^2 c}{R^2} \left(\frac{E}{m_o c^2}\right)^6 \left(\frac{r}{R}\right)^2 \quad (14)$$

r is the distance from the source to the observer.

The second part is contributed by the gravitational radiation proper. Previously a separation into two parts was not done correctly. We succeeded in doing this making use of the unique decomposition of the electromagnetic field into the radiative and nonradiative parts. In contrast to the electromagnetic case the gravitational radiation was shown to possess no maximum at

$\nu \sim \left(\frac{E}{m_o c^2}\right)^3$ the spectrum decreasing monotonously. The radiation intensity increases as the fourth power of $\left(\frac{E}{m_o c^2}\right)$ [8]:

$$I_{grav} = \frac{7}{3} \frac{G m_o^2 c}{R^2} \left(\frac{E}{m_o c^2}\right)^4 \tag{15}$$

REFERENCES

1. A.A. Sokolov, I.M. Ternov, V.G. Bagrow, D.V. Galtson, V. Ch. Zhukovskii, Zs. Phys. 211, 1, 1968; A.A. Sokolov, D. V. Galtsov, M.M. Kolesnikova, Comm. High School, phys. ser. (USSR), No. 4, 14, (1971).

2. K. C. Westfold, Astroph. J. 130, 241 (1959). V.L. Ginzburg et al. Progr. phys. Sci. (USSR) 87, 65 (1965), 94, 63 (1968).

3. D. Ivanenko, A. Sokolov, DAN (USSR) 59, 1551 (1948).

4. A.A. Sokolov, I.M. Ternov, Relativistic electron (USSR), 1974.

5. A.A.-Sokolov, I.M. Ternov, J. Exp. Theor. Phys. (USSR), 25, 698 (1953).

6. A.A. Sokolov, I.M. Ternov, DAN (USSR), 153, 1053 ; International Conf. on Accelerators, Dubna, 1963.

7. A.A. Sokolov, V. Ch. Zhukovskii, N.S. Nikitina, Phys. Lett., 43A, 85 (1973); A.A. Sokolov, I.M. Ternov, A.V. Borisov, V. Ch. Zhukovskii, Phys. Letts. 49A, 9 (1974).

8. D.V. Galtsov, Yu. M. Loskutov, A.A. Sokolov, V.R. Khalilov, Problems in the theory of gravitation , Moscow, 1975.

EDITORIAL COMMENTS

DEVELOPMENT OF SINGULARITIES

A number of theorems have been proven by Penrose and by Hawking concerning evolution of solutions of Einstein's Field Equations to a singularity. The work of Møller, presented here, displays this behavior in an unusually lucid way.

ON THE BEHAVIOUR OF PHYSICAL CLOCKS IN THE VICINITY OF SINGULARITIES OF A GRAVITATIONAL FIELD

C. Møller

The Niels Bohr Institute and NORDITA

Copenhagen

INTRODUCTION

In Einstein's theory of gravitation the proper time of an observer is defined by the length of the world line of the observer which is supposed to be given by the readings of a standard clock following the observer. The existence of standard clocks with this property is of principal importance because such clocks are necessary instruments in measuring the components of the metric tensor and other physical fields in any given system of coordinates[1]. An ideal standard clock in general relativity can also be defined as a clock that has the same rate in a local rest system of inertia as in a corresponding extended special relativity type system of inertia.

The working of a real physical clock of a definite construction is determined by the theory itself. Therefore it is possible to decide which conditions a physical clock must satisfy in order to be a good standard clock that measures the proper time with reasonable accuracy. This problem was treated in an old paper [2] by the present author for the case of a linear harmonic oscillator that represents the simplest conceivable model of a clock (the essential part of any clock, the balance is approximately a harmonic oscillator). It was shown that it is always possible to choose the parameters of the oscillator (the elastic constant, the mass and the amplitude of the oscillating particle) in such a way that it is a good standard clock for a given acceleration of the clock as a whole in a given non-singular gravitational field. In particular it was seen that oscillating systems with frequencies on the atomic scale are extremely good standard clocks under ordinary circumstances. This also justifies the assumption on which the usual derivation of the Einstein shift of spectral lines is based, viz. the assumption that the proper frequency of an atom, i.e. the number of oscillations pr. unit of proper time, is independent of the gravitational potential at the place where the atom is situated.

In the present paper we shall study the behaviour of physical clocks in the vicinity of singularities of the gravitational field, where the conditions set up in ref. 2 are not satisfied. It is one of the most surprising discoveries in the later years that such singularities are unavoidable in Einstein's theory, not only in the case of the universe as a whole but also for a sufficiently massive galaxy that is contracting under the influence of its own gravitational field. After a finite proper time, as measured on a standard clock following the matter the metric becomes singular. This is now generally recognized as a serious difficulty of Einstein's theory, primarily because this entrance into an unphysical state occurs after a finite proper time; for proper times in contrast with coordinate times are usually regarded as "physical" times measuring the natural rythms of actual events. As emphasized in particular by Charles Misner [3] it would be much more acceptable if the singularities occured either in the infinite past or in the infinite future and he put forward the idea that a finite proper time close to singularities might not correspond to a finite number of ticks of real physical clocks. In his speculations Misner based his challenge to the physical significance of proper time on global considerations, discarding the idea that considerations on a local basis could succeed, the argument being that locally special relativity is valid and that therefore sufficiently small physical clocks will measure proper time as in special relativity. However as we shall see in the following this is not quite correct, the reason being that a real physical clock, although arbitrarily small, cannot be a strict mathematical point since there must be something moving in a clock, otherwise it cannot measure time. As we shall see, this provides the possibility for "tidal forces" to influence the rate of physical clocks close to the singularities.

1. The Metric and the Equations of Motion of an Oscillator Particle in Fermi Coordinates

To start with let us consider a linear oscillator at rest in an extended system of inertia I in special relativity. It consists of a particle of proper mass m_0 oscillating around a fixed point 0 in I under the influence of an elastic force with the elastic constant k. If we assume that the velocity u of the particle is so small compared with the velocity c of light that terms of the order u^2/c^2 can be neglected, the relativistic equation of motion reduces to Newton's equation

$$\frac{d^2x}{dt^2} = -\frac{kx}{\omega_0} = -\omega_0^2 x(t), \qquad \omega_0 = \sqrt{k/m_0}. \tag{1.1}$$

With the initial conditions

$$x(0) = 0, \quad u(0) = \dot{x}(0) = \omega_0 A_0 \tag{1.2}$$

the solution of (1.1) is

$$x = A_0 \sin \omega_0 t \tag{1.3}$$

Since $u = |A_0 \omega_0 \cos \omega_0 t|$, the validity of (1.1) requires

$$\left(\frac{A_0 \omega_0}{c}\right)^2 \ll 1. \tag{1.4}$$

The number of passages of the particle through 0 is a measure of the Lorentz time t in I which is indentical with the proper time of 0. Thus, suitably calibrated, the oscillator represents a model of a standard clock at rest in I. Since the frequency ω_0 is independent of the amplitude we can make A_0 as small as we like (but not zero of course) so that we have a model of the "point like" clocks with which one operates in classical relativity theory. The special principle of relativity guarantees the validity of (1.1) and (1.3) in any system of inertia, so that the oscillator clock will show the correct proper time in any extended system of inertia.

However is it also a good standard clock when accelerated? Let us assume that the centre 0 is taken on a trip with a given world line

$$x^i = f^i(\tau), \quad i = 1, 2, 3, 4. \tag{1.5}$$

These equations express the Lorentz coordinates in I as given functions of the proper time τ of 0. In this case the motion of the oscillator particle around 0 is most conveniently described in a rigid non-rotational system of reference with coordinates

$$\check{x}^{i} = \{\check{x}^{\iota}, c\check{t}\} = \{\check{x}^{\iota}, c\tau\} \tag{1.6}$$

in which 0 is permanently situated at the origin $\check{x}^{\iota} = 0$.
In this accelerated system of coordinates the metric is simply[4)]

$$ds^2 = \delta_{\iota\kappa}\, d\check{x}^{\iota} d\check{x}^{\kappa} - c^2 d\check{t}^2 \left(1 + g_\iota \check{x}^\iota / c^2\right)^2. \tag{1.7}$$

Here the $g_\iota(\check{t})$ are functions of $\check{t}$ determined by the motion of 0 i.e. by the functions $f^i(\tau)$. Physically g_ι is at any time equal to the acceleration of 0 relative to a momentary inertial rest system.

The generally relativistic equations of motion corresponding to the metric (1.7) are

$$\frac{d^2 \check{x}^i}{d\tau_{(m)}^2} = -\check{\Gamma}^i_{kl} \frac{d\check{x}^k}{d\tau_{(m)}} \frac{d\check{x}^l}{d\tau_{(m)}} + \frac{\check{F}^i}{m_0} \tag{1.8}$$

where $\check{F}^i$ is the elastic four-force in this system and $\tau_{(m)}$ is the proper time of the oscillating particle. In the "Newtonian" approximation where the "amplitude" A and the velocity $\check{u}$ of the particle are so small that

$$g_\iota A / c^2 \ll 1 \,, \qquad \check{u}^2 / c^2 \ll 1 \,, \tag{1.9}$$

$\tau_{(m)}$ is equal to $\check{t} = \tau$ and the equations (1.8) with $i = \iota = 1,2,3$ reduce to

$$\frac{d^2 \check{x}^\iota}{d\check{t}^2} = \frac{c^2}{2} \check{g}_{44,\iota} + \frac{\check{F}^\iota}{m_0} \tag{1.10}$$

with

$$\check{g}_{44} = -\left(1 + g_\iota \check{x}^\iota / c^2\right)^2, \quad \frac{\check{F}^\iota}{m_0} = -\frac{k}{m_0} \check{x}^\iota = -\omega_0^2 \check{x}^\iota \tag{1.11}$$

or by (1.9)

$$\frac{d^2 \check{x}^\iota}{d\check{t}^2} = -\omega_0^2 \check{x}^\iota - g_\iota(\check{t}). \tag{1.12}$$

With the initial conditions (1.2) the solution of (1.12) is

$$\check{x}^\iota = A_0 \sin\omega_0 \check{t} + \frac{1}{\omega_0} \int_0^{\check{t}} g_\iota(t') \sin\omega_0 (t' - \check{t})\, dt' \tag{1.13}$$

which for arbitrary $g_L(\check{t})$ deviates essentially from the motion (1.3) of a harmonic oscillator.

However if the parameters k, m_0, A_0 are chosen such that at any time

$$\omega_0^2 A_0 \gg g_L \tag{1.14}$$

i.e. if the intrinsic acceleration of the oscillating particle is large compared with the acceleration g of the oscillator as a whole, the equations (1.12) and (1.13) are approximately equal to (1.1) and (1.3) respectively. Then the number of ticks of the clock are again a measure of the proper time $\tau = \check{t}$ of 0 and the oscillator is a good standard clock. In the case of an "atomic clock" the intrinsic accelerations are billions of times larger than the largest acceleration in the biggest accelerators, in accordance with the fact that one has never observed any influence of the external accelerations on spectra of accelerated atoms.

We shall now consider an oscillator clock moving through a (permanent) gravitational field corresponding the metric

$$ds^2 = g_{ik}\, dx^i dx^k \tag{1.15}$$

in an arbitrary system of coordinates. Let us first treat the simplest case of a freely falling clock in which case the world line of the centre 0 is a time-like geodesic C in 4-space (see the figure). Here the motion of the oscillator particle is most conveniently described in a system of Fermi coordinates (FC) with the base line C. The FC of an arbitrary event point Q are obtained by drawing a (space-like) geodesic from Q to a point P on C in such a way that its tangent unit vector μ^i at P is orthogonal to the tangent vector U^i of C in P.

Let τ be the measure of P_0P along C, i.e. the proper time of the centre 0 . Let $e_{(k)}^{\,i}(\tau) = e_{(k)}^{\,i}(\tau)$ (and $e_k^{(i)}(\tau)$) be a set of orthonormal tetrads (and their reciprocals) carried along C by parallel displacement. Since C is a geodesic we can choose the tetrads such that

$$e_{(4)}^{\,i}(\tau) = U^i/c \tag{1.16}$$

along the whole curve C. Further let σ be the distance PQ as measured along the geodesic PQ . Then the Fermi coordinates (χ^i) of Q are defined as

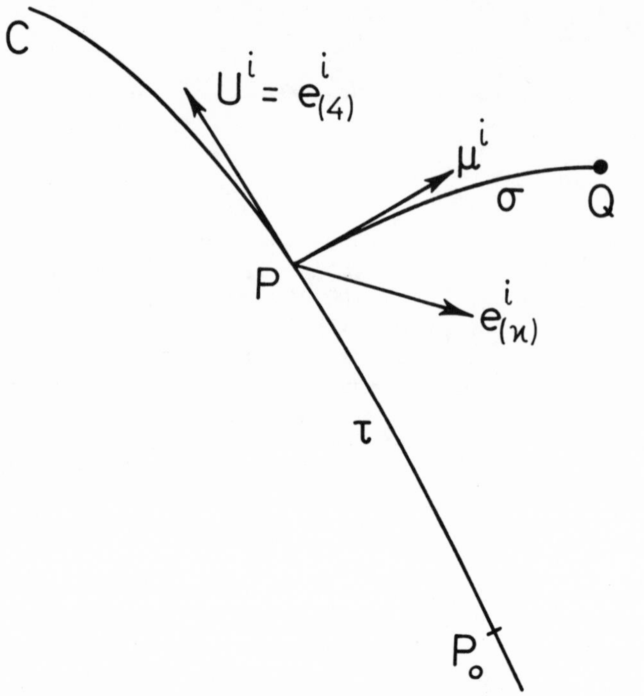

$$\overset{\circ}{x}^{\iota} = \mu^{(\iota)}\sigma \quad , \quad \overset{\circ}{x}^{4} = c\overset{\circ}{t} = c\tau \tag{1.17}$$

where

$$\mu^{(i)} = e^{(i)}_{k}\mu^{k} = \{\mu^{(\iota)}, 0\} \tag{1.18}$$

are the tetrad components of the vector μ^{i} . The FC of a point P on C are obviously

$$\overset{\circ}{x}^{i}_{P} = \delta_{i4}\, c\overset{\circ}{t} = \delta_{i4}\, c\tau . \tag{1.19}$$

For our purpose it is sufficient to know the metric tensor $\overset{\circ}{g}_{ik}$ in FC for points Q that are close to C . For such points the $\overset{\circ}{x}^{\iota}$ are small and using a Taylor expansion up to the second order in these quantitites, we obtain

$$\overset{\circ}{g}_{ik}(Q) = \overset{\circ}{g}_{ik}(P) + \frac{\partial \overset{\circ}{g}_{ik}(P)}{\partial \overset{\circ}{x}^{\lambda}}\overset{\circ}{x}^{\lambda} + \frac{1}{2}\frac{\partial^{2}\overset{\circ}{g}_{ik}(P)}{\partial \overset{\circ}{x}^{\lambda}\partial \overset{\circ}{x}^{\mu}}\overset{\circ}{x}^{\lambda}\overset{\circ}{x}^{\mu} , \tag{1.20}$$

From the definition of the FC one easily finds the following expressions for the coefficients in (1.20):

$$\overset{\circ}{g}_{ik}(P) = \eta_{ik} \qquad , \qquad \overset{\circ}{g}_{ik,l}(P) = 0 \tag{1.21}$$

where η_{ik} is the Minkowski matrix, and

$$\left.\begin{aligned} \overset{\circ}{g}_{ik,\lambda,\mu}(P) &= \frac{1}{3}\left[R_{(i\lambda\mu k)}(P) + R_{(i\mu\lambda k)}(P)\right] \\ \overset{\circ}{g}_{i4,\lambda,\mu}(P) &= -\frac{2}{3}\left[R_{(i\lambda\mu 4)}(P) + R_{(i\mu\lambda 4)}(P)\right] \\ \overset{\circ}{g}_{44,\lambda,\mu}(P) &= -\left[R_{(4\lambda\mu 4)}(P) + R_{(4\mu\lambda 4)}(P)\right] \end{aligned}\right\} \tag{1.22}$$

where $R_{(iklm)}$ are the invariant tetrad components of the curvature tensor defined by

$$R_{(iklm)} = e^{r}_{(i)}\, e^{s}_{(k)}\, e^{t}_{(l)}\, e^{u}_{(m)}\, R_{rstu} \, . \tag{1.23}$$

For a given world line C the right hand sides of (1.22) are known functions of $\tau = \overset{\circ}{t}$.

Thus (1.20) reduces to

$$\begin{aligned} \overset{\circ}{g}_{ik}(\overset{\circ}{x}) &= \eta_{ik} + h_{ik}(\overset{\circ}{x}) \\ h_{ik}(\overset{\circ}{x}) &= \frac{1}{2}\, \overset{\circ}{g}_{ik,\lambda,\mu}(\overset{\circ}{t})\, \overset{\circ}{x}^{\lambda} \overset{\circ}{x}^{\mu} \end{aligned} \tag{1.24}$$

with $\overset{\circ}{g}_{ik,\lambda,\mu}(\overset{\circ}{t})$ given by (1.22).

If we can assume that the world line of the oscillator particle is all the way close to the world line C of O and if the velocity $\overset{\circ}{u}$ is small compared with c it is easy to determine the motion of the particle. For in this "Newtonian" approximation, where

$$|h_{ik}| \ll 1 \quad , \quad \overset{\circ}{u}{}^{2}/c^{2} \ll 1 \; , \; \frac{\overset{\circ}{u}}{c}\sqrt{|h_{ik}|} \ll 1 \tag{1.25}$$

the generally relativistic equations of motion reduce to

$$\frac{d^2\overset{\circ}{x}^{\iota}}{d\overset{\circ}{t}^2} = \frac{c^2}{2}\overset{\circ}{h}_{44,\iota} - \frac{k}{m_0}\overset{\circ}{x}^{\iota} = -\omega_0^2\overset{\circ}{x}^{\iota} + c^2\overset{\circ}{g}_{44,\iota,\lambda}(\overset{\circ}{t})\overset{\circ}{x}^{\lambda} \tag{1.26}$$

on the analogy of (1.12). Here $\overset{\circ}{g}_{44,\iota,\lambda}(\overset{\circ}{t})$ is given by the last equation (1.22) and it is seen that the conditions for the oscillator clock to be a good standard clock is

$$\omega_0^2 \gg |c^2\overset{\circ}{g}_{44,\iota,\lambda}\overset{\circ}{x}^{\lambda}|. \tag{1.27}$$

As we shall see in the next section this condition is well satisfied under normal circumstances. However close to the singularities the time shown by the oscillator clock will deviate essentially from that of a standard clock.

2. The Rate of an Oscillator Clock Close to the Singularities in the Case of a Collapsing Galaxy

As a simple example we consider the collapse of a spherical distribution of incoherent matter originally at rest. In this case we can introduce a comoving and Gaussian system of space-time coordinates

$$x^i = \{r, \theta, \varphi, ct\} \tag{2.1}$$

for which

$$ds^2 = a(r,t)dr^2 + R(r,t)^2(d\theta^2 + \sin^2\theta d\varphi^2) - c^2dt^2. \tag{2.2}$$

In this case the world line of a fixed reference point is a geodesic described by

$$(r, \theta, \varphi) = \text{constants} \quad , \quad t = \tau \tag{2.3}$$

which simultaneously represent the equations (1.5) for the world line C of the centre O of a freely falling oscillator clock at a fixed reference point. For the tetrads $\overset{\circ}{e}{}^i_{(k)}$ that are carried along C by parallel transport we can take

$$\overset{\circ}{e}{}^i_{(1)} = \frac{\delta_{i1}}{\sqrt{a(r,t)}}, \quad \overset{\circ}{e}{}^i_{(2)} = \frac{\delta_{i2}}{R(r,t)}, \quad \overset{\circ}{e}{}^i_{(3)} = \frac{\delta_{i3}}{R\sin\theta}, \quad \overset{\circ}{e}{}^i_{(4)} = \delta_{i4}. \tag{2.4}$$

Inside homogeneously distributed matter the solution of Einstein's field equations gives[5)]

$$a(r,t) = \frac{S(t)^2}{1 - r^2/\ell^2} \quad , \quad R(r,t) = r\,S(t) \tag{2.5}$$

$$\frac{1}{\ell^2} = \frac{\kappa \mu_o c^2}{3} \quad , \quad S(t) = C(ct/\ell) . \tag{2.6}$$

Here μ_o is the original proper mass density and $C(u)$ is a function of $u = ct/\ell$ which describes a cycloid with

$$C(0) = 1 \quad , \quad C(\tfrac{\pi}{2}) = 0 . \tag{2.7}$$

For constant r the quantity R decreases from the value r at $t=0$ to zero at a time t_s given by

$$c t_s/\ell = \pi/2 . \tag{2.8}$$

Thus the metric becomes singular after a finite proper time t_s .

With the tetrads and the metric given by (2.4) - (2.6) it is straight forward to calculate the quantities (1.23), (1.22) and $h_{ik}(\overset{\circ}{x})$ in (1.24). Since $\overset{\circ}{t} = \tau = t$ we get

$$h_{ik} = - \frac{\overset{\circ}{x}{}^{\lambda} \overset{\circ}{x}{}^{\lambda}}{3\ell^2 S(\overset{\circ}{t})^3} \delta_{ik} + \frac{\overset{\circ}{x}{}^{i} \overset{\circ}{x}{}^{k}}{3\ell^2 S(\overset{\circ}{t})^3} \approx \frac{A^2}{\ell^2 S^3} \tag{2.9}$$

where A represents the maximum values of the $\overset{\circ}{x}{}^{i}$ or the amplitude in the oscillation. Further

$$h_{i4} = 0 \tag{2.10}$$

and

$$h_{44} = - \frac{\overset{\circ}{x}{}^{\lambda} \overset{\circ}{x}{}^{\lambda}}{2\ell^2 S(\overset{\circ}{t})^3} \approx \frac{A^2}{\ell^2 S^3} . \tag{2.11}$$

Thus the equations of motion (1.26) for the oscillator particle are of the form

$$\frac{d^2 \overset{\circ}{x}{}^i}{d\overset{\circ}{t}{}^2} = -\,\omega(\overset{\circ}{t})^2\, \overset{\circ}{x}{}^i \tag{2.12}$$

with

$$\omega(\overset{\circ}{t})^2 = \omega_0^2 + \frac{c^2}{2\ell^2 S(\overset{\circ}{t})^3} = \omega_0^2\left(1 + \frac{\lambda_0^2}{2\ell^2 S^3}\right) \tag{2.13}$$

where $\lambda_0 = c/\omega_0$ is the wave length corresponding to the frequency ω_0 of the oscillator in an extended system of inertia. According to (2.12), (2.13) the number of oscillations per unit proper time $\omega(\overset{\circ}{t})$ increases with $t = \tau = \overset{\circ}{t}$. The equations (2.12) are valid only as long as the conditions (1.25) are satisfied. According to (2.9)-(2.11) these conditions are

$$\frac{A^2}{\ell^2 S^3} \ll 1 \quad , \quad \frac{\overset{\circ}{u}{}^2}{c^2} \approx \frac{A^2\omega^2}{c^2} \ll 1 . \tag{2.14}$$

For a typical galaxy with mass M and radius r_b of the order

$$M \approx 10^{45}\ \mathrm{gm} \quad , \quad r_b \approx 10^{23}\ \mathrm{cm} \quad , \text{i.e. } \mu_0 \approx 10^{-25}\ \frac{\mathrm{gm}}{\mathrm{cm}^3} \tag{2.15}$$

we get from (2.6)

$$\ell \approx 10^{26}\ \mathrm{cm}. \tag{2.16}$$

Thus as long as S is of the order 1, the conditions (2.14) are extremely well satisfied, for even with $A \approx 1$ cm and $\omega_0 \approx 1\ \mathrm{sec}^{-1}$ we have

$$\frac{A^2}{\ell^2} \approx 10^{-52} \quad , \quad \frac{A^2\omega^2}{c^2} \approx 10^{-21} \quad , \quad \frac{\lambda_0^2}{\ell^2} \approx 10^{-31} \tag{2.17}$$

and by (2.13)

$$\omega \simeq \omega_0 . \tag{2.18}$$

A forteriori this is true for clocks with atomic frequencies and dimensions.

Since ω here is equal to ω_0 with a very high degree of accuracy, the oscillator clock is an extremely good standard clock during most of the contraction process. This is still true when the boundary of the matter sphere passes through the Schwarzschild radius

$$\alpha = \frac{\kappa M c^2}{4\pi} \tag{2.19}$$

which in our example is of the order

$$\alpha \approx 10^{17}\,\mathrm{cm} \tag{2.20}$$

This happens at a time t_α where

$$R_b = r_b\, S(t_\alpha) = \alpha \tag{2.21}$$

or

$$S = \frac{\alpha}{r_b} \approx 10^{-6} \tag{2.22}$$

and for this value of S we have even in the case (2.17)

$$\frac{\lambda_0^2}{\ell^2 S^3} \approx 10^{-13} \tag{2.23}$$

However, when t approaches the time t_S given by (2.8) the oscillator starts running faster than the standard clock and the times t_{osc} and t measured by the number of ticks of these clocks are approximately related by a differential equation of the type

$$d t_{osc} = \frac{\omega(t)}{\omega_0}\, dt = \sqrt{1 + \lambda_0^2/2\ell^2 S(t)^3}\; dt \tag{2.24}$$

In order to determine the exact relation between t_{osc} and the proper time t we have to solve the equation (2.12) which for a linear oscillator is of the form

$$\frac{d^2 \overset{\circ}{x}}{dt^2} + \omega(t)^2 \overset{\circ}{x} = 0 \tag{2.25}$$

with $\omega(t)$ equal to the steadily increasing function given by (2.13). Thereupon we have to determine the number of zero points of the solution $\overset{\circ}{x}(t)$ for a given interval of t . The exact solution of (2.25) is rather complicated. Therefore we replace the function $\omega(t)$ by a step-function $\rho(t)$ which is constant in successive intervals $\mathfrak{J}_0, \mathfrak{J}_1, \mathfrak{J}_2, \cdots\cdots, \mathfrak{J}_n, \cdots\cdots$ of decreasing lenghts and in each interval $\mathfrak{J}_n$ is equal to the value of ω at the lower end of $\mathfrak{J}_n$.

Let $t_0 = 0, t_1, t_2, \cdots\cdots, t_n, t_{n+1}, \cdots\cdots$ be the endpoints of these intervals, $\mathfrak{J}_n$ being the interval $t_n \leq t < t_{n+1}$. Then the value of ρ in the interval $\mathfrak{J}_n$ is

$$\omega_n = \omega(t_n) = \sqrt{\omega_0^2 + c^2/2l^2 S_n^3} \tag{2.26}$$

and we define the lengths of the intervals by

$$\omega_n (t_{n+1} - t_n) = 2\pi \quad \text{for } n = 0, 1, 2, \cdots\cdots. \tag{2.27}$$

The equation

$$\frac{d^2 y(t)}{dt^2} + \rho(t)^2 y(t) = 0 \tag{2.28}$$

obtained by replacing $\omega(t)$ in (2.25) by $\rho(t)$ has a solution of the form

$$y = A_n \sin \omega_n (t - t_n) \quad \text{in } \mathfrak{J}_n \tag{2.29}$$

where the amplitudes $A_0, A_1, A_2, \cdots\cdots A_n, \cdots\cdots$ in the different intervals are constants. On account of (2.27) the function $y(t)$ defined by the equations (2.29) is continuous at the endpoints of the interval, the t_n being zero points where

$$y(t_n) = 0 \tag{2.30}$$

The time derivative of y, the "velocity" u is in the arbitrary interval $\mathfrak{J}_n$

$$u(t) = \frac{dy}{dt} = \omega_n A_n \cos \omega_n (t - t_n). \tag{2.31}$$

Continuity of this function at the times $t_0, t_1, t_2, \cdots\cdots$ requires that $\omega_n A_n$ has the same value for all n, i.e.

$$\omega_n A_n = \omega_0 A_0 = u_0 \tag{2.32}$$

for all n, i.e. the amplitude A decreases as $1/\omega$ with increasing time. u_0 is the original velocity for which we have assumed $u_0^2/c^2 \ll 1$. Now it follows from (2.31) and (2.32) that the velocity $u \leq u_0$ so that the condition $u^2/c^2 \ll 1$ remains true for all $t < t_s$. Moreover we get from (2.9) - (2.11), (2.32) and (2.26)

$$h_{ik} \approx \frac{A_m^{\,2}}{\ell^2 S_m^{\,3}} = \frac{u_0^{\,2}}{\ell^2 S_m^3 \omega_m^{\,2}} = \frac{u_0^{\,2}}{\frac{c^2}{2} + \omega_0^2 \ell^2 S^3} \approx \frac{u_0^{\,2}}{c^2} \ll 1 .$$

Thus, provided the solution (2.29) of (2.28) is a reasonable approximation to the solution of (2.25), the conditions (1.25) or (2.14) will be satisfied all the way. Anyhow, since

$$\omega^2(t) \geq \rho^2(t) \tag{2.33}$$

for all $t < t_s$, the number of zero points of $\mathring{x}(t)$ cannot be smaller than the number of zero points of $\mathring{y}(t)$ in the same interval[6]. From (2.24) we get for the time t_{osc} measured by the oscillator clock in the proper time interval from 0 to t :

$$t_{osc} = \int_0^t \sqrt{1 + \lambda_0^2/2\ell^2 S(t')^3}\, dt' \tag{2.34}$$

which diverges logarithmically for $t \to t_s$.

Obviously this result is independent of the choice of the original amplitude A_0 and frequency ω_0. No matter how we choose these quantitites the oscillator will sooner or later cease to behave like a standard clock when approaching the singularity. Therefore it would seem that similar phenomena will occur for any real physical clock and one might conclude that it takes an infinite physical time to reach the singularity in accordance with Misner's suggestion. However, since the time measured be a physical clock in the region where it is not a good standard clock depends on the constitution of the clock it is perhaps more appropriate to say that no unique physical time exists close to the singularities. In the latter region the notions of proper time and of the metric field quantities themselves have no physical meaning since they are not measurable by real physical instruments.

This point of view is strengthened by considering more general types of singularities than those contained in (2.5). For instance in the empty space outside the matter sphere the solution of Einstein's field equations gives[4]

$$a(r,t) = \frac{R'(r,t)^2}{1-\alpha/r} \quad , \quad R(r,t) = r\,C\left(\sqrt{\frac{\alpha}{r^3}}\,ct\right) \tag{2.35}$$

where α is the Schwarzschild constant (2.19). Here the singularity occurs at a proper time t_s given by

$$\sqrt{\frac{\alpha}{r^3}}\;ct_s = \frac{\pi}{2}\,. \tag{2.36}$$

Consider again a freely falling oscillator clock with the centre 0 at a fixed reference point corresponding to constant (r, θ, φ). In this case we get instead of (2.9)-(2.11)

$$\left.\begin{aligned} h_{44} &= \frac{\alpha}{R^3}(\overset{\circ}{x}{}^1)^2 - \frac{\alpha}{2R^3}\left[(\overset{\circ}{x}{}^2)^2 + (\overset{\circ}{x}{}^3)^2\right] \\ h_{\iota 4} &= 0 \\ h_{\iota\kappa} &\approx \frac{\alpha A^2}{R^3}\,. \end{aligned}\right\} \tag{2.37}$$

Then (1.26) gives for $\iota = 1$

$$\frac{d^2\overset{\circ}{x}{}^1}{d\overset{\circ}{t}{}^2} = -\left(\omega_0^2 - \frac{\alpha c^2}{R(r,\overset{\circ}{t})^3}\right)\overset{\circ}{x}{}^1 = -\omega_0^2\left(1 - \frac{\alpha\lambda_0^2}{R^3}\right)\overset{\circ}{x}{}^1 \tag{2.38}$$

and for $\iota = 2,3$

$$\frac{d^2\overset{\circ}{x}{}^\iota}{d\overset{\circ}{t}{}^2} = -\omega_0^2\left(1 + \frac{\alpha\lambda_0^2}{2R(r,\overset{\circ}{t})^3}\right)\overset{\circ}{x}{}^\iota. \tag{2.39}$$

Thus the rate of a linear oscillator does not only depend on its constitution, i.e. on k, m_0 and A_0, but also on the direction of oscillation. It will for all directions of oscillation be a good standard clock as long as

$$\lambda_0^2 \ll \frac{R^3}{\alpha} = \frac{r^3}{\alpha}C^3\left(\sqrt{\frac{\alpha}{r^3}}\,ct\right). \tag{2.40}$$

This condition is amply satisfied most of the time also when the oscillator passes through the Schwarzschild radius α , since

$$\lambda_0^2 \ll \alpha^2 \tag{2.41}$$

in all practicle cases. However as t approaches t_S , where $\mathcal{C} \to 0$, a clock oscillating in the $\overset{\circ}{x}{}^1$ -direction, which is the radial direction, will vibrate with the decreasing frequency

$$\omega = \omega_0 \sqrt{1 - \alpha\lambda_0^2/R^3}$$

and with increasing amplitude until $\alpha\lambda_0^2/R^3$ has become larger than 1 after which the oscillator particle flies off and the clock goes to pieces. On the other hand a clock oscillating perpendicularly to the radial direction can be used as a time measuring instrument right up to the singularity, but its behaviour will be very much like that of clocks inside the matter, i.e. $t_{osc} \to \infty$ for $t \to t_S$.

Up to now we have only considered a freely falling oscillator clock. The case of an arbitrary motion of the centre O can be treated in a similar way by introducing Fermi coordinates with the world line C of O as base line. In this case C is not a geodesic and in constructing the FC the tetrads $e^i_{(k)}$ must be transported along C by a Fermi-Walker transport[7] instead of by parallel transport. Otherwise the construction of Fermi coordinates $\overset{\circ}{\chi}{}^i$ goes like before. These coordinates are a generalization of the coordinates $\overset{\vee}{x}{}^i$ defined by (1.7) in the case of arbitrary permanent gravitational fields. It is easy to calculate the motion of the oscillator particle in terms of these coordinates; but since it does not lead to any new results regarding our main problem, we shall not bring the calculations here. The only change in the earlier results is that the rate of the oscillator clock now depends also on the tetrad components of the four-acceleration of O besides on the tetrad components of the curvature tensor.

As a simple result of such calculations I shall only mention that an oscillator with the centre O kept at rest at a fixed reference point (R , θ , φ) = constant in the exterior Schwarzschild field outside the Schwarzschild radius α will have a proper frequency ω that depends on the direction of oscillation. If it is oscillating in a direction perpendicular to the radial direction the proper frequency is

$$\omega = \omega_0 \sqrt{1 + \alpha\lambda_0^2/2R^3} \tag{2.42}$$

Thus under the condition (2.41) the proper frequency ω in the field is equal to the frequency ω_0 in an extended system of inertia, i.e. the oscillator is a good standard clock. However if the oscillator is vibrating in the radial direction the proper frequency is

$$\omega = \omega_0 \sqrt{1 - (\alpha \lambda_0^2 / R^3) + \alpha^2 \lambda_0^2 / 4R^4 (1 - \alpha / R)} \tag{2.43}$$

Thus even under the condition (2.41) we can expect a considerable deviation of ω from ω_0 when R is sufficiently close to the Schwarzschild radius.

Conclusion

As a result of the investigation in the present paper we have come to the conclusion that the notion of proper time ceases to represent a physical quantity in the vicinity of singularities in the metric, since we cannot imagine any physical clock that can measure this quantity. This also means that the metric components themselves lose their physical meaning in this region and that Einstein's theory, which so admirably accounts for the gravitational phenomena inside our solar system, has to be changed in the case of super strong gravitational fields.

References

1) See f. inst.
C. Møller, Measurements in General Relativity and the Principle of Relativity. The publication of this paper, that was sent to a Russian journal in 1974, has been delayed for unknown reasons. A report on this paper was given also at the Erice meeting.

2) C. Møller, Dan. Mat. Fys. Medd. 30 No. 10 (1955);
Helv. Phys. Acta, Suppl. IV p. 54 (1956)

3) Charles W. Misner, Kip S. Thorne, John A. Wheeler, Gravitation, Freeman, New York (1973) § 30.7 p. 813

4) C. Møller, The Theory of Relativity, Clarendon Press, 2. Edition (1972), § 8.15 p. 287

5) C. Møller, Mat. Fys. Medd. Dan. Vid. Selsk. 39,7 (1975)

6) Riemann-Weber I, Differentialgleichungen der Physik, Vieweg Braunschweig 1925, p. 281

7) See f. inst.
J. L. Synge, Relativity, North-Holland Publishing Company, Amsterdam, 1960, Ch. II § 10

EDITORIAL COMMENTS

The Bimetric Theory of Gravitation proposed by Rosen appears to display the attractive features of General Relativity, without singularities.

BIMETRIC THEORY OF GRAVITATION

Nathan Rosen

Department of Physics, Technion

Haifa, Israel

I. FOUNDATIONS OF THE BIMETRIC THEORY OF GRAVITATION

In the general theory of relativity one regards space-time as a Riemannian manifold in which, at every point labelled by coordinates (x^0,x^1,x^2,x^3) there exists a metric tensor $g_{\mu\nu}$. This determines the interval ds between two neighboring points with coordinate differentials dx^μ according to the relation

$$ds^2 = g_{\mu\nu}dx^\mu dx^\nu \ , \tag{1-1}$$

and hence it describes the geometry of space-time. However, it is also considered to describe the gravitational field. The tensor $g_{\mu\nu}$ is determined by the Einstein field equations, involving $T_{\mu\nu}$, the energy-momentum density tensor of matter or other non-gravitational fields, together with initial and boundary conditions. In the presence of a gravitational field the space is in general curved, the curvature being described by the Riemann-Christoffel curvature tensor $R^\lambda_{\ \mu\nu\sigma}$ formed from $g_{\mu\nu}$. In the absence of gravitation, for example far from matter, the space is flat and the curvature tensor vanishes.

Some time ago it was proposed(1,2,3) to modify the formalism of general relativity by introducing at each point of space-time a second metric tensor $\gamma_{\mu\nu}$ corresponding to flat space, i.e., having a curvature tensor $P^\lambda_{\ \mu\nu\sigma}$ that vanishes everywhere. This means that between neighboring points one has a second interval $d\sigma$ given by

$$d\sigma^2 = \gamma_{\mu\nu}dx^\mu dx^\nu \ . \tag{1-2}$$

The presence of this flat-space tensor need not have any effect on the properties of the Riemannian space. One can regard $\gamma_{\mu\nu}$ as an auxiliary mathematical quantity determined by the choice of the coordinate system. We can now form two kinds of covariant derivatives, a g-derivative (denoted by ;) involving the Christoffel 3-index symbols $\left\{{\lambda \atop \mu\nu}\right\}$ formed from $g_{\mu\nu}$ and a γ-derivative (denoted by $|$) involving the symbol $\Gamma^{\lambda}_{\mu\nu}$ formed from $\gamma_{\mu\nu}$. Thus one has a tensor $g_{\mu\nu|\sigma}$, for example. There is also a new scalar, $\kappa=(g/\gamma)^{1/2}$.

It is natural to relate $g_{\mu\nu}$ to $\gamma_{\mu\nu}$ by requiring that, far from matter, where the space is flat, one should have $g_{\mu\nu} = \gamma_{\mu\nu}$. By a suitable choice of coordinates one can make $\gamma_{\mu\nu} = \eta_{\mu\nu} =$ diag.(+1, -1, -1, -1) , the Galilean metric tensor, but it may not always be convenient to do this.

With the help of the tensor $\gamma_{\mu\nu}$ it is found that one can improve the formalism of general relativity without changing its physical contents. For example, one can define a tensor to describe the energy-momentum density of the gravitational field, something which one cannot do in the conventional form of the general relativity theory.

The question I would now like to consider is whether, with the help of the two tensors $g_{\mu\nu}$ and $\gamma_{\mu\nu}$, one can set up a theory of gravitation different from the general relativity theory, a theory which I will call a bimetric theory of gravitation.

One may wonder why anybody would be interested in a theory different from general relativity. Several reasons can be given. Regarded as a system of differential equations, the Einstein field equations are rather complicated. They often give solutions having singularities, and a singularity must be regarded as representing a breakdown of the physical law described by the equations. The Einstein equations have solutions which are considered to describe "black holes". If these black holes are found in nature, this will be a great success of general relativity. Since there is no convincing evidence at present that black holes exist, one can take the standpoint that they represent a breakdown of the usual concepts of space-time and are therefore something unphysical. Then, too, in spite of the beauty of the general relativity theory and the conceptual framework it provides for the macroscopic phenomena of nature, it is interesting to explore other theories for comparison.

We shall look for a theory[4,5] which, like general relativity, satisfies the covariance and equivalence principles. For this purpose a variational principle will be used. The field variables will be the Riemannian metric tensor $g_{\mu\nu}$, the flat-space tensor $\gamma_{\mu\nu}$, and the vector ϕ_μ (it could be some other kind of tensor) describing the matter or other non-gravitational field. An attempt will be made to obtain a complete theory in the sense that all the equations of the theory will be derived from the variational principle. This will be taken to have the form:

$$\delta\int\mathcal{L}\,d\tau = 0\ , \qquad (1\text{-}3)$$

the variation arising from arbitrary variations of the field variables vanishing on the boundary of the integration region. Let us write

$$\mathcal{L} = \mathcal{L}_s + \mathcal{L}_m + \mathcal{L}_g\ , \qquad (1\text{-}4)$$

where $\mathcal{L}_s$ describes the flat-space properties of $\gamma_{\mu\nu}$, $\mathcal{L}_m$ the matter field, and $\mathcal{L}_g$ the gravitational field, each of these being a scalar density.

The form of $\mathcal{L}_s$ should be chosen so that the variational principle leads to the equation

$$P_{\lambda\mu\nu\sigma} = 0\ . \qquad (1\text{-}5)$$

In order to obtain this we need the variation of a fourth rank tensor. Let us therefore introduce another field variable $S^{\lambda\mu\nu\sigma}$ having the same symmetry properties as $P_{\lambda\mu\nu\sigma}$, and let us write

$$\mathcal{L}_s = \frac{1}{2}(-\gamma)^{1/2}S^{\lambda\mu\nu\sigma}P_{\lambda\mu\nu\sigma}\ . \qquad (1\text{-}6)$$

One finds that

$$(-\gamma)^{-1/2}\frac{\delta\mathcal{L}_s}{\delta S^{\lambda\mu\nu\sigma}} = \frac{1}{2}P_{\lambda\mu\nu\sigma}\ , \qquad (1\text{-}7)$$

$$(-\gamma)^{-1/2}\frac{\delta\mathcal{L}_s}{\delta\gamma_{\mu\nu}} = S^{\lambda\mu\nu\sigma}{}_{|\lambda\sigma} + \frac{1}{4}S^{\alpha\beta\gamma\delta}P_{\alpha\beta\gamma\delta}\gamma^{\mu\nu}\ . \qquad (1\text{-}8)$$

Since $S^{\lambda\mu\nu\sigma}$ is assumed to be present only in $\mathcal{L}_s$, we see from (7) that (3) leads to (5). If we write

$$W^{\mu\nu} = (-\gamma)^{-1/2}\frac{\delta\mathcal{L}_s}{\delta\gamma_{\mu\nu}}\ , \qquad (1\text{-}9)$$

then we now have

$$W^{\mu\nu} = S^{\lambda\mu\nu\sigma}{}_{|\lambda\sigma}\ . \qquad (1\text{-}10)$$

Since $S^{\lambda\mu\nu\sigma} = -S^{\lambda\mu\sigma\nu}$ and γ-differentiation is commutative, it follows that

$$W^{\mu\nu}{}_{|\nu} = 0\ . \qquad (1\text{-}11)$$

Next let us take $\mathcal{L}_m$ to be a function of ϕ_μ, $g_{\mu\nu}$ and their derivatives, but not of $\gamma_{\mu\nu}$. Since ϕ_μ is assumed to be present only in $\mathcal{L}_m$, Eq. (3) gives

$$\frac{\delta \mathcal{L}_m}{\delta \phi_\mu} = 0 , \tag{1-12}$$

and this represents the field equations of the matter. Let us also write

$$T_{\mu\nu} = \frac{1}{8\pi} (-g)^{-1/2} \frac{\delta \mathcal{L}_m}{\delta g^{\mu\nu}} . \tag{1-13}$$

This will be interpreted as the energy-momentum density tensor of the matter.

Let us now consider an infinitesimal coordinate transformation

$$x'^\mu = x^\mu + \xi^\mu(x) . \tag{1-14}$$

Then one finds

$$\delta g^{\mu\nu} = g^{\mu\alpha}\xi^\nu{}_{,\alpha} + g^{\nu\alpha}\xi^\mu{}_{,\alpha} - g^{\mu\nu}{}_{,\alpha}\xi^\alpha . \tag{1-15}$$

Taking into account the field equations (12), one has

$$\delta \int \mathcal{L}_m d\tau = 8\pi \int T_{\mu\nu}\delta g^{\mu\nu}(-g)^{1/2} d\tau . \tag{1-16}$$

Making use of (15), integrating by parts, and taking ξ^μ to vanish on the boundary, one finds

$$\delta \int \mathcal{L}_m d\tau = - 16\pi \int T_\mu{}^\nu{}_{;\nu}\xi^\mu(-g)^{1/2} d\tau . \tag{1-17}$$

Now the left side vanishes because the integral is invariant under a coordinate transformation. Since ξ^μ is arbitrary, one gets

$$T_\mu{}^\nu{}_{;\nu} = 0 . \tag{1-18}$$

Here and elsewhere indices are raised and lowered by means of $g_{\mu\nu}$ unless otherwise indicated.

Now let us consider $\mathcal{L}_g$. It will be assumed to be a function of $g_{\mu\nu}$, $\gamma_{\mu\nu}$ and $g_{\mu\nu|\lambda}$, that is of $g_{\mu\nu}$, $\gamma_{\mu\nu}$ and their first derivatives. Let us write

$$\frac{\delta \mathcal{L}_g}{\delta g^{\mu\nu}} = K_{\mu\nu}(-g)^{1/2} , \tag{1-19}$$

$$\frac{\delta \mathcal{L}_g}{\delta \gamma^{\mu\nu}} = H_{\mu\nu}(-\gamma)^{1/2} \ . \tag{1-20}$$

If we take again an infinitesimal coordinate transformation so that $\delta g^{\mu\nu}$ is given by (15) and $\delta\gamma^{\mu\nu}$ by a corresponding expression, then we find

$$\delta\int \mathcal{L}_g d\tau = -2\int\{K_\mu{}^\nu{}_{;\nu}(-g)^{1/2} + H_{\mu\underline{\nu}|\nu}(-\gamma)^{1/2}\}\xi^\mu d\tau \ , \tag{1-21}$$

where a line under an index means that it is to be raised or lowered with $\gamma_{\mu\nu}$. Again, since the integral on the left is invariant, we get the identity (with $\kappa = (g/\gamma)^{1/2}$)

$$\kappa K_\mu{}^\nu{}_{;\nu} + H_{\mu\underline{\nu}|\nu} \equiv 0 \ . \tag{1-22}$$

Let us now go back to (3). If we vary $g^{\mu\nu}$, then from (13) and (19) we get, as the gravitational field equations,

$$K_{\mu\nu} = -8\pi T_{\mu\nu} \ . \tag{1-23}$$

If we vary $\gamma^{\mu\nu}$ (noting that $W^{\mu\nu}\delta\gamma_{\mu\nu} = -W^{\underline{\mu\nu}}\delta\gamma^{\mu\nu}$), then from (9) and (20) we get

$$H_{\mu\nu} = W^{\underline{\mu\nu}} \ . \tag{1-24}$$

It follows from (11) that

$$H_{\mu\underline{\nu}|\nu} = 0 \ , \tag{1-25}$$

and hence from (22) that

$$K_\mu{}^\nu{}_{;\nu} = 0 \ . \tag{1-26}$$

It is interesting that this condition on $K_{\mu\nu}$ is imposed by the requirement that $\gamma_{\mu\nu}$ describe a flat space. We see that the divergence equations (18) and (26) are consistent with the field equations (23). Alternatively, one can say that (18) follows from (23) and (26) even without the considerations involving $\mathcal{L}_m$.

Let us go back to (19). One finds that it can be rewritten

$$\kappa K^{\mu\nu} = \left(\frac{\partial L_g}{\partial g_{\mu\nu|\sigma}}\right)_{|\sigma} - \frac{\partial L_g}{\partial g_{\mu\nu}} \tag{1-27}$$

where

$$L_g = (-\gamma)^{-1/2}\mathcal{L}_g \ . \tag{1-28}$$

Now Eq. (18) can be put into the form

$$(\kappa T_{\mu}{}^{\nu})_{|\nu} - \frac{1}{2}\kappa T^{\alpha\beta} g_{\alpha\beta|\mu} = 0 \; . \tag{1-29}$$

If we make use of (23) and (27), this becomes

$$(\kappa T_{\mu}{}^{\nu} + t_{\mu}{}^{\nu})_{|\nu} = 0 \; , \tag{1-30}$$

with

$$t_{\mu}{}^{\nu} = \frac{1}{16\pi}\left\{\frac{\partial L_g}{\partial g_{\alpha\beta|\nu}} g_{\alpha\beta|\mu} - \delta^{\nu}_{\mu} L_g\right\} . \tag{1-31}$$

This can be interpreted as the energy-momentum density tensor of the gravitational field. For $\gamma_{\mu\nu} = \eta_{\mu\nu}$, (30) represents a true conservation law.

Let us go back to Eq. (18), and let us apply it to the simple case of matter in the form of a pressure-less cloud of dust particles. The matter is characterized at each point by a scalar density ρ and a velocity vector u^{μ} which, for a particle with coordinates x^{μ} at this point, is given by

$$u^{\mu} = \frac{dx^{\mu}}{ds} \; , \tag{1-32}$$

so that

$$u_{\mu}u^{\mu} = 1 \; . \tag{1-33}$$

In this case one has

$$T^{\mu\nu} = \rho u^{\mu} u^{\nu} \; , \tag{1-34}$$

and Eq. (18) gives

$$(\rho u^{\nu})_{;\nu} u^{\mu} + \rho u^{\mu}{}_{;\nu} u^{\nu} = 0 \; . \tag{1-35}$$

Multiplying by u_{μ} and using (33), one gets

$$(\rho u^{\nu})_{;\nu} = 0 \; , \tag{1-36}$$

and (35) gives for $\rho \neq 0$

$$u^{\mu}{}_{;\nu} u^{\nu} = 0 \; , \tag{1-37}$$

which, in view of (32), is the equation of the geodesic. Since in this case every particle of dust is acted on only by a gravitational field, the conclusion to be drawn is that a test particle in a gravitational field moves on a geodesic.

The situation here is similar to that in the general relativity theory. Let us define a weak gravitational field by the condition

$$g_{\mu\nu} = \gamma_{\mu\nu} + h_{\mu\nu} , \tag{1-38}$$

where the $h_{\mu\nu}$ and their derivatives can be regarded as small quantities. In the case $\gamma_{\mu\nu} = \eta_{\mu\nu}$, if one considers the equation of the geodesic for a weak, static field and a slowly moving particle, we know from general relativity that, to get agreement with Newtonian mechanics, one must have

$$g_{00} = 1 + 2\phi , \tag{1-39}$$

where ϕ is the Newtonian gravitational potential satisfying

$$\nabla^2\phi = 4\pi\rho , \tag{1-40}$$

where ρ is the density of the matter producing the gravitational field.

Let us now go back to the variational principle and consider the form of $\mathcal{L}_g$ or L_g . It will be assumed that the field equations (23) are linear in the second derivatives of $g_{\mu\nu}$ and hence that L_g is a homogeneous quadratic function of $g_{\mu\nu|\lambda}$. There are many possible expressions of this kind. For example, to obtain the field equations of general relativity one takes for $\mathcal{L}_g$

$$\mathcal{L}_{GR} = (-g)^{1/2}\{ \frac{1}{4} g^{\lambda\rho}g^{\sigma\tau}g^{\mu\nu}g_{\lambda\sigma|\mu}g_{\rho\tau|\nu} - \frac{1}{2} g^{\lambda\rho}g^{\sigma\tau}g^{\mu\nu}g_{\lambda\mu|\sigma}g_{\rho\tau|\nu}$$
$$- \frac{1}{4} g^{\mu\nu}g^{\lambda\rho}g_{\lambda\rho|\mu}g^{\sigma\tau}g_{\sigma\tau|\nu} + \frac{1}{2} g^{\mu\nu}g^{\lambda\rho}g_{\lambda\rho|\mu}g^{\sigma\tau}g_{\sigma\nu|\tau}\} . \tag{1-41}$$

One can choose arbitrarily a particular form for L_g and justify it subsequently by the results obtained from it. However, one can try to set up some general criteria to help in making the choice. Let us require (a) that L_g be homogeneous of degree zero in $g_{\mu\nu}$ and its derivatives, so that it is invariant under a constant change of scale of $g_{\mu\nu}$, and (b) that $\gamma^{\mu\nu}$ will be present only in the contractions with all the differentiation indices, so that in the gravitational field equations there will be no free differentiation indices. One finds that

$$L_g = ag^{\lambda\rho}g^{\sigma\tau}g_{\lambda\sigma|\alpha}g_{\rho\tau|\underline{\alpha}} + bg^{\lambda\rho}g_{\lambda\rho|\alpha}g^{\sigma\tau}g_{\sigma\tau|\underline{\alpha}} , \tag{1-42}$$

where a and b are arbitrary constants.

To determine the constants, one considers the case $\gamma_{\mu\nu} = \eta_{\mu\nu}$ with coordinates $(t,x,y,z) = (x^0,x^1,x^2,x^3)$, for a weak, static field produced by a static matter density ρ . The constants in (42) are chosen so that the field equations (23) have a solution of the form

$$ds^2 = (1+2\phi)dt^2 - (1-2\phi)(dx^2+dy^2+dz^2) , \tag{1-43}$$

when ϕ satisfies Eq. (40), so that the theory agrees in this case with general relativity, and hence with Newtonian mechanics. One obtains:

$$a = \frac{1}{4} , \qquad b = -\frac{1}{8} , \tag{1-44}$$

so that

$$\mathcal{L}_g = (-\gamma)^{1/2}\{ \frac{1}{4} g^{\lambda\rho}g^{\sigma\tau}g_{\lambda\sigma|\alpha}g_{\rho\tau|\underline{\alpha}} - \frac{1}{8} A_\alpha A_{\underline{\alpha}} \} , \tag{1-45}$$

where

$$A_\alpha = g^{\lambda\rho}g_{\lambda\rho|\alpha} = 2 \frac{\kappa_{,\alpha}}{\kappa} . \tag{1-46}$$

One finds

$$\kappa K_{\mu\nu} = N_{\mu\nu} - \frac{1}{2} g_{\mu\nu}N , \tag{1-47}$$

with

$$N_{\mu\nu} = \frac{1}{2} g_{\mu\nu|\alpha\underline{\alpha}} - \frac{1}{2} g^{\lambda\sigma}g_{\mu\lambda|\alpha}g_{\nu\sigma|\underline{\alpha}} , \tag{1-48}$$

and

$$N = g^{\lambda\sigma}N_{\lambda\sigma} = \frac{1}{2} A_{\alpha|\underline{\alpha}} . \tag{1-49}$$

The field equations (23) take the form

$$N_{\mu\nu} - \frac{1}{2} g_{\mu\nu}N = - 8\pi\kappa T_{\mu\nu} , \tag{1-50}$$

so that in empty space

$$N_{\mu\nu} = 0 . \tag{1-51}$$

Eqs. (50) can be rewritten

$$N_{\mu\nu} = - 8\pi\kappa(T_{\mu\nu} - \frac{1}{2} g_{\mu\nu}T) . \tag{1-52}$$

If, for simplicity, we take $\gamma_{\mu\nu} = \eta_{\mu\nu}$, then for a given $T_{\mu\nu}$ satisfying (18) , these represent a set of hyperbolic partial differential equations for $g_{\mu\nu}$. If at $t = 0$, $g_{\mu\nu}$ and $g_{\mu\nu,0}$ are given at all points of space, the equations can be integrated to give $g_{\mu\nu}$ for any $t > 0$.

Using (31) one finds that the gravitational energy-momentum density tensor is given by

$$t_{\mu\nu} = t_\mu^{\ \nu} = \frac{1}{32\pi}\{g^{\lambda\rho}g^{\sigma\tau}g_{\lambda\sigma|\mu}g_{\rho\tau|\nu} - \frac{1}{2}A_\mu A_\nu - 2\gamma_{\mu\nu}L_g\} . \qquad (1\text{-}53)$$

We see that it is symmetric, so that, with $\gamma_{\mu\nu} = \eta_{\mu\nu}$, we have angular momentum conservation, at least in the absence of matter.

II. STELLAR STRUCTURE AND COLLAPSE

Let us now apply the bimetric gravitation theory to the case of a static spherically symmetric body the interior of which is characterized by a density ρ and a pressure p .

Using spherical polar coordinates $(x^0,x^1,x^2,x^3) \equiv (t,r,\theta,\Phi)$ with the origin at the center of the body, so that

$$d\sigma^2 = dt^2 - dr^2 - r^2d\theta^2 - r^2\sin^2\theta d\Phi^2 , \qquad (2\text{-}1)$$

one can write[5]

$$ds^2 = e^{2\phi}dt^2 - e^{2\psi}(dr^2+r^2d\theta^2+r^2\sin^2\theta d\Phi^2) \qquad (2\text{-}2)$$

with $\phi = \phi(r)$, $\psi = \psi(r)$. The energy-momentum density tensor $T_\mu^{\ \nu}$ has as non-vanishing components

$$T_0^{\ 0} = \rho(r), \quad T_1^{\ 1} = T_2^{\ 2} = T_3^{\ 3} = -\, p(r) , \qquad (2\text{-}3)$$

and the energy-momentum equation (1-18) gives

$$p' + (\rho + p)\phi' = 0 , \qquad (2\text{-}4)$$

where a prime denotes an r-derivative.

The field equations (1-52) can be written

$$N_\mu^{\ \nu} = -\, 8\pi\kappa(T_\mu^{\ \nu} - \frac{1}{2}\delta^\mu_\nu T) . \qquad (2\text{-}5)$$

Inside the body, i.e., for $r \leq R$, where R is its radius, these take the form

$$\phi'' + \frac{2}{r}\phi' = 4\pi\kappa(\rho+3p) , \qquad (2\text{-}6)$$

$$\psi'' + \frac{2}{r}\psi' = -\, 4\pi\kappa(\rho-p) , \qquad (2\text{-}7)$$

where

$$\kappa = e^{\phi+3\psi} . \tag{2-8}$$

Outside of the body for $r > R$, one has the above equations with $\rho = p = 0$. Hence the solution can be written

$$\phi = -\frac{M}{r} , \qquad \psi = \frac{M'}{r} , \tag{2-9}$$

where the primary mass M and the secondary mass M' are given by

$$M = 4\pi \int_0^R \kappa(\rho+3p) r^2 dr , \qquad M' = 4\pi \int_0^R \kappa(\rho-p) r^2 dr , \tag{2-10}$$

so that $M' \leqslant M$. The primary mass, which has been shown to be equal to the total energy of the body[6], corresponds to the Newtonian mass and determines the motion of a particle at a large distance from the body. The secondary mass influences the deflection of light and of a particle near the body (i.e., having a high velocity).

In the case of a body like the sun where p is negligible compared to ρ (in general relativity units) one can write $M' = M$, and Equation (2) takes the form

$$dS^2 = e^{-2M/r} dt^2 - e^{2M/r}(dr^2+r^2 d\theta^2+r^2 \sin^2\theta d\phi^2) . \tag{2-11}$$

If one compares this with the line element of the isotropic form of the Schwarzschild solution of general relativity, it is found that g_{oo} agrees to second-order terms in M/r, while the other components of $g_{\mu\nu}$ agree to first order terms. It follows therefore that the bimetric theory gives the same agreement with present-day observations as the general relativity theory.

Let us go back to the field equations in the general case. They can be used to determine the structure of a star, for example. However, for this we need to know the equation of state connecting p and ρ. This, together with (4), enables us to express p and ρ in terms of ϕ, so that we are in a position to solve Equations (6) and (7). For this purpose, we impose the boundary conditions:

$$r = 0 : \qquad \phi' = \psi' = 0 , \qquad \phi < 0 , \qquad \psi > 0 , \tag{2-12}$$

$$r = R : \qquad R\phi' = -\phi, \qquad R\psi' = -\psi . \tag{2-13}$$

The conditions (13) are required in order that the solutions of (6) and (7) should join smoothly to the exterior solutions given by (9).

If one takes a reasonable equation of state, and obtains solutions of the field equations, one finds that the primary mass M increases as the central density $\rho(0)$ increases, reaching a maximum value M_{max} for a certain value of this density. As the central density increases further, M undergoes "damped oscillations", approaching a limiting value $M_\infty (< M_{max})$ as $\rho(0) \to \infty$. It appears then that, as the mass of the star M is gradually increased, the star will remain in an equilibrium state until the value M_{max} is reached. For larger values of M no equilibrium is possible.

The same problem of stellar structure can be treated in the framework of the general theory of relativity. Since the procedure is well known, it will not be discussed here. What is important is that the mass of the star M (in general relativity there is no secondary mass M' different from M), as a function of $\rho(0)$ behaves qualitatively in the same way as in the bimetric case, i.e., as $\rho(0)$ increases, M reaches a maximum value M_{max} and then undergoes damped oscillations[7], approaching the limit M_∞ as $\rho(0)$ goes to infinity. However the values of M_{max} and M_∞ are different from those in the bimetric theory.

In order to compare the predictions of the two theories some calculations were carried out together with Joe Rosen on the basis of several simple equations of state.

Let us take as the equation of state

$$p = \frac{1}{3}(\rho - \rho_0) \qquad (\rho \geq \rho_0) , \tag{2-14}$$

where ρ_0 is the density at the surface of the star $(r = R)$. This is not a very realistic equation of state since it gives $dp/d\rho = 1/3$, which is much too large for small values of ρ and may be too small for large values of ρ. However it has the advantage of leading to simple calculations.

Thus with the help of (14), Eq. (4) leads to the relations

$$\rho = \frac{1}{4}\rho_0(3e^{-4\phi-4\alpha} + 1) , \tag{2-15}$$

$$p = \frac{1}{4}\rho_0(e^{-4\phi-4\alpha} - 1) , \tag{2-16}$$

where $\alpha = -\phi(R)$. Putting (15) and (16) into the field equations (6) and (7), one can integrate the latter numerically. The results can be expressed in terms of dimensionless quantities. Let us write

$$m = (8\pi\rho_0)^{1/2} M , \qquad a = (8\pi\rho_0)^{1/2} R . \tag{2-17}$$

Corresponding to M_{max} one has m_{max}. Numerical calculations give $m_{max} = 0.44 (m' = 0.21, a = 0.81)$.

Calculations carried out according to general relativity for the same equation of state give $m_{max} = 0.26$ $(a = 0.96)$.

The choice of ρ_o is arbitrary. Let us take $\rho_o = 2\times10^{14}$ gm/cm^3 . Then one gets according to general relativity theory $M_{max} = 2.9\ M_\odot$ (R = 16 km). On the other hand, the bimetric theory gives $M_{max} \cong 4.9\ M_\odot$ ($M' = 2.3\ M_\odot$, R = 13 km).

We see that the ratio of M_{max} in the bimetric theory to that in general relativity is about 1.7, and this is independent of the choice of ρ_o .

For comparison, let us now take, instead of the equation of state (14),

$$p = \rho - \rho_o \, . \tag{2-18}$$

This represents the most extreme equation of state permitted by relativity theory. For the acoustic velocity is given by $(dp/d\rho)^{1/2}$, and according to the theory of relativity this velocity cannot exceed that of light, which is equal to unity (in our units); here we have the limiting case: $dp/d\rho = 1$. In this case Eq. (4) leads to the relations

$$\rho = \rho_o e^{-2\phi-2\alpha} \, , \tag{2-19}$$

$$p = \rho_o(e^{-2\phi-2\alpha} - e^{-\phi-\alpha}) \, . \tag{2-20}$$

If these are substituted into the field equations (6) and (7), numerical integration gives the results $m_{max} = 7.3$ ($m' = 0.72$, $a = 3.3$).

Corresponding calculations in general relativity give $m_{max} = 0.43$ $(a = 1.2)$.

Let us now take $\rho_o = 4\times10^{14}$ gm/cm^3 (in order that in the present case, as in the previous one, M_{max} according to general relativity should be of the order of 2-3 solar masses). It is found that general relativity gives $M_{max} = 3.3\ M_\odot$ (R = 14 km). On the other hand the bimetric theory gives $M_{max} \cong 57\ M_\odot$ ($M' = 5.7\ M_\odot$, R = 38 km).

In the present case the ratio of M_{max} in the bimetric theory to that of general relativity is about 17 (for any value of ρ_o) .

It is noteworthy that the maximum mass is larger in the bimetric theory than in general relativity. However to make this point more convincing, one should use a more realistic equation of state than those given previously. Let us now take as the equation of state

$$p = \rho - \rho_o + (2\lambda-1)\rho_o\{1 - [1 + \frac{1}{\lambda}(\frac{\rho}{\rho_o} - 1)]^{1/2}\} . \qquad (2\text{-}21)$$

This equation has the property that, as $\rho \to \infty$, $dp/d\rho \to 1$. However it is characterized by two parameters, ρ_o , the density for which $p = 0$, and $\lambda(> 1/2)$ such that

$$\left(\frac{dp}{d\rho}\right)_{\rho_o} = 1/2\lambda(< 1) . \qquad (2\text{-}22)$$

In this case Eq. (4) leads to

$$\rho = \rho_o[\lambda e^{-2\phi-2\alpha} + 1 - \lambda] , \qquad (2\text{-}23)$$

$$p = \rho_o[\lambda e^{-2\phi-2\alpha} + (1-2\lambda)e^{-\phi-\alpha} + \lambda - 1] . \qquad (2\text{-}24)$$

It is necessary to choose the parameters. If one takes (arbitrarily) $\rho_o = 5\times10^{13}$ gm/cm^3, one finds that taking $\lambda = 40$ gives a reasonable fit to existing equations of state[8] based on the properties of nuclear matter.

This choice of parameters gives according to general relativity the results: $M_{max} = 1.5\ M_\odot$ (R = 10 km). The bimetric theory gives $M_{max} = 8.1\ M_\odot$ ($M' = 1.0\ M_\odot$, R = 8.5 km).

In this case the ratio of M_{max} in the two theories is about 5.6. Evidently the large ratio of the masses here (and in the previous case) is associated with the fact that the central density of the star according to the bimetric theory is much larger than that according to general relativity. In the present case one finds that for $M = M_{max}$ one has, according to general relativity, $\rho(0)/\rho_o = 70$, while according to bimetric relativity $\rho(0)/\rho_o = 1.8 \times 10^5$.

The fact that M_{max} is so much larger in bimetric gravitation than in general relativity allows one to interpret the dark compact celestial objects observed up to now (such as Cygnus X-1) as being neutron stars rather than black holes.

Since, in the bimetric theory, there is a maximum mass for the star (larger than that in general relativity) beyond which no equilibrium states can exist, it follows that, for a mass greater than M_{max} , the star must undergo collapse. However, one would expect that the nature of the collapse would be quite different from that in general relativity because of the apparent absence of black holes in the present theory.

In the case of gravitational collapse according to general relativity theory, one can get a qualitative picture of what happens by considering the case of zero pressure. Then each particle moves on a geodesic. A particle on the surface of the star moves on a geodesic of the Schwarzschild solution. The time required for the particle to reach the Schwarzschild sphere $(r = 2M)$, for an observer moving with the particle (i.e., the proper time of the particle), is finite. Once the surface of the star has entered the Schwarzschild sphere, one has a black hole. It is believed[9] that the matter of the star and its radiation are then permanently trapped in the black hole.

However, from the standpoint of a distant observer at rest in the Schwarzschild coordinate system, the time (i.e., the coordinate time) required for the stellar surface to contract to the Schwarzschild sphere is infinite, so that he will never see the star in the state of a black hole. As the surface approaches the Schwarzschild sphere, the light emitted will be red-shifted more and more, and the star will look darker and darker. Eventually the star will appear to the observer as a dark body of about the size of the Schwarzschild sphere.

From the standpoint of the bimetric theory, in the case of zero pressure one can follow the motion of an inward moving particle inside the star until it reaches the center of the star. A rough estimate indicates that the time required to reach the center is finite, both from the standpoint of a co-moving observer and from that of a distant observer at rest in the coordinate system. This particle, on reaching the center, could begin to move outward on the opposite side. Hence there is the possibility that the inner part of the star will first contract and then expand. If the star were composed of an ideal fluid, it could oscillate. In the case of real matter the behavior could be more complicated: the expanding inner part could collide with the contracting outer part and eject part of the mass. Considerable energy emission in various forms could take place.

As a rough way of comparing the process of collapse in the two theories one can examine a very simple cosmological model, a homogeneous isotropic universe, spatially flat and containing only pressureless matter characterized by a density $\rho = \rho(t)$.

Let us consider first the general theory of relativity, and let us take the line element in the form

$$ds^2 = dt^2 - e^{2\psi}(dx^2+dy^2+dz^2) , \qquad (2\text{-}25)$$

with $\psi = \psi(t)$. The Einstein field equations then have as their solution

$$e^{\psi} = A(t-t_0)^{2/3} , \tag{2-26}$$

with A (> 0) and t_0 arbitrary constants, and

$$\rho = \frac{1}{6\pi(t-t_0)^2} . \tag{2-27}$$

We see from (26) that for $t < t_0$, e^{ψ} decreases as t increases, so that the universe is contracting, while for $t > t_0$, e^{ψ} grows with t , so that the universe is expanding. However at $t = t_0$ there is a singularity ($e^{\psi} = 0, \rho = \infty$) , and this precludes any causal relation between the contraction and the expansion. That is, Eq. (26) describes two unrelated processes, a contraction ending in a singularity and an expansion beginning from a singularity ("the big bang"). The first one can be regarded as analogous to the gravitational collapse of a star.

Now let us consider the same cosmological model in bimetric gravitation theory. We take as line element

$$ds^2 = e^{2\phi}dt^2 - e^{2\psi}(dx^2+dy^2+dz^2) , \tag{2-28}$$

with $\phi = \phi(t)$, $\psi = \psi(t)$, and we choose $\gamma_{\mu\nu} = \eta_{\mu\nu}$. The field equations (5) are found to have the form (a dot denoting a t-derivative)

$$\ddot{\phi} = -4\pi\rho e^{\phi+3\psi} , \tag{2-29}$$

$$\ddot{\psi} = 4\pi\rho e^{\phi+3\psi} , \tag{2-30}$$

so that

$$\ddot{\phi} + \ddot{\psi} = 0 . \tag{2-31}$$

As in general relativity, one obtains from Eq. (1-18) (for $p = 0$) the relation

$$\rho = \rho_0 e^{-3\psi} \qquad (\rho_0 = \text{const.}) . \tag{2-32}$$

Eq. (29) can then be integrated to give

$$e^{\phi} = \frac{\alpha^2}{2\pi\rho_0 \cosh^2\alpha(t-t_0)} \qquad (\alpha, t_0 = \text{const.}). \tag{2-33}$$

From (31) one gets

$$\psi = - \phi + At + B , \tag{2-34}$$

with A,B arbitrary constants.

For the present purpose let us take $A = B = 0$. Then

$$e^{\psi} = \frac{2\pi\rho_o}{\alpha^2}\cosh^2\alpha(t-t_o) \; . \tag{2-35}$$

We see that here we also have a contraction for $t < t_o$ and an expansion for $t > t_o$. In order to compare with the general relativity solution let us take a new time variable

$$T = \int_{t_o}^{t} e^{\phi}dt = \frac{\alpha}{2\pi\rho_o}\tanh\alpha(t-t_o) \; . \tag{2-36}$$

Then

$$ds^2 = dT^2 - e^{2\psi}(dx^2+dy^2+dz^2) \; , \tag{2-37}$$

where, as a function of T,

$$e^{\psi} = \frac{1}{2\pi\rho_o(T_o{}^2-T^2)} \; , \tag{2-38}$$

with $T_o = \alpha/2\pi\rho_o$. From (32) we can write

$$\rho = (2\pi)^3\rho_o{}^4(T_o{}^2-T^2)^3 \; . \tag{2-39}$$

We see that for $-T_o < T < 0$ there is a contraction and for $0 < T < T_o$ an expansion. However at $T = 0$ there is no singularity, so that the contraction goes over smoothly into the expansion.

This suggests by analogy that, in the bimetric theory, a collapsing star will go down to some small, finite size and then expand. Further calculations for the case of the star are required.

III. GRAVITATIONAL WAVES

It will be recalled that the Einstein field equations of the general theory of relativity have the form

$$G_{\mu\nu} \equiv R_{\mu\nu} - \frac{1}{2}g_{\mu\nu}R = -8\pi T_{\mu\nu} \; , \tag{3-1}$$

where $T_{\mu\nu}$ is the energy-momentum density tensor of the matter generating the gravitational field. The case of a weak field can be characterized by the fact that, in a suitable coordinate system, one can write

$$g_{\mu\nu} = \eta_{\mu\nu} + h_{\mu\nu} , \tag{3-2}$$

where $\eta_{\mu\nu}$ is the Galilean metric tensor, and the components $h_{\mu\nu}$ are small quantities. If we define

$$\bar{h}_{\mu\nu} \equiv h_{\mu\nu} - \frac{1}{2} \eta_{\mu\nu} h , \tag{3-3}$$

with

$$h \equiv h_{\alpha\underline{\alpha}} , \tag{3-4}$$

an underlined index being raised (or lowered) with $\eta_{\mu\nu}$, the field equations in the first-order approximation take the form

$$\frac{1}{2} (\bar{h}_{\mu\nu,\alpha\underline{\alpha}} - \bar{h}_{\mu\alpha,\underline{\alpha}\nu} - \bar{h}_{\nu\alpha,\underline{\alpha}\mu} + \eta_{\mu\nu} \bar{h}_{\alpha\beta,\underline{\alpha\beta}}) = - 8\pi T_{\mu\nu} . \tag{3-5}$$

Let us now consider an infinitesimal coordinate transformation

$$x^{\mu} = x'^{\mu} + \xi^{\mu} , \tag{3-6}$$

where ξ^{μ} are small quantities. In the primed coordinate system one has

$$g'_{\mu\nu} = g_{\alpha\beta} \frac{\partial x^{\alpha}}{\partial x'^{\mu}} \frac{\partial x^{\beta}}{\partial x'^{\nu}} , \tag{3-7}$$

and if one writes, corresponding to (2) ,

$$g'_{\mu\nu} = \eta_{\mu\nu} + h'_{\mu\nu} , \tag{3-8}$$

one finds that in first order

$$h'_{\mu\nu} = h_{\mu\nu} + \xi^{\underline{\mu}}{}_{,\nu} + \xi^{\underline{\nu}}{}_{,\mu} . \tag{3-9}$$

Since ξ^{μ} are arbitrary we can choose them to satisfy four conditions in the primed system. Dropping the primes, let us take the conditions in the form

$$\bar{h}_{\mu\nu,\underline{\nu}} = 0 . \tag{3-10}$$

Eq. (5) then becomes

$$\bar{h}_{\mu\nu,\alpha\underline{\alpha}} = - 16\pi T_{\mu\nu} . \tag{3-11}$$

The field equations now consist of Eqs. (10) and (11). Since the matter tensor satisfies the equation

$$T_{\mu}{}^{\nu}{}_{;\nu} = 0 , \tag{3-12}$$

which, in first order, can be written

$$T_{\mu\underline{\nu},\nu} = 0 \;, \tag{3-13}$$

Eq. (11) gives

$$\bar{h}_{\mu\underline{\nu},\nu\alpha\underline{\alpha}} \equiv \Box\bar{h}_{\mu\nu,\underline{\nu}} = 0 \;, \tag{3-14}$$

so that Eqs. (10) and (11) are consistent. For a suitable matter tensor $T_{\mu\nu}$ the field equations have solutions describing gravitational waves(10). As a consequence of Eq. (10) the gravitational waves are found to be transverse.

Now let us compare the situation in the bimetric theory. Here the field equations are given by (1-50). The case of a weak field is described by

$$g_{\mu\nu} = \gamma_{\mu\nu} + h_{\mu\nu} \;, \tag{3-15}$$

and if we write

$$\bar{h}_{\mu\nu} = h_{\mu\nu} - \frac{1}{2}\gamma_{\mu\nu}h \qquad (h \equiv h_{\alpha\underline{\alpha}}) \;, \tag{3-16}$$

then the field equations take the form in the first-order approximation

$$\bar{h}_{\mu\nu|\alpha\underline{\alpha}} = -\,16\pi T_{\mu\nu} \;. \tag{3-17}$$

If we take $\gamma_{\mu\nu} = \eta_{\mu\nu}$, which is always possible, then we have

$$\bar{h}_{\mu\nu,\alpha\underline{\alpha}} = -\,16\pi T_{\mu\nu} \;. \tag{3-18}$$

The field generated by the matter is given by the "retarded potential" solution of Eq. (18), i.e.

$$\bar{h}_{\mu\nu}(x,y,z,t) = -\,4\int \frac{1}{r}\,T_{\mu\nu}(x',y',z',t-r)\,d^3x' \;, \tag{3-19}$$

with $(t,x,y,z) = (x^0,x^1,x^2,x^3)$ and $r = \{\Sigma_{k=1}^{3}(x'^k - x^k)^2\}^{1/2}$.

In the present case the matter satisfies Eq. (12) and hence (13). From (19) together with (13) a calculation gives

$$\bar{h}_{\mu\nu,\underline{\nu}} = 0 \;. \tag{3-20}$$

Thus we finally have Eqs. (18) and (20), which are the same as the weak-field equations of general relativity, Eqs. (11) and (10), although the steps that lead to them are different in the two theories. Hence one can conclude that in the weak-field case the gravitational waves emitted by a physical system according to the bimetric theory will be the same as those emitted according to general relativity. Furthermore, since in both theories the motion of a particle in a gravitational field is described by the geodesic equation, it follows that the detection of these waves will be described in the bimetric theory in the same way as in general relativity[10].

The bimetric formalism is somewhat different from that of general relativity, if we take $\gamma_{\mu\nu}$ different from $\eta_{\mu\nu}$. Since one can arrive at any flat-space $\gamma_{\mu\nu}$ by a coordinate transformation from $\eta_{\mu\nu}$, we see that in the general case, one can replace (20) by

$$\bar{h}_{\mu\nu|\underline{\nu}} = 0 \, , \tag{3-21}$$

and the field equations are given by (17) and (21). If the $\gamma_{\mu\nu}$ field is also taken to be weak, i.e.

$$\gamma_{\mu\nu} = \eta_{\mu\nu} + \theta_{\mu\nu} \, , \tag{3-22}$$

where $\theta_{\mu\nu}$ are small, then (15) can be written

$$g_{\mu\nu} = \eta_{\mu\nu} + \psi_{\mu\nu} \, , \tag{3-23}$$

where

$$\psi_{\mu\nu} = h_{\mu\nu} + \theta_{\mu\nu} \, . \tag{3-24}$$

If we now carry out an infinitesimal coordinate transformation, both $\psi_{\mu\nu}$ and $\theta_{\mu\nu}$ are transformed, but $h_{\mu\nu}$ as given by (24) remains unchanged. The field equations (17) and (21) in first order again have the form (18) and (20), but $h_{\mu\nu}$ as defined by (15) is "gauge invariant", i.e., invariant under an infinitesimal coordinate transformation.

Let us now consider the case of weak gravitational waves on a background of a strong gravitational field which is slowly varying in space and time. Let us take the bimetric field equations for empty space:

$$N_{\mu\nu} = 0 \, . \tag{3-25}$$

Denoting the metric tensor describing the background field (the field in the absence of gravitational waves) by $\overset{\circ}{g}_{\mu\nu}$, let us write

$$g_{\mu\nu} = \overset{\circ}{g}_{\mu\nu} + h_{\mu\nu} \, , \qquad g^{\mu\nu} = \overset{\circ}{g}{}^{\mu\nu} - h^{\mu\nu} \, , \tag{3-26}$$

where $h_{\mu\nu}$, describing the gravitational waves, is assumed to be small, and indices are raised and lowered with $\overset{\circ}{g}_{\mu\nu}$. Correspondingly let us write

$$N_{\mu\nu} = \overset{0}{N}_{\mu\nu} + \overset{1}{N}_{\mu\nu} . \tag{3-27}$$

In the absence of the waves we have as field equations, by (1-48),

$$\overset{0}{N}_{\mu\nu} \equiv \frac{1}{2} \overset{\circ}{g}_{\mu\nu|\alpha\underline{\alpha}} - \frac{1}{2} \overset{\circ}{g}^{\lambda\rho} \overset{\circ}{g}_{\lambda\mu|\alpha} \overset{\circ}{g}_{\rho\nu|\underline{\alpha}} = 0 . \tag{3-28}$$

With waves present we have Eq. (25), which with the help of (27) and (28) gives in first order

$$\overset{1}{N}_{\mu\nu} \equiv \frac{1}{2} h_{\mu\nu|\alpha\underline{\alpha}} + \frac{1}{2} h^{\lambda\rho} \overset{\circ}{g}_{\lambda\mu|\alpha} \overset{\circ}{g}_{\rho\nu|\underline{\alpha}}$$

$$- \frac{1}{2} \overset{\circ}{g}^{\lambda\rho} (h_{\lambda\mu|\alpha} \overset{\circ}{g}_{\rho\nu|\underline{\alpha}} + \overset{\circ}{g}_{\lambda\mu|\alpha} h_{\rho\nu|\underline{\alpha}}) = 0 . \tag{3-29}$$

However, with the assumption that the background field is slowly varying (compared to the gravitational wave) we can neglect all the terms compared to the first. Hence (29) gives

$$\bar{h}_{\mu\nu|\alpha\underline{\alpha}} = 0 . \tag{3-30}$$

We see that the propagation of the gravitational wave is determined by the flat-space $\gamma_{\mu\nu}$ and not by the background $g_{\mu\nu}$. On the other hand, the propagation of light is determined by $g_{\mu\nu}$ (or, in the present case, $\overset{\circ}{g}_{\mu\nu}$) , so that in general the gravitation cone is different from the light cone. This is in contrast to the situation in the general relativity theory where the two cones are identical, and it leads to some predictions differing from those of general relativity:

1. In the case of waves going through a nearly static gravitational field, the light velocity will be smaller than the gravitation velocity. This can be seen by taking $\gamma_{\mu\nu} = \eta_{\mu\nu}$ and noting that for a weak static gravitational field

$$g_{00} = 1 + 2\phi , \quad g_{jk} = - \delta_{jk}(1-2\phi) \quad (j,k = 1,2,3) , \tag{3-31}$$

where $\phi(< 0)$ is the Newtonian potential, so that the gravitation velocity is unity, while the light velocity is $1 + 2\phi < 1$.

It follows that if an event takes place (such as the explosion of a supernova) which generates both electromagnetic and gravitational radiation, the gravitational wave should reach us before the electromagnetic wave, the interval between them depending on the distance from the event to us and on the gravitational fields in the intervening space[11] .

2. A beam of gravitational radiation passing close to a star will not be deflected. According to general relativity it will be deflected in the same way as a light beam.

Let us now consider the general case of strong fields. Instead of the linear approximation equations (17) we now have the exact equations (1-50). The question arises: what, if any, equation is to replace the divergence condition, Eq. (21) ? In general relativity, instead of the divergence condition of Eq. (10) in the weak field case one often takes the De Donder condition

$$(\sqrt{-g}\, g^{\mu\nu})_{,\nu} = 0 . \qquad (3\text{-}32)$$

The left-hand member is not a vector. In the bimetric theory there is a vector

$$\kappa S^{\mu} = (\kappa g^{\mu\nu})_{|\nu} , \qquad (3\text{-}33)$$

which, for $\gamma_{\mu\nu} = \eta_{\mu\nu}$, is equal to the left side of (32) .

It is natural to try to impose the condition

$$S^{\mu} = 0 , \qquad (3\text{-}34)$$

since this goes over into Eq. (21) in the weak-field case. However, it is found that in the general case Eq. (34) leads to an inconsistency. One finds that the energy-momentum relation, Eq. (12) , can be rewritten with the help of the field equations in the form

$$S_{\mu|\alpha\underline{\alpha}} = (g^{\alpha\lambda}{}_{|\alpha} g_{\mu\lambda|\underline{\nu}} - g^{\alpha\lambda}{}_{|\underline{\nu}} g_{\mu\lambda|\alpha} + 16\pi\kappa T_{\mu}{}^{\nu})_{|\nu} , \qquad (3\text{-}35)$$

with $S_{\mu} = g_{\mu\lambda} S^{\lambda}$. In general the right side of Eq. (35) is different from zero, and one cannot impose the condition (34) . It is not clear whether, in the general case, no conditions (other than the field equations) can be imposed, or whether some condition different from (34) is possible. Further work is needed on this point.

An exception arises in the case $\gamma_{\mu\nu} = \eta_{\mu\nu}$, for empty space ($T_{\mu}{}^{\nu} = 0$) , if $g_{\mu\nu}$ is diagonal, for then the right side of Eq. (35) vanishes and Eq. (34) can be imposed. To be sure, one does not know whether this condition is consistent with the equations of motion of the source of the wave, but let us ignore this point. Consider for example, the case of a plane wave propagating in the positive direction of the x-axis. The field equations are satisfied if one takes

$$g_{\mu\nu} = g_{\mu\nu}(t-x) . \qquad (3\text{-}36)$$

Let us take

$$g_{\mu\nu} = \eta_{\mu\nu} e^{2\psi_\mu (t-x)} \tag{3-37}$$

Imposing Eq. (34) , one finds that, to within an additive constant,

$$\psi_o = \psi_1 \quad , \tag{3-38}$$

$$\psi_2 = -\psi_3 \ . \tag{3-39}$$

If one takes, $\psi_2 = \psi_3 = 0$, one obtains a solution for which $R^\lambda{}_{\mu\nu\sigma} = 0$ and which can therefore be transformed into $g_{\mu\nu} = \eta_{\mu\nu}$ by a coordinate transformation. On the other hand, for $\psi_o = \psi_1 = 0$, the line element, which can be written (with $\psi = \psi_2$)

$$ds^2 = dt^2 - dx^2 - e^{2\psi}dy^2 - e^{-2\psi}dz^2 \ , \tag{3-40}$$

describes a plane transverse gravitational wave. The gravitational energy-momentum density tensor is found to have components

$$t_{oo} = t_o{}^1 = t_{11} = \frac{1}{4\pi} (\psi_{,o})^2 \ . \tag{3-41}$$

A rotation of 45° about the x-axis puts the line element into the form

$$ds^2 = dt^2 - dx^2 - \cosh 2\psi(dy^2+dz^2) - 2\sinh 2\psi dydz \ . \tag{3-42}$$

REFERENCES

[1] N. Rosen, Phys. Rev. 57, 147 (1940).

[2] A. Papapetrou, Proc. Roy. Irish Acad. 52A, 11 (1948).

[3] N. Rosen, Ann. Phys. (N.Y.) 22, 1 (1963).

[4] N. Rosen, GRG Jour. 4, 435 (1973).

[5] N. Rosen, Ann. Phys. (N.Y.) 84, 455 (1974).

[6] I. Goldman, GRG Jour. (to appear).

[7] B.K. Harrison, Phys. Rev. 137B, 1644 (1965).

[8] V. Canuto, Ann. Rev. Astron. Ap. 12, 167 (1974).

[9] C.W. Misner, K.S. Thorne and J.A. Wheeler, Gravitation (San Francisco, 1973), p. 846.

[10] J. Weber, General Relativity and Gravitational Waves (New York, 1961).

[11] This was pointed out by Dr. T. Damour.

SPECULATIVE REMARKS ON PHYSICS IN GENERAL AND RELATIVITY IN PARTICULAR

André Mercier

Institut für theoretische Physik

Universität Bern

SUMMARY

Perhaps gravitation is, strictly speaking, not a "field" of the same nature as the fields which propagate interaction, and hence not quantizable, at least not in the usual sense. Signals are perhaps not transmitted by gravitation. Space-time is the generalization of (Newtonian) time, not of space. But (already Newtonian) Time is not a measurable quantity like all other properties of systems, so space-time is that on which all physical phenomena depend, and since space-time renders gravitation by means of its particular Riemannian structure, the very concept of gravitational waves or signals and that of gravitons rest perhaps on a misunderstanding, which may call for a radical reinterpretation of Relativity Physics.

1. The following dilemma arose and will presumably not be solved before the end of the nineteen seventies:

On the one hand, lots of good physicists are engaged in the observation of what they call "gravitational signals". Even if no one can say that such signals have been registered with certainty, the arguments in favour of a satisfactory observation in due time seem extremely convincing.

The same text has been published in DIALECTICS AND HUMANISM, No. 3/1975, p. 125 under the heading: Philosophical Problems of Physics

On the other hand, General Relativity strictly speaking accounts neither for a clear notion of energy, nor for a rigorous propagation of waves. Even the analogy between gravitation and electromagnetism is utterly criticable.

What is it then we are talking about when we call it gravitational signal?

2. There are periods in the development of physics when the received views do not quite fit into the correct description of Nature. Perhaps we are in such a period, because we are still used to interpret the observations we make by means of concepts and representations proceeding from a philosophical background elaborated by a generation of physicists who were faced with problems different from ours and issued from difficulties that we are not expected to overcome any more.

So, maybe it is not all too "mal placé" if I permit myself to behave like a naughty boy and try to contend many a received view and see what happens. Probably the reader will find me crazy, but Niels Bohr was once of the opinion that one never may be crazy enough, except very exceptionally.

For Bohr knew that a time would come when all our opinions would have to be in their turn completely revised - as completely indeed as he himself and ALBERT EINSTEIN had revolutionized the current *Weltbild*.

3. 1974 in Erice, I gave three seminar lectures on Cosmology from an epistemological view point[1]. In these Erice lectures, I started from the fact that cosmology intends to be the theory of the world as a whole and that according to EINSTEIN its only truly universal property is gravitation.

If we then make a geometrical picture of the world, the fundamental structure of its geometry must allow for our interpretation as gravitation.

I have also given the reason why I think that gravitation is not an interaction like the "other" interactions. One problem arises here specifically. When fields are quantized, their particles

[1]These lectures are in print under the title: "Epistemological Questions concerning Cosmology and Gravitation", GRG Journal, Vol. 6, No. 5 (1975).

are found to be the bearers or transmitters - one might say even the propagators - of the interactions between other fields, and a systematic classification can be drawn from their consideration. For that to be successful, however, the quantization must assume the linearity of the field theory and imply consequently a superposition principle for the corresponding waves.

4. In a certain approximation, there are gravitational waves. At least this is a possible interpretation of the formal aspect of the Theory. Hence, in so far as gravitation is interpreted as a field in the usual sense, there should - people say - be gravitons by straightforward quantization within the said approximation. Since gravitation is the most general, i.e. truly universal property of matter, gravitons would then be the most fundamental particles.

However, apart from their being practically unobservable, such gravitons are very much different from the particles of other fields; the only similar property they have to other particles is a spin: a spin of 2. But a spin of *two* seems to me highly improbable for the particle which ought to be *the* most fundamental of all particles, or rather: the fact that will-be gravitons have a spin of 2 is only the consequence of an artificial construction of those particles, but their construction is unrealistic, at most formal, and perhaps we should do better to forget about them, i.e. to refrain from quantizing an artificially linearized, assumed field of gravitation. This would mean that the so-called field of gravitation is not of the same nature as other fields like the electromagnetic one and that all the analogies used in reasonings are no other than formal analysis.

Another reason to doubt that gravitons exist is that if they did exist, they would have, on the ground of EINSTEIN's non-approximate, i.e. complete equations, to themselves emit and absorb gravitons, precisely because of the non-linearity of these equations, and I find this as hard to believe as it was difficult for NEWTON's contemporaries to believe in the immediate, infinitely quick propagation of the effects of gravitation through arbitrarily long distances in space.

5. One has been much influenced by the idea that physics is fundamentally a quantum physics. However, the original part of physics which was submitted to quantization was electromagnetism alone, first by means of PLANCK's oscillators leading to EINSTEIN's photons, as well as in the form of old quantum theory, then in the form of classical quantum theory and, finally, as quantum electrodynamics. Only after this, did quantum field theory actually arise

in the mind of people as *the* general form of physics and as yielding the possible definition of particles as quanta of the various fields. This situation was similar to that of Newtonian mechanics becoming the accepted form of physics after having been conceived by NEWTON originally as a theory of gravitation only.

Indeed, the success of Newtonian and post-Newtonian mechanics was such that physicists came to believe that every physical theory has to be elaborated in the same form. The only exception to that belief was thermodynamics, and there is no doubt that CARNOT, who must be considered as the founder of thermodynamics, even if he was partly inspired by comparisons between the flow of heat and the flow of fluids, did realise that a completely different approach to Reality was necessary if he was to explain successfully the dissipation of energy (though he did not use this modern terminology). And statistical mechanics did not reduce thermodynamics to mechanics, it only showed how to introduce micro-mechanisms into the description of (macroscopic) engines.

Newtonian mechanics was an atomic view on reality, since it started with point masses. Only by the consideration of EULER and LAGRANGE on fluids as quasi-continuous assemblies of particles - or rather conversely the quasi-atomic analysis of fluids as distributions of mass, the infinitesimal elements of which behave like point masses - was it possible to develop a mechanics of continuous media similar to what later has been called field. However, these media were not endowed with independent properties that would make out of them entities of their own, i.e. proper fields. The only necessary supplementary postulate was the assumption that their stress tensor is symmetric, else the conservation theorems cannot be demonstrated. (Originally, conservation principles were admitted and the symmetry of the stresses deduced). Only when FARADAY conceived of the magnetic and electric fields as independent physical realities was the concept of a field made available to physics.

6. If we now look at Relativity theory, we notice that two view points may be taken, as I have repeatedly explained. Either things are defined as world-lines, or there are no objects but fields. I am tempted to say that there is a kind of complementarity between world-lines and fields, and that consequently it does not make good sense to consider world-lines embedded in a field.

6.a In the world-line picture, as I have explained in the Landé Jubilee Volume, the objects of physics are subject to what I have called super-determinism, for since an object is identical with its world-line and each object simply is as it is by its own right so to speak, nobody disposes of that object, for its world-line is a necessary part of the world which itself consists of nothing but invariable world-lines and it makes no sense to pretend that we are able to let any world-line go through a particular point of space-time, which is an event, chosen in the sense of an initial condition, for we do not have any power or disposal over world events; whereas in Newtonian mechanics, we are allowed to choose, - we must even give, - the initial conditions, which is a certain and precise, though limited, freedom of the will, the limitations of which lie in the second order of the differential equation of Newtonian dynamics. There is no such free will of Newtonian kind in Relativity Theory, unless it be made completely artifical. LAPLACE, conversely, made Newtownian mechanics artificially completely determined, but Laplacian absolute determinism does not follow from Newtonian mechanics; it is valid only under the further assumptions, first that all bodies without exception (i.e. including people like physicists) evolve according to Newtonian mechanics only and second that the world was created once for all, whereas the relativistic picture of world-lines includes the invariability of the set of all world-lines and is therefore the picture of a huge four-dimensional fossil that has always been a fossil. This picture does not fit very well in our physics. Indeed, does not our physics rest on the assumption that observers receive signals if they choose to stand at the right place with the right instruments to capture them ? Therefore, (6.b), the other alternative, i.e. the picture of fields is presumably the more suitable one. This is exactly what relativistic electrodynamics revealed in its special relativistic form. The interesting thing is that after its generalization, Relativity Theory did not attribute any more to the electromagnetic field the nature of the fundamental field. Actually, this is not so very astonishing, for the fundamental field (I call it still a field for want of another better name) should not be one which transmits a particular interaction, especially if other fields are at hand in the usual sense, capable of a quantization and of the attribution of various kinds of charge. The fundamental field is gravitation and has no specific property other than that of being the fundamental field. As a matter of fact, the only features of that field are first its admitting of null-lines and hence null-cones locally, and second its being the consensus of all matter.

7. At the time of NEWTON's and during a couple of centuries since observed signals were always and on a world-wide scale only light signals, or eventually sound where there was an elastic medium, but that latter case is of minor importance and even not relevant, since in the atomic picture there is only vacuum between

the point masses. In principle, all celestial bodies would have to immediately change their position and momentum if I chose to move my hand even just a little bit, and this does not provide for any practicable signalisation. On the contrary it produces a permanent disturbance.

Indeed, are signals which are instantly transmitted real signals ? Can they be anything else than the built-in synchronism of ideal clocks devoid of reality at the various places of a Newtonian world, i.e. the material aspect of the assumption of a universal and absolute time ? So the question arose, whether there can be gravitational signals.

This question is not only a question of Newtonian mechanics where it always sounded more or less meaningless or unbelievable, but also one of General Relativity. In General Relativity physics, such signals would of course travel with the limit velocity c, which was found very satisfactory and helped overcome the old difficulty. In that respect, the limit c needs of course not be just the velocity of light; on the contrary, it is primarily the limit velocity which defines the null-lines, i.e. - experimentalists might say - the velocity of any signals properly speaking, though I personally would rather not look at it in that experimental perspective, but preferably as the value of a universal constant to be necessarily used as natural unit of all measurements of velocities, which is important enough. From that view point, it does not appear necessary that there should be gravitational signals, especially if gravitation, - as I have explained in the Erice lectures, - is nothing but the material aspect of time, for, as I argued, space-time is not to be considered as a space, but as time itself endowed with a richer structure than Newtonian time. In that case, the so-called gravitational field would not be a field proper, i.e. a function of space-time, but space-time itself, whereas proper fields would arise from the specific behaviour of matter due to certain of its properties, and only such specific fields could be used for the transmission of signals, because they would admit of some quantization and eventually transport well-defined amounts of energy in some ordered shape, which are the conditions for signals first to be paid for, second to be understandable.

8. So the question should be asked at this point whether assumed gravitational signals would only be noticeable or rather make sense if and only if some sort of quantization of gravitation would succeed. Yet serious doubts may arise in that respect, for if that quantization succeeds, the next question is: would it be comparable with ordinary field quantization ? Precisely this can be doubted,

for apart from the fact that it should be done with full respect due to the nonlinearity of the field equations, a further difficulty seems to arise from the fact that the very gravitational picture of modern physics is not really a local picture, but a global one (though EINSTEIN's original application of differential geometry was only local). The necessity of a global picture was realized, I believe, by various scientists after the second world war, e.g. by MICHEL KERVAIRE and myself when we prepared together the Golden Jubilee of the Theory of Relativity in 1955 and discussed the matter then. The reason I would give for that is, that Relativity physics really is to be a cosmology, not a local engineering-like attempt at getting at individual things like test particles or singularities, which are both rather unphysical, though they do work very well in cases like the motion of the perihelion or the description of black holes. The fact that Relativity theory is a cosmolgoy, however, makes, it seems to me, a quantization in BOHR's sense illusory, for there is no observer outside the cosmos and hence no quantum system to which a ψ-function can be attributed in such a way that an observation would fix its state at the event of the observation, and this seems to me in a way pretty evident: How could we indeed look at the cosmos from outside since we are inside ?

9. At the 1955 Jubilee Conference, the possibilities and difficulties of quantizing gravitation were already under discussion, especially in a paper by EUGENE WIGNER. Twenty years of effort have not led to much progress. For the problem solved in the meantime, mainly by LICHNEROVICZ, viz. to express quantization recipes in a generally covariant manner, is not the proper question to be really put. The problem is to quantize somehow the seemingly unquantizable gravitation. At a time, much attention was given by DIRAC, BERGMANN and others, to the homogeneous canonical formalism because, fundamentally, SCHROEDINGER's equation is nothing but the analogue, by the principle of correspondence, of the so-called accessory condition of that homogeneous formalism. However, could such consideration lead to more than the unnecessary quantization of the state of a non-realistic test particle ?

According to recent attempts at a global approach by means of symplectic mechanics using space and co-tangent space, it seems that some sort of quantization may succeed. But I wonder whether it will then be one similar to the quantization yielding a spectrum of possible states in which the system under consideration can be found by observation, or whether it will rather be the imposition upon the world of something like an inherently discontinuous structure endowed with minimum cell measure of a lattice or the like, comparable to cells of measure h^n in phase-space. This would be very different from field quantization, and at the same time very nice in two or three respects, for it would yield the missing elementary universal constant to be used as a natural unit, and possibly define at the same time a universal clock built in the universe and available "at every event-point". And that is what is desirable, since space-time is in my view time itself and should not depend upon the choice between different kinds of clocks based either on gravitation like old kitchen clocks and the motion of the Earth on the ecliptic, or on electro-magnetism like wrist-watches and atomic clocks, or on the life-time of radioactive substances etc. Finally, it could perhaps explain why gravitation appears to be a macroscopic effect analogous to statistical description of nature.

However, this would not at all be a quantization delivering quanta of a gravitational field - gravitons - and therefore it would neither support nor contradict the idea that there are gravitational signals similar to waves and packets of such. Waves would then at most be lattice waves of the 4-dimensional space-time, i.e. something like fluctuations in time itself, whose observation could only be made eventually in comparison with real clocks, but not with ideal clocks.

The fact that so far no gravitational waves or signals have with any certainty been observed does not prove that there are no such signals. For these signals, if they exist, may be too weak to be noticed by means of our devices. Yet, it may be an indication that gravitation is not that what some think it to be. I should like to insist however that even if the attempts at observing these signals have so far not led to conclusive results, the idea behind them is perhaps still more important than was imagined by those who started them.

10. Indeed, we can never know what are the most important or crucial problems of the physics just ahead of us. If we remember for example the origin of quantum theory, who would have dreamed in 1900 that the solution to the difficulties to explain the energy distribution of the black body would open such a revolution ? or that the fact of MAXWELL's equations not being GALILEI-covariant would start Relativity physics ? In both cases, the available theory was to become obsolete or eventually a mere approximation: that is quite a warning. Perhaps either Quantum theory even in its most advanced form of quantum field theory, or General Relativity, or both, are to become obsolete or approximate. Indeed, why should a theory that was made for the electromagnetic quantum field be the right one for other objects, especially since it already is unable to satisfactorily get rid of its infinities ? And why should a

theory meant to reintroduce the description of gravitation lost in Special Relativity, be exactly the correct grasp of the very fundamentals of time, i.e. of the independent parameter upon which all further functional description of the properties of the world and its constituents depend ?

The sort of argumentation I have used could also be applied to a discussion of the question whether a unitary field theory makes sense. I shall not touch on it. It would lead to far-reaching considerations about the nature of physical interactions as well as about the very question: what is meant by the unification aimed at - a question upon which I have dwelt elsewhere. So let me rather conclude.

11. I know that I have presented no single positive result, and that I have been speculating in a way that will sound very unphysical to many a physicist ! Actually, I have done something quite different: This paper has been conceived in its very methodology as a continuous questioning. I have successively questioned the received views, especially on the existence or reality of conceived physical entities. This procedure of systematically doubting propositions just uttered in order to get at successive insights into the nature of things is typical of a philosophical method known as phenomenology. The idea of systematic doubt goes of course back to DESCARTES; but there it was just doubting the received views, whereas phenomenology to-day as it was conceived by HUSSERL is a methodology of its own. In a way, if one pleases to say so, I have developed these considerations precisely in the form of an exercise in phenomenology, with the purpose of finding out, how far a philosophical method like that can be seriously applied to a scientific field sufficiently familiar to me: an experiment in philosophy, if you will. Looking back at it, one can say, it seems to me, that the experiment has been successful in its own right. For it has shown that it makes perfectly good sense to analyse an important (though not uncontroversial) piece of physics by means of the phenomenological method.

GRAVITATION AND TACHYONS[(°)]

Erasmo RECAMI

Istituto di Fisica Teorica, Università di Catania

57 Corso Italia, I-95129 Catania, Italy

1. - INTRODUCTION. EXTENDED RELATIVITY IN TEN POINTS

Recently, special relativity has been extended[1-4] to Superluminal inertial frames and faster-than-light objects (tachyons). It yielded the framework for building up a classical, relativistic theory of tachyons[1,2]. Let us outline the logical scheme that guided that extension of special relativity:

1) The very (Einsteinian) principle of relativity refers to inertial frames with constant velocity u, without any a priori restriction on the value of ($u \gtrless c$). We wish to express the relativity principle (RP) in the form << physical laws of mechanics and electromagnetism are required to be co-variant when passing from an inertial frame f_1 to another frame f_2 moving with constant relative velocity u, where $-\infty<u<+\infty$>>. We complete this postulate by adding the requirement that <<physical signals are transported only by positive-energy objects>> (cf. ref.(1)), i.e. the Principle of Retarded Causality or "Reinterpretation Principle".

2) We assume i) the RP and ii) the following postulate: <<space-time is homogeneous and space is isotropic>>, besides the Principle of retarded causality[(1)].

3) From the previous assumptions 2), the existence of an invariant speed v follows, e.g. by generalizing the procedures in ref.[(5)]; and experience shows that such a velocity **is** the speed of light: $v\equiv c$. Notice that the only invariant speed will be that of light: the divergent speed, e.g., will not be invariant.

4) From points 2) and 3), there follows a <<duality principle>> (DP) [1–4,6]: <<the terms bradyon (*) (B), tachyon (T), subluminal (s), Superluminal frame (S) do not have an absolute meaning, but only a relative one. Light speed invariance allows an exhausite partition [7] of all inertial ($u \lessgtr c$) frames in two sets {s},{S}, which are expected to be such that a (subluminal) Lorenz transformation (LT) maps {s},{S} separately into themselves, and a Superluminal <<Lorentz transformation>> (SLT) maps {s} into {S} and vice versa>>. A one-to-one correspondence may be set between frames s(u) and S(U), with $\vec{u} \| \vec{U}$, where $u \rightleftarrows U=c^2/u$, such a mapping being an <<inversion>> (i.e. a particular conformal mapping).

5) From RP, light speed invariance and DP it follows [1–4,8] that transformations between two frames f_1, f_2 must be linear and such that, for every tetravector (four-position, four-momentum, four-velocity,...),

$$c^2t^2-\vec{x}^2=\pm\,(c^2t'^2-\vec{x}'^2) \qquad (u \lessgtr c). \tag{1}$$

The metric (+---) will be adopted throughout this work. Natural units (c=1) will be used, when convenient.

6) In eq. (1), the sign <u>plus</u> holds for $u < c$, and <u>minus</u> for $u > c$. In fact, when going from a frame s to a frame S, the type of four vectors associated with the same observed object changes from timelike to spacelike, or vice versa, as follows from point 4).

For instance, in the four-momentum space, since for bradyons

$$E^2 - p^2 \equiv p^2 = m_o^2 > 0 \qquad (\beta^2 < 1), \tag{2a}$$

then for a tachyon, i.e. after a SLT, we shall have [1]

$$E'^2 - p'^2 \equiv p'^2 = -m_o^2 < 0 \qquad (\beta^2 > 1), \tag{2b}$$

where of course m_o is always real.

7) It is easy to verify [1–3] that, in the simple case of Superluminal, collinear, relative motion along the x-axis (†), eq. (1) breaks up into the two requirements

$$\left\{\begin{array}{l} c^2t'^2+(ix')^2=(ict)^2+x^2\,, \\ \\ (iy')^2+(iz')^2=y^2+z^2 \end{array}\right. \qquad (u^2 > c^2), \tag{1bis}$$

where use has been made of $g_{\mu\nu} = \delta_{\mu\nu}$ and of Einstein's notation. By the way, we shall always avoid explicit use of a metric tensor by writing the generical chronotopical vector as

$$x \equiv (x_o, x_1, x_2, x_3) = (ct, ix, iy, iz).$$

8) Transformations satisfying eq. (1), in the case of collinear (subluminal or Superluminal) motion with velocity $\vec{u}$ along the x-axis, can be shown([2]) to be the <<generalized Lorentz transformation>> (GLT): (^)

$$(3)\quad \begin{cases} x' = \pm \dfrac{x-ut}{\sqrt{|1-\beta^2|}}, & t' = \pm \dfrac{t-ux/c^2}{\sqrt{|1-\beta^2|}}, \\ & \qquad (-\infty < \beta \equiv \dfrac{u}{c} < \infty), \\ y' = \pm y\sqrt{\dfrac{1-\beta^2}{|1-\beta^2|}}, & z' = \pm z\sqrt{\dfrac{1-\beta^2}{|1-\beta^2|}}, \end{cases}$$

which obviously reduce to the standard (subluminal) ones when $\beta^2 < 1$. All the GLT's form a new group G of transformations ([1-3,9]). All discussions relative to these GLT's (in particular, relative to the SLT's) may be found in ref. ([1,2]). For instance, from eqs. (3) the <<generalized velocity composition law>> may be easily derived ([3]).

In a more compact form, eqs. (3) may be written as follows([1,3]):

$$(4)\quad \begin{cases} x' = \eta\gamma(x - ct\cdot tg\phi), & t' = \eta\gamma(t - \dfrac{x}{c}\, tg\phi), \\ & \qquad (\beta^2 \gtrless 1), \\ y' = (\eta\delta)y, & z' = (\eta\delta)z, \end{cases}$$

where we set

$$(5)\quad \begin{cases} \beta \equiv tg\phi, & \gamma \equiv +(|1-tg^2\phi|)^{-1/2} \\ \eta \equiv \eta(\phi) \equiv \dfrac{\cos\phi}{|\cos\phi|}\,\delta^2, & \delta \equiv +[(1-tg^2\phi)/|1-tg^2\phi|]^{1/2} \end{cases}$$

In eqs. (4), the angle ϕ runs over the whole round angle. For instance ([1]), usual subluminal, homogeneous, orthochronous, proper LT's are got for ϕ running from $-\pi/4$ to $+\pi/4$; and the corresponding nonorthochronous ones for $3\pi/4 < \phi < 5\pi/4$.

Form (4) parametrizes the GLT's in a <<continuous>> fashion, where our parameter ϕ runs (with continuity) from 0 to 2πrad. In particular, form(4) allows a straightforward (continuous) geometrical interpretation of generalized Lorentz transformations, for $\beta^2 \gtrless 1$ (see, e.g., Fig. 5,8 of ref.([2]) . For example, in the bidimensional case we have

$$(6)\qquad \begin{cases} x=C(x'\cos\phi+ct'\sin\phi)\,, \\ \qquad\qquad\qquad\qquad (0\leq\phi\leq 2\pi)\,, \\ t=C(t'\cos\phi+\dfrac{x'}{c}\sin\phi)\,, \end{cases}$$

where

$$C\equiv+\;[(1+tg^2\phi)/|1-tg^2\phi|]^{1/2}\;,$$

and where the physical meaning is quite clear. The SLT's corresponding to $\phi_1 = \pi/2$ and $\phi_2=(3/2)\pi$ will be called the <<trascendent transformations>> (1) K_+ and K_-, respectively.

10) Let us assume that we know, besides the class A of usual physical laws (of mechanics and electromagnetism) for bradyons, also the class B of the physical laws for tachyons(and antitachyons (1-3,10)). When we pass from a subluminal frame s to a Superluminal one S, class A--because of point 4)--will transform into class B, and vice versa (2,9,4). In this sense, the totality of physical laws ($A\cup B$) will be covariant under the whole group G, i.e. <<G-covariant>>(3). And in this sense inertial frames (with relative velocities $|u| \gtrless c$) are all equivalent (1).

From the previous considerations, the <<rule of tachyonization>> (TR) immediately follows, which is just a consequence of point 4): <<the relativistic laws (of mechanics and electromagnetism, at least) for tachyons follow by applying a SLT--e.g. the transcendent transformation K--to the corresponding laws of bradyons>>.

Moreover, it resulted (1,3) that physical laws (of special relativity) may be written in a (universal) form valid for both B's and T's, a form obviously coinciding with the usual one in the bradyonic case. For example, the G-covariant expression (1,3)

$$(7)\qquad m=\frac{m_o}{\sqrt{|1-\beta^2|}}\qquad (\beta^2\gtrless 1;\ m_o\ \text{real})$$

in such a form has <<universal>> validity (see also the following).

2. - SUPERLUMINAL SOURCES AND DOPPLER EFFECT

Once the <<Superluminal Lorentz transformations>> (SLT) are given for $u^2>c^2$, it is immediate to get the generalization (") of the Doppler-effect formula from the time transformation law. Namely, in the case of relative collinear motion along the x-axis, one has (11) for $u^2 \gtrless c^2$

$$(8)\qquad \nu=\nu_o\,(|1-\beta^2|)^{1/2}/(1+\beta\cos\alpha)\,,\qquad (u^2\lessgtr c^2)\,,$$

where $u \equiv \beta c$ is the relative speed and $\alpha \equiv \widehat{\vec{u}\vec{\ell}}$, the vector $\vec{\ell}$ being directed from the observer to the source.

In the particular case of relative motion (strictly) along the observation ray, since

$$[\mathrm{sign}(u)].[\mathrm{sign}(\cos\alpha)] = \begin{cases} - & \text{corresponds to } \underline{\text{approach}} \\ + & \text{corresponds to } \underline{\text{recession}}, \end{cases}$$

we obtain the behavior represented in Fig. 1, where the dashed curve refers to <<approach>> and the solid one to <<recession>>.

It is interesting that, in the particular case $\sin\alpha = 0$, our formula (3) has been obtained - independently by us ([2,3,11]) - also by GREGORY([12]), through heuristical considerations.

The first important point we want to stress is the following. In Fig. 1, the two solid curves (recession) are one the conformal correspondent of the other, as expected, in the sense that the same frequency will be obtained both for $u = \bar{v} < c$ and $U = c^2/\bar{v} > c$. Therefore, an astrophysical source receding with Superluminal velocity U is expected to exhibit a Doppler effect identical to the one of a usual source traveling away at velocity $u = c^2/U$.

The second result to be emphasized is the following. The above-mentioned conformal correspondence holds even for the two dashed curves (approach), except for the sign. Precisely, in the case of Superluminal approach, eq. (3) yields a negative sign ([1,2,11,12]) (Fig. 1). Such an occurrence represents the fact that a (subluminal) observer receives the radioemission of an approaching Superluminal source in the reversed chronological order, as is made clear in Fig. 2.

Therefore, if a macroscopic phenomenon is known to produce a radioemission obeying a certain chronological law, and one happens to detect the reversed radioemission, the observed source should be considered as a Superluminal, approaching object.

3.- RADIOCONTACT POSSIBILITIES, QUASARS (AND PULSARS)

Let us now examine an observer s, and the Minkowski space-time as <<seen>> by him. Namely, let us study ([3]) the relative position of the world-lines of both bradyons and tachyons, and of the light-cones associated with one or more events of their history. Notice ([3]) that--in a three-dimensional space-time, for simplicity-- the light-cones springing from a bradyonic world-line are strictly one inside the other, since the locus of the vertices passes inside the cones. This is not true for the light-cones springing from a

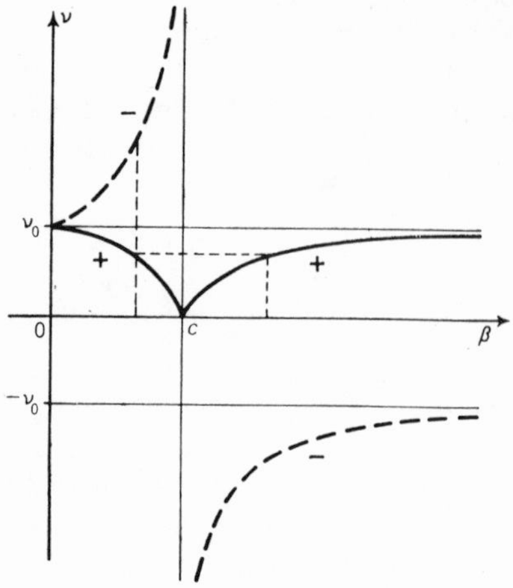

Fig. 1. – Doppler-effect extension: observed frequency *vs.* relative velocity for motion along the x-axis. The sign *minus* refers to approach (dashed line) and the *plus* to recession (solid line). The interpretation of the negative values appearing for the Superluminal approach is given in Fig. 2.

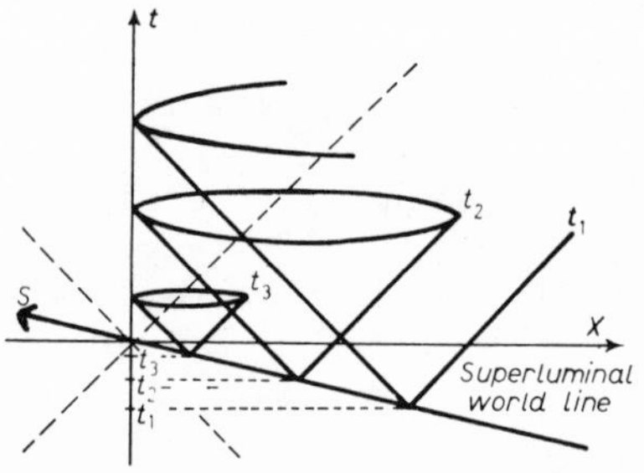

Fig. 2. – The radioemission of a Superluminal source, approaching the observer along the x-axis, will be received in reversed chronological order. This is the meaning of the negative frequencies entering Fig. 1. The line S is the Superluminal world-line.

tachyonic world-line. In the second case, their envelope is constituted by two planes: the <<retarded light-cones>> occupy entirely one dihedron with angle larger than 90°. We can easily get what follows (in partial correction of what appeared about this point in Refs. [3,2]):

A) Case of subluminal receiver B. The detector B may receive radio-signals (RS) from both bradyonic and tachyonic emitters.

B) Case of a Superluminal receiver T. Due to the Principle of Relativity (and to the Duality Principle), also T must be able to receive RS both from bradyons and from tachyons, even if great care must be used when describing such radio-contact processes from the initial (subluminal) observer.(^)

Here we want to add only the following observations:

(i) As any free bradyon, in a particular frame (its rest-frame), is represented by a point in space extended "along a line" in time, so any free tachyon -- in a particular frame (its "transcendent" frame) -- is represented by a "point" in time extended along a line in space. In other words, an infinite-speed (electromagnetic) source will appear as an "instantaneous" line [or, better, to optical observation it will appear as a couple of objects, departing from the same point H and then (symmetrically, simultaneously) moving far apart along a straight line ℓ, with an apparent relative speed initially very large and asymptotically tending to 2c]. In the case of a tachyonic source T traveling, with infinite or finite speed, along a (straight) line ℓ very far from our observation point O, when T is around the point H, where $OH \perp \ell$, then we shall observe the same phenomenon so as when getting radiations from two sources practically coinciding, but one receding and one approaching: i.e. we shall observe optical beats.

(ii) When a tachyonic (electromagnetic) source T, moving with constant speed V along a line ℓ is seen in a position P such that

$$V \sin\alpha = c$$

where α=POH, then we shall be struck by an "optical bang," somewhat similar to the well-known "sound-bang". Might quasars be far, tachyonic sources observed under the "optic-bang" conditions? It would then be easy, incidentally, to get statistical information, about the distributions of tachyonic sources in the universe, from observations about quasars.

Causality problems connected with these situations have been solved, e.g., in ref. ([1,2]) and references therein.

4. - FASTER-THAN-LIGHT OBJECTS IN THE GRAVITATIONAL FIELD

4.1 - Introduction

First of all, let us remember that, even in the Galilean-relativity case, the universal-gravitation law may be written ([13]) in a form similar to that of Einstein's gravitational equations. Namely, given a certain space-time, let us call $\Gamma^{\mu}_{\rho\sigma}(x)$ the <<affinity>> (or affine connection) components defined on that space-time ([13]). The Galilei-Newton space-time has a very particular affine geometry: the <<flat affine geometry>>, where $\Gamma^{\mu}_{\rho\sigma}(x) \equiv \overset{\circ}{\Gamma}{}^{\mu}_{\rho\sigma}$ vanishes at every point of the space-time manifold.

Then, the Newtonian equations of motion of a particle experiencing a gravitational force F^{μ} may read([13])

$$\text{(9)} \qquad \frac{d^2x^{\mu}}{dt^2} + \overset{\circ}{\Gamma}{}^{\mu}_{\rho\sigma} \frac{dx^{\rho}}{dt} \frac{dx^{\sigma}}{dt} = F^{\mu}/m_o \qquad \text{(Newton's case).}$$

4.2 Tachyons and gravitational field in special relativity.

Also in special relativity theory (Minkowski space-time), it is possible to introduce the gravitational field[13], and to build up a theory apparently explaining all observed gravitational phenomena (advances of planetary perihelia, bending of light rays, and so on), and deducible by a variational principle[13]. We are referring essentially to the formulation by BELINFANTE[14] and by FIERZ[14]. By the way, such a theory may be improved[15], so as to lead to the same field equations of general relativity.

In Belinfante's theory [13,14] (which, e.g., succeeds also in being associated with a definite spin value, namely 2), the gravitational field is described by a symmetric second-rank tensor $h_{\mu\nu}$. The equations of motion for a bradyon essentially result to be [13,14,15]

$$\frac{d^2x^\mu}{ds^2} + \Gamma^\mu_{\rho\sigma} \frac{dx^\rho}{ds}\frac{dx^\rho}{ds} = 0 \qquad \text{(ds timelike)}, \tag{10}$$

where the quantities $\Gamma^\mu_{\rho\sigma}$ are the components of a certain<<symmetric affinity>>[13], i.e. are the Christoffel symbols formed from a certain tensor[13] $f_{\mu\nu}$ (built up by means of the Minkowskian metric tensor $g_{\mu\nu}$ and of $h_{\mu\nu}$). For our purposes, it is noticeable that Christoffel symbols behave as (third-rank) tensors with respect (only) to linear transformations of the co-ordinates. This is our case, since all GLT's are linear transformations of Minkovsky space-time.

We are of course assuming that the gravitational interaction is relativistically covariant. Equations (10) tell us that the considered bradyon suffers the gravitational 4-force:

$$F^\mu = - m_o \Gamma^\mu_{\rho\sigma} \frac{dx^\rho}{ds}\frac{dx^\rho}{ds} \qquad (\beta^2 < 1). \tag{10 bis}$$

From the <<tachyonization rule>> we may immediately derive the gravitational 4-force suffered by a tachyon, by applying a SLT (e.g. the transcendent transformation $K_{\pm}$) to eq. (10 bis). In the simple case of collinear motion along the x-axis, transformations $K_{\pm}$ have been shown in ref. [1,2] to be represented by the 4 x 4 matrices

$$K_{\pm} = \begin{pmatrix} \pm\sigma_2 & 0 \\ 0 & -i\sigma_o \end{pmatrix} \equiv -(\mp\sigma_2 \oplus i\sigma_o), \tag{11}$$

$$\text{where} \quad \sigma_o \equiv \begin{pmatrix} 1 & 0 \\ 0 & 1 \end{pmatrix} \quad \text{and} \quad \sigma_2 \equiv \begin{pmatrix} 0 & -i \\ i & 0 \end{pmatrix}$$

are Pauli Matrices.

By the way, we eliminated the distinction between <<covariant>> and <<contravariant>> components, by means of our convention of suitably introducing imaginary units in the spatial-component definition. A summation is understood over repeated indices.

In conclusion, for $\beta^2 > 1$ we get

$$F'^{\mu} = + m_o \Gamma'^{\mu}_{\rho\sigma} \frac{dx'^{\rho}}{ds'} \frac{dx'^{\sigma}}{ds'} \qquad (\beta^2 > 1), \tag{12}$$

where now m_o is the (real) proper mass <u>of the tachyon</u> considered.

In other words, a tachyon is seen to experinece a gravitational repulsion (and not attraction!). However, since the fundamental equation of tachyon dynamics (see ref. ([1])):

$$F^{\mu} = - c \frac{d}{ds} \left(m_o c \frac{dx^{\mu}}{ds} \right) \qquad (\beta^2 > 1)$$

brings about another sign change, the equations of motion for a tachyon in a gravitation field will still read

$$\frac{d^2x'^{\mu}}{ds'^2} + \Gamma'^{\mu}_{\rho\sigma} \frac{dx'^{\rho}}{ds'} \frac{dx'^{\sigma}}{ds'} = 0 \qquad (ds' \text{ spacelike}). \tag{10'}$$

Therefore, eqs. (10) and (10') are already <u>G-covariant</u>; they hold for <u>both</u> bradyons <u>and tachyons</u>. They may be (still <u>G-covariantly</u>) written

$$a^{\mu} + \Gamma^{\mu}_{\rho\sigma} u^{\sigma} = 0 \qquad (\beta^2 \gtrless 1), \tag{13}$$

where a and u are four-acceleration and four-velocity respectively. Notice that, whilst dx/ds is a four-vector only with respect to the proper Lorentz group, on the contrary $u \equiv dx/d\tau_o$ is a G-four-vector.

As regards tachyons, let us, e.g., consider a tachyon going towards a gravitational-field source. Since it will experience a repulsive force, its energy ([1-3])

$$E = \frac{m_o c^2}{\sqrt{\beta^2 - 1}} \qquad (\beta^2 > 1;\ m_o \text{ real})$$

will decrease (contrary to the bradyonic case); however (see ref. ([1-3]) its velocity will increase. Therefore, we spoke about <<gravitational repulsion>> since the gravitational force applied to the tachyon is centrifugal (and not centripetal); however, as a result, the tachyon will accelerate towards ([17]) the gravitational field source (similarly to a bradyon). In conclusion, under our hypotheses it follows that:

a) From the dynamical (and energetical) point of view, tachyons appear as gravitationally repulsed. They might be considered to couple negatively to the gravitational field (the source velocity being inessential!), i.e. to be the <<antigravitational particles>>.

b) From the kinematical viewpoint, however, tachyons appear as <<falling down>> towards([17]) the gravitational-field source (the source velocity being inessential, as well as for electromagnetic field).

Let us explicitly remember that usual antiparticles do couple positively to the gravitational field. There is no gravitational difference between objects and their antiobjects (in accordance with the <<reinterpretation principle>> : see ref. ([1 3,18]) and the first point in Sect. 1)).

For tachyons, in the <<non-relativistic limit>> (speed near $c\sqrt{2}$: cf. eq. (7)) we would find for the three-force magnitude no longer Newton's law $F=-Gm_{o1}m_{o2}/r^2$, but

$$F=+G\,\frac{m_{o1}\,m_{o2}}{r^2} \qquad (\beta^2 > 1;\ m_{o1}, m_{o2} \text{ real}),$$

since when passing to tachyons G transforms into $-G$.

All the foregoing accords with the principle of duality.

In fact, given observer O, all objects that appear as bradyons to him will also appear to couple positively to the gravitational field; on the contrary, all objects that appear as tachyons to him will also appear to couple negatively to the gravitational field. In any case, bradyons, tachyons and luxons (*) will all bend towards the gravitational-field sources, even if with different curvatures.

5. - ON THE SO-CALLED <<GRAVITATIONAL ČERENKOV RADIATION>> AND TACHYONS. CASE OF THE VACUUM.

All that was previously said holds under the assumption of negligible gravitational-wave radiation.

If we improved eq.(10), for bradyon in a gravitational field, by allowing gravitational-wave emission, then the tachyonization rule (in the spirit of the duality principle) would still permit us to derive immediately the corresponding law for tachyons (now allowed to emit or absorb gravitational waves when interacting with that gravitational field).

But here we want only to deal with the problem of the so called <<gravitational Čerenkov radiation>> (GCR).

A usual bradyon (e.g. without charges, but of course massive) will generally radiate gravitational waves[19] as follows:

A) First of all, *when* (and if) it happens to enter a *sensu lato* medium (i.e. a force field), with speed larger than the gravitational-wave speed in that field, then it is expected to radiate a cone of gravitational radiation (somewhat similarly to the sound wave cone emitted by an ultrasonic airplane in air). Such a cone should not be confused with gravitational Čerenkov radiation, which on the contrary has the following features[20]: i) it is emitted by the bodies (or particles) of the material medium and not by the traveling object itself, ii) it is a function of the velocity (and not of the acceleration) or the traveling object.

B) Secondly, when the considered bradyonic object (e.g. a galaxy) B happens for example to enter a cluster of uniformly distributed galaxies, with a speed larger than the gravitational-wave speed inside that <<material medium>> (the cluster), then B will cause the cluster galaxies to emit gravitational waves such that possibly they will coherently sum up to form a gravitational Čerenkov cone[20,21].

Now, we want to investigate if and when tachyonic objects can emit cones of gravitational waves, and in particular cause emission of gravitational-radiation Čerenkov cones.

The first result is that *free tachyons in vacuum will not emit gravitational Čerenkov radiation, or gravitational-wave cones, or any gravitational radiation*.

This result may be immediately obtained -- see ref. [1,2,9] -- by using the tachyonization rule, i.e. by applying a SLT to the behaviour of a free bradyon (e.g. at rest) in vacuum.

As regards the gravitational Čerenkov effect, let us consider an object that appears to us (frame s_o) as a T moving at constant velocity in vacuum. Such an object will appear as a B with respect e.g. to its rest frame S; according to frame S, therefore, the (bradyonic) differential energy loss through GCR in vacuum is obviously zero(22):

$$\left(\frac{dE}{ds}\right)_o = 0 \qquad (\beta^2 < 1). \tag{14a}$$

If we transform such a law by means of the SLT from S to s_o, we get the differential energy loss through GCR for a tachyon in vacuum(2,22):

$$\left(\frac{dE'}{ds'}\right)_o = 0 \qquad (\beta^2 > 1). \tag{14b}$$

Moreover, an object uniformly moving in vacuum does not emit any radiation both in the subluminal case and in the superluminal one, as required by (extended) relativity.

Let us repeat that it is necessary not to confuse GCR with the radiation that an object can indeed emit in vacuum when it is accelerated.

Lastly, let us again emphasize that GCR comes out from the objects of the material medium, and not from the <<radiating>> object itself(20,21). Therefore, the very expression <<GCR in vacuum>> would have been meaningless, unless one simultaneously provided a suitable theory about a <<vacuum structure>>.

6.- CASE OF MATERIAL <<MEDIA>> .

Since the previous case A), considered in Sect. 5, is substantially not different from case B)(13,23), we shall confine ourselves to the study of Čerenkov radiation cones caused by tachyonic objects. The principle of duality, together with tachyonization rule, allows us to solve also the general problem of GCR from tachyons in a material medium(22).

Let us first define the proper <<gravitational refraction index>> N of a medium M_1 with respect to a second medium M_2 as the ratio of the gravitational-wave (GW) speed in M_2 over the GW speed in M_1. Let us now generalize (Fig. 3) the <<drag effect>>(8) for the Superluminal velocities, i.e. extend(22) the calculation of the apparent speed v of gravitational waves in a moving medium, with respect to a given observer 0, for tachyonic media.

Figure 3 just plots the (extended) <<gravitational refraction index>> N, for moving media, vs. the medium speed u, derived from mere consideration of the generalized velocity composition law (see ref.([2])):

$$(15) \qquad N(u) \equiv N_c \frac{N_o c+u}{c+N_o u} \qquad (u^2 \gtrless c^2);$$

the quantity $v=c/N$ will give the GW speed in the moving medium considered, with respect to the observer 0. The GW speed in vacuum is of course assumed to be equal to the light speed c.

It is easy to observe that i) the GW speed v in a bradyonic medium appears to a bradyonic observer 0 always slower than c; ii) the GW speed v in a hypothetical, ideal <<luminal medium>> (u = c) would appear to 0 as always equal to c; iii) the GW speed in a tachyonic medium will always appear to 0 as larger than c.

It is immediate to deduce that

i) a bradyon B can emit gravitational "Cerenkov" only in a subluminal medium (when it happens to travel with speed w larger than the apparent GW speed v in that bradyonic medium). From duality and the tachyonization rule it is then straightforward to get that

ii) a tachyon T can emit GCR only in a Superluminal medium, and namely when it happens to travel in the (tachyonic) medium with speed w slower than the apparent GW speed v in that medium.

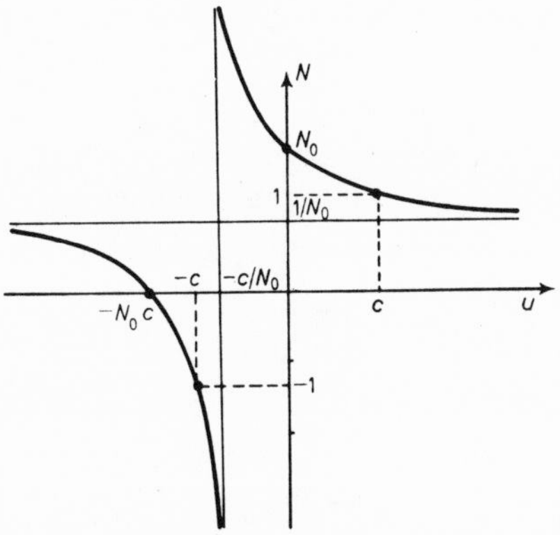

Fig. 3. – « Drag effect » extension: generalization of the formula for the observed « gravitational *refraction index* » N of a medium *vs.* its relative velocity u. The medium is supposed to move collinearly with respect to the observer. N_0 is the *proper* « gravitational refraction index ». See the text.

Therefore, a tachyon will not emit gravitational Čerenkov in usual, subluminal media. Conversely, it is possible that a tachyon appears to move in a tachyonic medium with velocity slower than the apparent GW velocity there: see Fig. 3.

From the analytical point of view, the formula (°) of the standard, differential energy loss (per unit path length) through GCR from a bradyon in a medium may be written[22]

$$\text{(16a)} \qquad \frac{dW}{dZ} \propto \frac{4\pi^2 Gm^2}{c^2} \int_{(\beta N_o > 1)} \left(1 - \frac{1}{\beta^2 N_o^2}\right) \nu d\nu \; ; \quad \left(\beta \equiv \frac{w}{c},\ w^2 < c^2\right),$$

where now $\beta \equiv w/c$, the quantity w being the bradyon speed relative to 0, an N_o is the proper «gravitational refraction index» of the medium. In particular, for $N_o = 1$ (i.e. for the vacuum) one gets

$$\text{(17a)} \qquad \frac{dW}{dZ_o} = 0 \qquad\qquad (w^2 < c^2).$$

Then, by applying the transcendent Superluminal Lorentz transformation K_+ to eq. (16a), we find the analogous formula of the differential energy loss (per unit path length), through GCR, from a a tachyon in a medium[22]:

$$\text{(16b)} \qquad \frac{dW}{dZ} \propto \frac{4\pi^2 Gm^2}{c^2} \int_{(N_o/\beta > 1)} \left(1 - \frac{\beta^2}{N_o^2}\right) \nu d\nu ; \quad \left(\beta \equiv \frac{w}{c};\ w^2 > c^2\right),$$

where N_o is still the proper «gravitational refraction index» of the medium, and w the tachyon speed (relative to 0). In particular, for N_o 1 (i.e. for the vacuum and for bradyonic media), one gets

$$\text{(17b)} \qquad \left(\frac{dW}{dZ}\right)_o = 0 \qquad\qquad (w^2 > c^2).$$

The Čerenkov relation giving the cone angle in the bradyonic case[21]

$$\text{(18a)} \qquad \cos\theta = \frac{1}{\beta N_o} \qquad\qquad (\beta^2 < 1)$$

transformsunder the same SLT into the following relation, giving the cone angle in the tachyonic case:

$$\cos\theta = \frac{\beta}{N_o} \qquad (\beta^2 > 1) \tag{18b}$$

All the previous papers([24 27]) dealing with GCR, in our opinion, are not correct, since they are based on an uncritical extension of the usual procedure for bradyons to the tachyons' case. They, e.g., concluded about a GCR from tachyons in vacuum. We have seen from (extended) relativity, on the contrary, that Superluminal objects do not emit GCR in bradyonic media (e.g. in bradyonic galaxy-clusters).

* * *

The present work has been done in collaboration with R. MIGNANI

The authors are grateful to Profs. A. AGODI, M. BALDO, P. CALDIROLA and V. DE SABBATA, and dr. E. MASSARO for useful discussions, and to Mr. F. ARRIVA, Dr. L.R. BALDINI and Mrs. G. GIUFFRIDA for their kind collaboration. Finally, one of them, (E.R.) wishes to thank Professors N. DALLAPORTA, V. DE SABBATA, J. WEBER for their very kind interest.

References

(0) Work performed in collaboration with R. MIGNANI

(1) E. RECAMI and R. MIGNANI: Riv. Nuovo Cimento, 4, 209,398 (1974).

(2) R. MIGNANI and E. RECAMI: Nuovo Cimento, 14A, 169 (1973); Erratum, 16A,208(1973); E. RECAMI: I tachioni, in Annuario Scienza e Tecnica 73, Enciclopedia EST-Mondadori (Milano, 1973), p. 85.

(3) E.RECAMI and R.MIGNANI: Lett. Nuovo Cimento, 4,144 (1972);8, 110(1973); and references therein.

(4) L.PARKER: Phys. Rev., 188, 2287(1969).

(5) See, e.g. V. GORINI and A. ZECCA, J Phys. 11, 2226 (1970); V. BERZI and V. GORINI: J. Math. Phys., 10, 1518 (1969). These papers are to be generalized.

(6) See also F. ANTIPPA: Nuovo Cimento, 10A, 389 (1972).

(*) Slower-than light particle.

(7) See, e.g., A. AGODI: Lezioni di fisica teorica, Catania, 1972 (unpublished).

(8) See, e.g., W. RINDLER: Special Relativity (Edinburgh, 1966).

(†) We always neglect space-time translations.

(^) For problems connected to eqs. (3) see also R. MIGNANI and E. RECAMI: Lett. Nuovo Cimento, 9,357 (1974); H.C.CORBEN: Lett. Nuovo Cimento, 11 , 533 (1974).

(9) R. MIGNANI, E. RECAMI and U. LOMBARDO: Lett. Nuovo Cimento, 4, 624 (1972).

(10) See M. BALDO, G. FONTE and E. RECAMI: Lett. Nuovo Cimento 4, 241 (1970), and references therein.

('') In ref. (1) the generalization of Maxwell equations for Superluminal sources is given, too. See as well R. MIGNANI and E. RECAMI; Lett. Nuovo Cimento, 9, 367 (1974).

(11) R. MIGNANI and R. RECAMI: General Relat. Gravit. 5, 615 (1974).

(12) C. GREGORY: Nature Phys. Sci., 239, 56 (1972).

(13) J.L. ANDERSON: Principles of Relativity Physics, Chapters 1, 2, 5, 2,8,10,10,6 (New York, N.Y. 1967)

(14) F.J. BELINFANTE: Phys. Rev. 89, 914 (1953); M. FIERZ: Helv. Phys. Acta 12,3 (1939). See also ANDERSON: ref. (13), Chap. 6 and 8.10.

(15) S.N. GUPTA: Phys. Rev., 96, 1683 (1954); R.M. KRAICHNAN: Phys. Rev. 98, 1118 (1955)

(16) See, e.g., L.D. LANDAU and E.M. LIFSHITZ: Theorie du champ (Moscow, 1966)

(17) See also R.O. HETTEL and T.M.HELLIWELL: Nuovo Cimento, 13B, 82 (1973). In this paper the problem of tachyon trajectories in a Schwartzschild field has been considered within general relativity. Cf. also R.W. FULLER and J.A. WHEELER: Phys. Rev., 128, 919 (1962).

(18) O.M.P. BILANIUK, V.K. DESHPANDE and E.C. SUDARSHAN: Am. Journ. Phys., 30, 718 (1962)

($^{*}_{*}$) Objects traveling at the light speed, like photons and neutrinos.

(19) See e.g., J. WEBER; General Relativity and Gravitational Waves (New York, N.Y., 1961); V.DE SABBATA: L'Elettrotecnica, 57 (9), 1 (Milano, 1970).

(20) See, e.g., L.D. LANDAU and E.M. LIFSHITZ: Electrodynamics of Continuous Media (Oxford, 1960), p. 357.

(21) Cf. J. V. JELLEY: Čerenkov Radiation and Its Applications (London, 1958).

(22) Cf. also R. MIGNANI and E. RECAMI: Lett. Nuovo Cimento, 7, 388 (1973); 9 362 (1974).

(23) See, e.g., W. PAULI, Relativitatstheorie (Leipzig, 1921).

(o) As follows by analogy with the electromagnetic case. Cf. ref. (22)

(24) A. PERES: Phys. Lett. 31A, 361 (1970).

(25) L.S. SCHULMAN: Nuovo Cimento, 2B, 38 (1971).

(26) A.S. LAPEDES and K.C. JACOBS: Nature Phys. Soc., 234, 6 (1972).

(27) H.K. WIMMEL: Nature Phys. Sci., 236, 79 (1972). See also R. GOTT III: preprints (to be published).

GENERAL RELATIVITY AND QUANTUM THEORY

E. SCHMUTZER

Sektion Physik, Friedrich-Schiller-Universität
DDR-69 Jena, Max-Wien-Platz 1

ABSTRACT

The fact, that the traditional quantum mechanics does not fulfill the requirement of Einstein's General Principle of Relativity is the reason for presenting this paper. Here a new foundation of quantum theory is given [1]. The fundamental quantum laws proposed are:

1. time-dependent simultaneous laws of motion for the operators, general states and eigenstates,
2. commutation relations,
3. time-dependent eigenvalue equations.

The main results are:

1. Meta-quantum theory including macroscopic phenomena (thermodynamics etc.), which can be specialized to usual quantum mechanics.
2. The fundamental laws are picture-free and valid for arbitrary frames of reference (covariant).
3. New method of calculation for time-dependent problems.

1. INTRODUCTION

To promote the problem of field quantization in a curved space-time on the basis of Einstein's theory of gravitation it is necessary to understand the relationship between General Relativity and quantum theory (in non-relativistic quantum mechanics with restricted coordinate transformations) fully. We believe that quantum theory in its present formulation is not adequate to the requirements of General Relativity (in the limiting case of arbitrarily moving frames of reference with non-relativistic velocities).

Let us sketch our motivation for a change of the traditional basis of quantum theory in the following.

1. The traditional formulation of quantum mechanics is based on the Heisenberg picture

$$\text{a)}\ \frac{dF^{(H)}}{dt} = \frac{\partial F^{(H)}}{\partial t} + \frac{i}{\hbar}\left[H^{(H)}, F^{(H)}\right], \quad \text{b)}\ \frac{d|\Psi\rangle^{(H)}}{dt} = 0 \qquad (1)$$

or on the Schrödinger picture

$$\text{a)}\ \frac{dF^{(S)}}{dt} = \frac{\partial F^{(S)}}{\partial t}, \quad \text{b)}\ \frac{d|\Psi\rangle^{(S)}}{dt} = \frac{1}{i\hbar} H^{(S)} |\Psi\rangle^{(S)}, \qquad (2)$$

which are connected by a unitary transformation. All other pictures obtained by a unitary transformation are equivalent. Usually inertial frames of reference are taken as a basis.

As is well known the covariance (forminvariance) of the above equations within a certain picture is destroyed by a unitary transformation. Is it possible to find a covariant form of quantum theory valid in any frame of reference ab initio, i.e. to have a picturefree formulation ab initio as in classical mechanics, electromagnetism, etc.? If such a form of the fundamental laws of quantum theory were to exist, they would then have to be considered as the proper primary basis - in accordence with General Relativity.

2. With respect to the time the formulation (1) and (2) are asymmetric and unaesthetic. We doubted that such fundamental laws of nature exhibit asymmetries of such kind.

3. In the Heisenberg picture as well in the Schrödinger picture there are strange differencies concerning the laws of motion for the general state and the eigenstate. Should not an eigenstate as a specialization of a general state obey the same law?

4. The form-invariance of (1a) leads, if we transform the operator $F^{(H)}$ according to

$$\bar{F}^{(H)} = U^{(H)} F^{(H)} U^{(H)+} , \tag{3}$$

to the following transformation law of the Hamiltonian

$$\bar{H}^{(H)} = U^{(H)} H^{(H)} U^{(H)+} - i\hbar \frac{dU^{(H)}}{dt} U^{(H)+} . \tag{4}$$

The form-invariance of (2b) has a different transformation law, namely

$$\bar{H}^{(S)} = U^{(S)} H^{(S)} U^{(S)+} + i\hbar \frac{dU^{(S)}}{dt} U^{(S)+} , \tag{5}$$

as a consequence, if we transform the general state like

$$\overline{|\Psi\rangle}^{(S)} = U^{(S)} |\Psi\rangle^{(S)} . \tag{6}$$

Our question is: which of the two is the true Hamiltonian after the transformation? Because the Hamiltonian is a basic quantity, we were not satisfied with this ambiguity.

From all these considerations as well as several others - including philosophical ones - we were led to the following generalization of Einstein's idea of the Principle of General Relativity:

Principle of Fundamental-Covariance

"Fundamental physical laws are form-invariant:

1. with respect to coordinate transformations of space-time (Principle of General Relativity)
$\longrightarrow$ (Principle of coordinate-covariance),
2. with respect to operator-state-transformations of the Hilbert space
$\longrightarrow$ (Principle of operator-covariance)."

In other words:

"Fundamental physical laws have the same form for two observers of different states of motion, independently of the coordinates or operators and states being used."

2. FUNDAMENTAL LAWS OF QUANTUM MECHANICS

We consequently follow the line of our Principle of Fundamental-Covariance and construct the laws accordingly. The elements of construction are the position operators $Q_K(t)$ and the canonical momentum operators $P_K(t)$, from which we are led to the general operator

$$F = F(Q_K(t), P_K(t), t). \tag{7}$$

Consequently we generalize the usual concept of the state $|\Psi\rangle^{(S)} = |\Psi(t)\rangle^{(S)}$ of the traditional theory as follows:

$$|\Psi\rangle = |\Psi(Q_K(t), P_K(t), t)\rangle. \tag{8}$$

While the algebraic rules of calculation are not

changed, the last notion induces a change of the rules of the differential calculas.

Our proposed basic laws are:

<u>Fundamental laws I (Heisenberg's commutation rules):</u>

a) $[Q_K, Q_L] = 0$, b) $[P_K, P_L] = 0$, c) $[Q_K, P_L] = i\hbar\delta_{KL}$. (9)

<u>Fundamental law II (Heisenberg's equation of motion of an operator):</u>

$$\frac{dF}{dt} = \frac{\partial F}{\partial t} + \frac{i}{\hbar}[H, F] , \tag{10}$$

where $H(Q_K, P_K, t)$ is the Hamiltonian of the quantum mechanical system. The partial derivative denotes the explicit time dependence.

<u>Fundamental law III (equation of motion of a general state):</u>

$$\frac{d|\Psi\rangle}{dt} = \frac{\partial|\Psi\rangle}{\partial t} + \frac{i}{\hbar} H|\Psi\rangle \tag{11}$$

Our coining of the concept of the state denoted in (8) allows us to distinguish between the total and the partial time derivative of the state. The explicit time dependence corresponds to external temporal influences which are interpreted later.

One should realize that as well as introducing the term with the partial time derivative we also changed the sign of the second term compared with the Schrödinger equation. This step had to be taken to obtain a unique transformation law of the Hamiltonian.

Furthermore, we stress that (10) and (11) form the simultaneous system of the fundamental equations of motion.

Fundamental law IV (eigenvalue equation):

In addition to the basic laws formulated above the eigenvalue equation

$$F|f_\sigma\rangle = f_\sigma|f_\sigma\rangle \tag{12}$$

for an operator F plays an important role, too. In the time-dependent case the eigenstates $|f_\sigma\rangle$ and the eigenvalues f_σ in general are time-dependent.

Fundamental law V (equation of motion for an eigenstate):

In accordance with (11) we postulate the following law of motion for the eigenstates

$$\frac{d|f_\sigma\rangle}{dt} = \frac{\partial|f_\sigma\rangle}{\partial t} + \frac{i}{\hbar} H|f_\sigma\rangle \;. \tag{13}$$

For application the following eigenvalue problems are of great importance:

Position operator problem:

$$Q_K|q_K\rangle = q_K|q_K\rangle \;. \tag{14}$$

Because the position operator is not explicitly time-dependent, equation (13) takes the form

$$\frac{d|q\rangle}{dt} = \frac{i}{\hbar} H|q\rangle \qquad \left(|q\rangle = |q_1\rangle \cdots |q_{3N}\rangle\right). \tag{15}$$

For the Schrödinger representation we notice the relation

$$\langle q|F(Q_K,P_K,t)|\Lambda\rangle = F_D\left(q_K,\frac{\hbar}{i}\frac{\partial}{\partial q_K},t\right)\Lambda \;, \tag{16}$$

where F_D is the differential operator corresponding to F and $\Lambda = \langle q|\Lambda\rangle$.

Momentum operator problem:

$$P_K|p_K\rangle = p_K|p_K\rangle \;. \tag{17}$$

Formulae similar to the case of the position operator are valid.

Hamiltonian problem:

$$H|h_\nu\rangle = h_\nu |h_\nu\rangle \quad . \tag{18}$$

From (13) we obtain the following law of motion for the eigenstates of the Hamiltonian:

$$\frac{d|h_\nu\rangle}{dt} = \frac{\partial|h_\nu\rangle}{\partial t} + \frac{i}{\hbar} H|h_\nu\rangle = \frac{\partial|h_\nu\rangle}{\partial t} + \frac{i}{\hbar} h_\nu |h_\nu\rangle \quad . \tag{19}$$

3. DEFINITIONS OF THE PARTIAL DERIVATIVES OF OPERATORS AND STATES WITH RESPECT TO THE POSITION AND MOMENTUM OPERATORS

First we define

$$\text{a)} \quad \frac{\partial F}{\partial Q_K} = \frac{1}{i\hbar}[F, P_K] \quad , \text{ b)} \quad \frac{\partial F}{\partial P_K} = \frac{1}{i\hbar}[Q_K, F] \tag{20}$$

The dependence of the state (8) from the Q_K and P_K according to our conception forces us to answer the question after the partial derivatives of the state with respect to the Q_K and P_K. We define

$$\text{a)} \quad \frac{\partial|\Psi\rangle}{\partial Q_K} = -\frac{1}{i\hbar} P_K|\Psi\rangle \quad , \text{ b)} \quad \frac{\partial|\Psi\rangle}{\partial P_K} = \frac{1}{i\hbar} Q_K|\Psi\rangle \quad . \tag{21}$$

4. OPERATOR-COVARIANCE OF THE PROPOSED THEORY (TRANSFORMATION THEORY)

The main argument for postulating the above fundamental laws as simultaneous laws consists in their operator-covariance which leads to the validity of these laws for arbitrary (non-relativistic) frames of reference. As it can be shown in detail the whole scheme is covariant, if we perform unitary transformations according to the rules:

a) $\overline{Q_K} = U Q_K U^+$, b) $\overline{P_K} = U P_K U^+$ (22)

or in general

a) $\overline{F} = F(\overline{Q}_K, \overline{P}_K, t) = U F U^+$, b) $\overline{\frac{\partial F}{\partial t}} = U \frac{\partial F}{\partial t} U^+$; (23)

further

$$\overline{H} = U H U^+ - i\hbar \frac{dU}{dt} U^+ , \tag{24}$$

a) $\overline{|\Psi\rangle} = U|\Psi\rangle$, b) $\frac{\partial \overline{|\Psi\rangle}}{\partial t} = U \frac{\partial |\Psi\rangle}{\partial t}$, (25)

a) $\overline{|f_\sigma\rangle} = U|f_\sigma\rangle$, b) $\frac{\partial \overline{|f_\sigma\rangle}}{\partial t} = U \frac{\partial |f_\sigma\rangle}{\partial t}$. (26)

5. TRANSITION TO A DISTINGUISHED SPECIAL CASE ("STATIC ASPECT")

Because our basic laws (10) and (11) are operator-covariant, there does not exist a unitary transformation which destroys the form of these two equations, i.e. the symmetry of these laws cannot be broken by a unitary transformation. The consequence of this fact is the impossibility of passing over to the Heisenberg picture or to the Schrödinger picture. But there exists a unitary operator N (U→N), defined by

$$i\hbar \frac{\partial N}{\partial t} = HN \tag{27}$$

which simplifies the shape of the laws of motion (we denote the N-transformed quantities by a hook).

Namely in this case from (24)

$$\check{H} = 0 \quad , \text{ i.e. } \quad \check{h}_\nu = 0 \tag{28}$$

results.

Now the basic laws of motion (10), (11) and (13) read

$$\text{a) } \frac{d\check{F}}{dt} = \frac{\partial \check{F}}{\partial t} \quad , \text{ b) } \frac{d|\check{\Psi}\rangle}{dt} = \frac{\partial|\check{\Psi}\rangle}{\partial t} \quad , \text{ c) } \frac{d|\check{f}_\nu\rangle}{dt} = \frac{\partial|\check{f}_\nu\rangle}{\partial t} \ . \tag{29}$$

Further the constancy of the transformed position operators and momentum operators results:

$$\text{a) } \check{Q}_K = Q_{K0} \quad , \text{ b) } \check{P}_K = P_{K0} \ . \tag{30}$$

Therefore this special case of presentation of the quantum laws could be called "static aspect".

6. DIFFERENTIAL EQUATION FORMALISM (SCHRÖDINGER REPRESENTATION)

Using the scalar products

$$\text{a) } \Psi(q,t) = \langle q|\check{\Psi}\rangle \quad , \text{ b) } \hat{\Psi}(q,t) = \langle q|\Psi\rangle \ , \tag{31}$$

$$\text{a) } \varphi_\nu(q,t) = \langle q|\check{h}_\nu\rangle \quad , \text{ b) } \phi_\nu(q,t) = \langle q|h_\nu\rangle \tag{32}$$

we find from (11)

$$\frac{\partial \Psi}{\partial t} = \left(\frac{\partial \Psi}{\partial t}\right)_{expl} + \frac{1}{i\hbar} H_D \Psi \qquad \text{(generalized Schrödinger equation),} \tag{33}$$

where

$$\left(\frac{\partial \Psi}{\partial t}\right)_{expl} = \langle q| \frac{\partial|\check{\Psi}\rangle}{\partial t} \ ; \tag{34}$$

from (19)

$$\frac{\partial \varphi_\nu}{\partial t} = \left(\frac{\partial \varphi_\nu}{\partial t}\right)_{expl} + \frac{1}{i\hbar} H_D \varphi_\nu$$ (wave equation for the eigenfunctions), (35)

where

$$\left(\frac{\partial \varphi_\nu}{\partial t}\right)_{expl} = \langle q | \frac{\partial | \check{h}_\nu \rangle}{\partial t} ; \qquad (36)$$

and from (18)

$$H_D \phi_\nu = h_\nu \phi_\nu$$ (eigenvalue equation) (37)

or in another form

$$(NHN^+)_D \varphi_\nu = h_\nu \varphi_\nu . \qquad (38)$$

From (33) the balance equation for the probability

$$\frac{\partial(\Psi^* \Psi)}{\partial t} = \Psi^* \left(\frac{\partial \Psi}{\partial t}\right)_{expl} + \Psi \left(\frac{\partial \Psi^*}{\partial t}\right)_{expl} + \frac{1}{i\hbar}\left[\Psi^*(H_D \Psi) - \Psi (H_D \Psi)^*\right] \qquad (39)$$

results, a consequence of which the relation

$$\int \left[\Psi^* \left(\frac{\partial \Psi}{\partial t}\right)_{expl} + \Psi \left(\frac{\partial \Psi}{\partial t}\right)^*_{expl}\right] dq = 0 \qquad (40)$$

is.

7. PHYSICAL INTERPRETATION

When we postulated the equation of motion (11) for general states, we stated that the partial derivative corresponds to external temporal influences. Only after precribing these explicit effects the equation (11) is fully determined. Up to now we have left this question open.

Because (39) takes the form of a continuity equation, i.e. of a conservation law for the probability, if

$$\left(\frac{\partial \Psi}{\partial t}\right)_{expl} = 0 \quad , \quad i.e. \quad \frac{\partial |\Psi\rangle}{\partial t} = 0 \quad , \tag{41}$$

we realize that this condition leads to the level of usual quantum mechanics, where a quantum mechanical system and electromagnetism are coupled (Schrödinger theory).

The equation

$$\left(\frac{\partial \Psi}{\partial t}\right)_{expl} \neq 0 \quad , \quad i.e. \quad \frac{\partial |\Psi\rangle}{\partial t} \neq 0 \tag{42}$$

surpasses the level of quantum mechanics and could be taken as a basis of a meta-quantum theory. This line of development offers a possibility for taking into account macroscopic phenomena and the measuring process.

On the basis of Schrödinger theory the expectation value is defined by

$$\bar{f} = \langle \check{\Psi} | F | \check{\Psi} \rangle = \int \Psi^* F_D \Psi \, dq \quad , \tag{43}$$

in agreement with the usual definition. Then for the expectation values Ehrenfest's theorem and Heisenberg's uncertainty relation are guaranteed.

8. NEW METHOD OF CALCULATION FOR TIME-DEPENDENT HAMILTONIANS

Method of solving the time-dependent eigenvalue equation

The Hamiltonian is decomposed in the usual way into a time-independent and a time-dependent part:

$$H_D = \overset{\circ}{H}_D + \overset{t}{H}_D(t) \quad , \quad h_\nu = E_\nu + \overset{t}{h}_\nu \quad . \tag{44}$$

We use the time-independent eigenvalue equation

$$\overset{\circ}{H}_D \chi_\sigma = E_\sigma \chi_\sigma \tag{45}$$

and expand

$$\phi_\nu(q,t) = \sum_\lambda c_{\nu\lambda}(t)\,\chi_\lambda(q) \quad . \tag{46}$$

Inserting into (37) we are left with the linear system

$$\sum_\lambda c_{\nu\lambda}\left[\delta_{\sigma\lambda}(E_\lambda - E_\nu) + H_{\sigma\lambda} - \overset{t}{h}_\nu\,\delta_{\sigma\lambda}\right] = 0 \tag{47}$$

for the coefficients $c_{\nu\lambda}$, where

$$H_{\sigma\lambda}(t) = \int \chi_\sigma^*\,\overset{t}{H}_D(t)\,\chi_\lambda\,dq \quad . \tag{48}$$

The secular equation corresponding to (47) reads

$$\det\left|H_{\sigma\lambda} + \delta_{\sigma\lambda}(E_\lambda - E_\nu - \overset{t}{h}_\nu)\right| = 0 \quad . \tag{49}$$

Hence the quantities $\overset{t}{h}_\nu$ can be calculated.

<u>Method of solving the Schrödinger equation</u>

In contrast to Dirac's perturbation method we expand with respect to the time-dependent eigenfunctions ϕ_σ :

$$\Psi = \sum_\sigma S_\sigma(t)\,\phi_\sigma(q,t) \quad . \tag{50}$$

In the usual way we find the system of linear differential equations

$$\dot{S}_\mu - \frac{1}{i\hbar}\,h_\mu S_\mu + \sum_\sigma S_\sigma\,\varphi_{\mu\sigma} = 0 \tag{51}$$

with

$$\varphi_{\mu\sigma} = \int \phi_{\mu}^{*} \frac{\partial \phi_{\sigma}}{\partial t} \, dq = - \varphi_{\sigma\mu}^{*} . \tag{52}$$

For application of this new method on concrete time-dependent problems one should consider paper [2].

We renounce to present the matrix formalism and the generalization to quantum field theory in a metric field. This subjects are treated in detail in [1].

REFERENCES

1. Schmutzer, E. (1975). Nova Acta Leopoldina (Halle), Suppl. Nr. 8, Vol. 44
(1976). Exp. Technik d. Physik, 24, 131

2. Schmutzer, E. (1976). Acta Physica Austriaca (in press)

AUTHOR INDEX

SUBJECT INDEX